Thermal Infrared Remote Sensing

This book systematically and comprehensively introduces concepts, theories, principles, and applications of Thermal Infrared (TIR) imaging spectral techniques and TIR remote sensing. It introduces in detail algorithms, techniques, and methods for processing and analyzing TIR data, including radiometric calibration and atmospheric correction in TIR region, retrieving land surface emissivity and temperature (LST) from various TIR data, and downscaling LST and TIR radiance spatiotemporally. Focusing on TIR-RS application orientation, this book introduces state-of-the-art techniques and methods applied in areas such as hydrology, biomass burning, urban thermal environment, and other areas of interest.

Features:

- Presents a complete coverage of thermal infrared remote sensing, from concepts to applications.
- Provides a thorough survey of characteristics and types of airborne and spaceborne TIR sensors, systems, and missions.
- Discusses principles and methods of radiometric calibration and atmospheric correction for correcting TIR data.
- Includes case studies that demonstrate the applicability and potential of a variety of techniques and methods to process TIR data.
- Offers a comprehensive review of TIR remote sensing applications.

This book is a great resource for professionals in the academic and corporate world and an insightful reference for graduate and senior undergraduate students studying geology, geography, geophysics, and environments, as well as natural resources and survey management.

Thermal Infrared Remote Sensing

Principles and Applications

Ruiliang Pu and Qiuyan Yu

CRC Press
Taylor & Francis Group
Boca Raton London New York

CRC Press is an imprint of the
Taylor & Francis Group, an **informa** business

Designed cover image: iStock

First edition published 2026
by CRC Press
2385 NW Executive Center Drive, Suite 320, Boca Raton FL 33431

and by CRC Press
4 Park Square, Milton Park, Abingdon, Oxon, OX14 4RN

CRC Press is an imprint of Taylor & Francis Group, LLC

ISBN: 978-1-032-23180-8 (hbk)
ISBN: 978-1-032-23182-2 (pbk)
ISBN: 978-1-003-27617-3 (ebk)

DOI: 10.1201/9781003276173

Typeset in Times
by Apex CoVantage, LLC

Contents

Preface

This book is the culmination of years of dedicated research, scholarship, and passion for the field of Thermal Infrared Remote Sensing. It was primarily authored by my esteemed advisor, Dr. Ruiliang Pu, a distinguished scholar and professor at the University of South Florida, School of Geosciences. Tragically, Dr. Pu passed away in 2023, just before completing the final touches on this manuscript. It is with deep respect, admiration, and a profound sense of responsibility that I have undertaken the task of bringing his work to completion.

Dr. Pu was a pioneer in thermal infrared remote sensing, and his contributions to the field were both foundational and transformative. Throughout his career, he was deeply committed to advancing our understanding of remote sensing applications, from urban heat island analysis, environmental monitoring, and land surface temperature retrieval. His ability to bridge theoretical principles with real-world applications made his research invaluable to both academia and industry.

This book is structured to provide a comprehensive exploration of Thermal Infrared Remote Sensing, covering its principles, methodologies, sensor technologies, and applications across various scientific domains. Beginning with an overview of fundamental concepts, the chapters delve into radiative properties of materials, atmospheric corrections, land surface temperature retrieval, and downscaling techniques, among other key topics. The later chapters highlight its applications in hydrology, biomass burning, urban climate studies, and environmental monitoring, reflecting the wide-reaching impact of thermal remote sensing technology.

Completing this book posthumously has been both an honor and a challenge. While I have remained as faithful as possible to Dr. Pu's vision, certain editorial adjustments were necessary to ensure clarity, coherence, and completeness. I extend my heartfelt gratitude to the colleagues, peers, and students who contributed their expertise, insights, and encouragement throughout this process. Their support has been instrumental in finalizing this book and preserving the legacy of an exceptional scholar.

It is my sincere hope that this book will serve as both a tribute to Dr. Ruiliang Pu's enduring contributions and a valuable resource for students, researchers, and professionals in remote sensing and related fields. May it inspire new explorations, discoveries, and innovations in the ever-evolving landscape of thermal infrared remote sensing.

With deep respect and gratitude,

Qiuyan Yu, PhD
Exponent Inc.
2/9/2025

About the Authors

Ruiliang Pu was Professor at the School of Geosciences at the University of South Florida (USF). He received his MSc (forest management) from Nanjing Forestry University (NFU), China in 1985, and earned a PhD (cartography and geographic information systems) in 2000, conducted at University of California (UC) at Berkeley, from Chinese Academy of Sciences. He has published more than 138 journal papers, book chapters, and segments in English. He also published more than 26 journal papers in Chinese. In addition, he published a *Hyperspectral Remote Sensing: Fundamentals and Practices* (2017, CRC Press, Taylor & Francis Group), which was translated into Chinese and published in 2020 by Higher Education Press, Beijing, China. He has also co-written two books in Chinese with his Chinese colleagues. His publications have appeared in top- or higher-tier journals.

Qiuyan Yu is Senior Scientist at Exponent Inc. in the biological and ecological sciences practice. Before joining Exponent, she served as Research Assistant Professor at New Mexico State University. She earned her PhD in geography and environmental science and policy from the University of South Florida in 2018 under the mentorship of Dr. Ruiliang Pu and also holds an MSc in cartography and geographic information systems from Beijing Normal University, China, as well as a BSc in geographic information systems from Lanzhou University, China. Dr. Yu has led NASA-funded research projects, served on NASA and NSF review panels, and published more than 45 journal articles. She has contributed to over 100 industry-focused projects spanning environmental science, ecology, civil engineering, and mechanical engineering, applying remote sensing, geographic information systems, data science, machine learning, and artificial intelligence-based modelling. Beyond her scientific work, Dr. Yu remains dedicated to education, having founded platforms that promote scientific innovation and science education.

1 Overview of Thermal Infrared Remote Sensing

Thermal infrared (TIR) remote sensing, while less explored than optical and microwave counterparts, offers unique insights into the thermal properties of the Earth's surface. This chapter delves into the fundamentals of TIR energy and remote sensing, contrasting it with optical and microwave techniques. We trace the historical evolution of TIRRS, highlighting key missions and sensors across three distinct stages of development. A comprehensive overview of TIRRS applications in diverse fields is provided, followed by a forward-looking perspective on its future potential. By understanding the distinct advantages and applications of TIRRS, we aim to stimulate further research and innovation in this promising field.

1.1 CONCEPTS OF TIR REMOTE SENSING

1.1.1 TIR Energy and Radiation

Any objects on the Earth's surface and in the atmosphere (e.g., rock, soil, vegetation, water, clouds, and even people) at a temperature above absolute zero in degrees Kelvin (i.e., 0 K; −273.16°C; −459.69°F) can emit thermal radiation due to kinetic energy of molecules within objects. The electromagnetic radiation emitted or reflected from an object is called radiant flux while the concentration of the amount of the radiant flux exiting an object is its radiant temperature (Jensen, 2007). Therefore, all the features we encounter on the Earth's surface and in the atmosphere on a typical day enable emitting or reflecting TIR electromagnetic radiation.

Thermal energy or radiation refers to electromagnetic waves between the 3 and 20 μm portion of the spectrum (Figure 1.1). Most thermal remote sensing applications make use of the 8 to 14 μm thermal spectral range. Human eyes cannot identify differences of objects in TIR spectral range because they are only sensitive to visible light from 0.4 to 0.7 μm, not shortwave infrared from 0.7 to 3.0 μm or TIR energy (3–14 μm). The Earth's surface and atmosphere radiate TIR energy outward owing to heating by solar irradiation and by Earth's internal heat flow. The Earth's internal energy is originally absorbing solar shortwave energy and then converting longwave energy. By measuring and recording such emitted TIR radiation in parts of the TIR region of the spectrum, TIR sensors/systems can produce very informative data that provide both distinctive signatures and/or indirect indications of TIR properties of Earth's surface materials and atmospheric components. An object's internal kinetic heat enables to convert to radiant energy, often called external or apparent energy, which allows us to measure it by using TIRRS sensors. The radiant energy acquired by the TIRRS usually has a high positive correlation with the object's true kinetic temperature. This is a basis for TIRRS enabling one to detect and map the TIR properties of objects on Earth's surface and in the atmosphere.

DOI: 10.1201/9781003276173-1

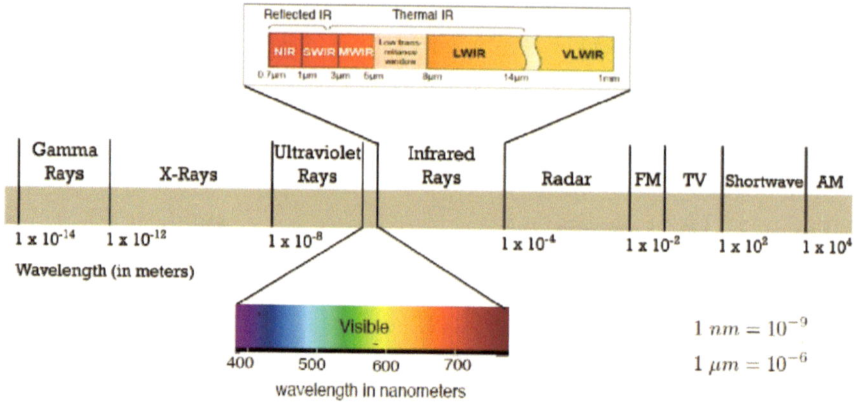

FIGURE 1.1 Electromagnetic spectrum (modified from Kaya, 2014).

1.1.2 TIR Imaging and Acquisition

Thermal imagery is produced by imaging sensors and their recording the TIR energy through its radiation emitted or reflected from objects that are sensed (Figure 1.2). The TIR radiation measurement or imaging obeys Planck's function, which addresses the relationship between the object (blackbody) absolute temperature and its spectral radiant exitance. A blackbody is an object that is a perfect emitter and absorber. Theoretically, three physical laws can be used to describe the behaviors of the TIR radiation: Stefan-Boltzmann law, Wien's displacement law, and Kirchoff's radiation law. The Stefan-Boltzmann law actually is related to the Planck's function and is used to calculate the maximum spectral radiant exitance with a known blackbody temperature; the Wien's displacement law describes the relationship between a true temperature of a blackbody in degree Kelvin and its peak spectral radiant exitance or dominant wavelength; and Kirchoff's law addresses, per the infrared portion of the spectrum, that the spectral emissivity of an object generally equals its spectral absorptance. However, no objects in nature are true blackbodies but selective bodies (Jensen, 2007), such as vegetation, soil, rock, water body, etc. Therefore, when applying the three physical laws to characterize the TIR radiation, an emission efficiency factor, called emissivity (ε), should be considered. The ε is generally a function of wavelength and surface conditions of objects, and it is a factor that denotes how efficiently an object radiates energy compared to a blackbody.

The TIR imagery can be acquired by two types of remote sensing sensors onboard airborne and space-borne platforms: across-track thermal scanners and pushbroom linear and area-array charge-coupled-device (CCD) detectors (Jensen, 2007). The thermal radiation scanner is an optical-electronic scanning system. It works on the principle of direct interaction between photons of radiation incident on them and the energy levels of electrical charge carriers within the detector material. Earlier working models directly recoded thermal energy on the film to form an image, but the modern scanning systems are capable of imaging temperature resolution on an order of 0.1°C. Currently, most thermal image data are collected with a single detector and

FIGURE 1.2 A thermal image sample (https://www.animations.physics.unsw.edu.au/jw/thermal-radiation.htm, accessed on May 2, 2022).

a scanning mirror and also with making use of both linear and area-array multiple CCD detectors to acquire thermal imagery. The individual CCD detectors also work on the same principle as that a single detector operating for the across-track thermal scanner. However, since the CCD detectors can view the ground resolution element for a longer time, each of which is fixed in an unchanged position relative to other elements, the quality of the TIR imagery acquired is improved significantly in terms of both radiometric and geometric resolutions.

1.1.3 TIR Remote Sensing

TIR remote sensors can measure TIR energy emitted from the Earth's surface and atmosphere in single wavelength bands (broadband), tens of bands (multiband), and in hundreds of bands (hyperspectral). Today, broadband surface temperature mapping is mostly done with airborne and satellite sensors. Thermal inertia mapping is done using broadband measurements taken during the day and night. Emissivity mapping is done using multiband and hyperspectral sensors, which are capable of measuring small changes in radiant flus emitted from objects. Figure 1.3 shows a comparison between daytime and nighttime TIR images. Currently, most popularly used airborne and satellite TIR sensors include the Thermal Infrared Multispectral Scanner (TIMS); the Moderate Resolution Imaging Spectroradiometer (MODIS); Advanced Very High-Resolution Radiometer (AVHRR); Advanced Spaceborne Thermal Emission and Reflection Radiometer (ASTER); Landsat series with TIR band/TM, ETM+, and TIRS; Geostationary Operational Environmental Satellite (Imager-GOES); and Meteosat Second Generation (SEVIRI-MSG), etc. The goal of TIRRS is to obtain the thermal spectral energy for each pixel in an image of a scene, with the purpose of finding TIR properties of objects, identifying materials, or detecting processes. Thermal data prove to be complementary to other remote sensing data: optical and microwave remote sensing data. Since TIRRS is developed for improving identification of materials and quantitatively determining thermal

Day Night

FIGURE 1.3 A sample of comparison between daytime and nighttime of TIR images. Day image: contrasts between heated buildings and streets and areas in shadow create a scene that resembles an aerial photo. Night image: many streets are evident because, being asphalt paved, they absorb more heat and remain warmer through the night (courtesy D. Quattrochi and J. Luvall, 1999, Project ATLANTA; NASA Marshall Space Flight Center, Global Hydrology Climate Center, Huntsville, AL).

properties on targets of interest, such as minerals, water body, vegetation, soils, and man-made materials, the TIRRS technology is most interested in remote sensing as a tool for many application fields, such as hydrology, plant ecology, limnology, pedology, climatology, and atmospheric sciences.

Although the TIRRS technology has many advantages over optical and microwave remote sensing, such as acquiring nighttime thermal imagery, there are several challenges in TIRRS. For example, in TIRRS, we must be care of interpreting interfaces that are boundaries between different features/objects, which, sometimes, are not clear as on optical and microwave images depending on objects' temperature and the temperature difference between the two neighbor objects. Usually, an object with high temperature in a thermal image may visually move its 'boundary' outgoing. Thermal energy transports often in three modes: convection, conduction, and radiation. TIRRS only measures and records the thermal energy in a radiation mode. Indirect measurements of thermal radiance are useful for estimating the temperature of complex surfaces, such as soils, rocks, and vegetation canopies. The 'surface' temperature is determined by how homogeneous surface materials and, per TIRRS, how homogeneous a surface is within a field-of-view (IFOV) in some finite wavelength band. Norman and Becker (1995) pointed that, per homogeneous surfaces in thermal equilibrium, the surface temperature in an IFOV area can be sensed and is not ambiguous, while, per heterogeneous surfaces with uneven thermal distribution within an IFOV area, the determination of surface temperature in an IFOV is ambiguous and needs to be careful of analyzing thermal measurements. This is because the heterogeneous surface temperature within an IFOV area measured by a TIR sensor (only measuring thermal energy in a radiation mode) may be affected by an energy transport in a conduction and/or convection mode, such as a case of a bare soil surface and water surface mixed within an IFOV area. Therefore, a relationship between the temperature measured by a thermal sensor (radiometer) over a homogeneous surface and its kinetic temperature is different from that between the temperature measure by a thermal sensor over a heterogenous surface and its kinetic temperature, which need to be noted when we analyze the thermal image acquired over a heterogenous landscape.

1.1.4 Differences among Optical, TIR, and Microwave Remote Sensing

TIR remote sensing (TIRRS), as a form of remote sensing, concentrates on the use of TIR bands to sense temperature emitted from targets, and it is only the activity of sensing energy remotely and processing that energy into information in an image form. Comparing all three remote sensing categories: optical, TIR, and microwave remote sensing, there are five major differences among the three different remote sensing categories, which are summarized in Table 1.1. The following are brief summaries of their differences.

TABLE 1.1
Differences Among Optical, TIR, and Microwave Remote Sensing

Characteristics	Optical remote sensing	TIR remote sensing	Microwave remote sensing
Spectral range	Including visible: 0.4–0.7 μm, NIR: 0.7–1. 0 μm, & SWIR: 1.0–3. 0 μm	3–14 μm, typically 8–14 μm	Typically 1 cm–100 cm
Imaging and working mode	Imaging energy reflected from targets, photogrammetry, and optical-machine scanning	Imaging energy emitted from targets, across-track scanning and charge-coupled-device detectors (CCDs) linear and area-array pushbroom	Active remote sensing by self-emitting pulse and then receiving backscatter energy
Time and weather condition to acquire images	Clear sky and daytime	Day and nighttime	All weather conditions and day and nighttime
Image and data (interpretation)	Easy to interpret, analyze and understand	Not easy to interpret and target boundaries not clear and moving depending on targets' temperature	More complicated and complex and difficult to interpret and understand
Cost	Cost-effective as it is easy to carry and operate	Cost higher compared to optical remote sensing	High cost of obtaining data

First of all, the spectral wavelength ranges covered by the three categories of remote sensing are different (Figure 1.1). Optical remote sensing covers visible: 0.4–0.7 μm, NIR: 0.7–1. 0 μm, and SWIR: 1.0–3. 0 μm; TIRRS covers 3–14 μm, typically 8–14 μm; and microwave remote sensing typically operates within 1 cm–100 cm. Second, their imaging principles and working modes are different. The traditional optical remote sensing acquires imagery by measuring the solar energy reflected from targets (e.g., soil, vegetation, water, man-made materials, etc.) using photogrammetry (most for visible and NIR ranges) and optical-machine scanning (most for SWIR range); TIRRS images the TIR energy emitted from targets using across-track thermal scanners and pushbroom linear and area-array charge-coupled-device detectors (CCDs); the microwave remote sensing is active remote sensing by self-emitting pulse and then receiving backscatter energy to form RADAR imagery. Third, the imaging time and required sky and weather conditions are different. The optical remote sensing needs clear sky conditions and operates in the daytime only because it relies on, records, reflected energy of solar light from targets during the daytime; most TIRRS can operate at nighttime to measure the TIR energy emitted from targets, but it can also record reflected TIR energy from targets during the daytime; active microwave remote sensing can work during both day- and nighttime and under almost all weather conditions because the microwave energy/pulse has a capability to penetrate a thick cloud layer, dense vegetation canopy and surface soil, etc. Fourth, the image appearance and data interpretation are also different among the remote sensing categories. Acquiring data and imagery using optical remote sensing is smooth looking, and a pixel value on the image mainly denotes the reflection of material(s) covered by the pixel, so the image is relatively easy to interpret, analyze, and understand; a TIR image created by TIRRS is not easy to interpret and understand as boundaries between different targets are not clear and may be 'moved' depending on target's temperature; RADAR or microwave imagery is more complicated and complex and thus difficult to interpret and understand compared to optical and TIR images. Finally, the cost to collect imagery is very different among the three remote sensing categories. Acquiring optical data is cost-effective as it is easy to carry and operate; the cost of collecting TIR data, due to higher requirements for TIR sensors, is relatively higher than that using optical sensors; active microwave remote sensing is highly disadvantaged due to its high cost of obtaining data, and the use of passive sensors may mitigate the cost, so that most countries heavily rely on the passive microwave sensors.

1.2 A BRIEF HISTORY OF TIR REMOTE SENSING

Based on the history and characteristics of representative TIR instruments and sensors, a brief history of TIRRS, by referring to Jensen (2007), could be chronologically introduced in three rough development stages: Stage 1 (early), initially discovering TIR radiation and measurement (1738–1900); Stage 2 (middle), TIRRS technology preliminarily used in the military (World Wars I and II); Stage 3 (modern), TIRRS technology developed and used for both military and civilian purposes (after 1960). However, there was not an exact chronological timetable that could clearly be divided the three stages. It is just for convenience in reviewing the history and development stages of TIRRS.

Thermal remote sensing at its very early stage is related to an astronomer. Sir Frederick William Herschel (1738–1822) discovered the TIR portion of the electromagnetic spectrum in 1800, which was described in his famous paper entitled 'Investigations of the Powers of the Prismatic Colors to Heat and Illuminate Objects: with Remarks.' In 1800, he produced a spectrum of colors using a prism and measured different parts of the spectrum with a thermometer to discover the modern so-called TIR portion. However, accurate temperature at 1/10,000°C could not be measured until 1880 by S. P. Langley in 1889, who developed a device that could measure temperature variations with such an accuracy.

Primitive radiation thermometers were employed during World War I, denoted by S. O. Hoffman who could detect men and eventually aircraft at 120 m. In the 1930s, Germany developed the Kiel system for discriminating between bombers and night fighters. A single most important development in TIR surveillance technology was the development of detector elements by nations at war during World War II, including Great Britain and the United States. Detectors could convert incoming radiant temperature and thermal energy into an analog electrical signal, which was a vast progress in the history of TIRRS development. Early TIR detectors were lead sulfide photodetectors and were the first widely used detectors in industrial radiation thermography then. Now we have very fast detectors to replace them with mercury-doped germanium (Ge:Hg), indium antimonide (InSb) and other substances that are very responsive and sensitive to TIR radiation energy. Per the TIR technology for military application, very sensitive satellite and aircraft remote sensing systems and techniques were initially developed in the 1960s.

In 1968, the U.S. government declassified the production of TIR remote sensing systems and started to use them for civilians, but as with those prior to 1968, thermal data could only be collected at a very coarse resolution. The coarse resolution TIR data were ideal for monitoring regional cloud patterns and frontal movement. For example, the first civilian satellite remote sensing data (including TIR data at a very coarse resolution) were acquired for weather observation in 1960, which TIR sensor was onboard the U.S. Television IR Operational Satellite (TIROS), a weather satellite. Then, the U.S. National Aeronautics and Space Administration (NASA) launched the Heat Capacity Mapping Mission (HCMM) in 1978 for Earth remote sensing observations. HCMM obtained 600 m × 600 m spatial resolution TIR (10.5–12.6 μm) scenes during both day- and nighttime. This was one of the first scientifically oriented (geology) TIR systems. In the same year, NASA launched the Coastal Zone Color Scanner (CZCS), which included a TIR sensor, used for monitoring sea-surface temperature. Beginning with TIROS-N in 1978, the U.S. National Oceanic and Atmospheric Administration (NOAA) also began acquiring TIR data with the Advanced Very High-Resolution Radiometer (AVHRR) onboard the Polar Orbiting Environmental Satellites (POES), in five channels with typical TIR channels 3–5. The latest generations of the AVHRR sensor still collects TIR data at spatial resolutions of 1.1 km × 1.1 km local area coverage (LAC) data and of 4 km × 4 km global area coverage (GAC) data. In 1999, NASA's EOS (Earth Observing System) mission included sensor Moderate-resolution Imaging Spectroradiometer (MODIS) carried on Terra and Aqua (2002) satellites. The MODIS has two typical TIR channels 31 and 32 at spatial resolution of 1 km × 1km. At the modern stage, perhaps the most widely used TIR data at moderate resolution include those collected from satellites for Earth observations: Landsat series TIR data. The Landsats 4 and 5 carrying TM sensors were launched by NASA in 1982 and 1984 respectively and collected TIR data at 120 m × 120 m spatial resolution over the TIR spectral range 10.4–12.5 μm. Landsat 7 was launched in 1999, carrying the Enhanced Thematic Mapper Plus (ETM+) sensor with a TIR band spatial resolution increased to 60 m × 60 m. Landsat 8, launched in 2012, carried a TIRS sensor, with two TIR bands covering TIR spectral wavelengths, 10.60–11.19 μm, and 11.50–12.51 μm, at a spatial resolution 100 m × 100 m. In addition, NASA launched the Advanced Space-borne Thermal Emission and Reflection Radiometer (ASTER) on the Terra satellite. ASTER can collect multispectral TIR data at a 90 m × 90 m spatial resolution over five bands between 8.12 and 11.65 μm.

Geostationary orbits TIR sensors can provide TIR image data within the same region (dish area) of the Earth at high temporal resolution (e.g., 15 min) (Mao et al., 2021). They include the NOAA Geostationary Operational Environmental Satellite (GOES) carrying Imager beginning 1975, SEVIRI onboard MSG (Meteosat Second Generation) beginning 2005, and AGRI onboard Fengyun-4A in 2016, etc. They can provide TIR data at 3 km to 8 km resolutions.

At this stage, in addition to the development of satellite TIR sensors and systems, NASA has also developed a number of airborne TIR instruments and scanners that have been used for Earth science research. For example, the Thermal Infrared Multispectral Scanner (TIMS) sensor is a six-band instrument that collects TIR data within the thermal spectral range from 8.2 to 11.60 μm. NASA's Advanced Thermal and Land Applications Sensor (ATLAS) was an advanced version of TIMS, which covers the same six bands as TIMS but also includes nine spectral bands in visible and middle infrared reflectance ranges. The German Aerospace Center (DLR) developed a 79-channel Digital Airborne Imaging Spectrometer (DAIS 7915). This sensor covers the spectral range from the visible to the TIR spectral ranges. It has six TIR channels ranging from 8.5 to 12.3 μm and one channel at 4.3 μm. More recently, the DLR has developed the Airborne Reflective Emissive Spectrometer (ARES), which has 30 multiple TIR bands. Airborne TIR sensors have some advantages over satellite sensors. They may include—but are not limited to—that airborne TIR sensors can collect TIR data at different altitudes and thus are capable of providing TIR data at different spatial resolution to study specific Earth processes, such as monitoring geological phenomena, biological and physiological interactions, landscape and agricultural characteristics, forest energy exchanges, and a variety of energy flux interactions (Quattrochi and Luvall, 1999) and that the TIR data collected by airborne sensors may be used for different calibrations for developing satellite sensors and systems.

1.3 OVERVIEW OF TIR REMOTE SENSING APPLICATIONS

Based on properties of thermal imagery, since the advent of the modern TIRRS technology in 1960s, a variety of TIR sensors' data have been investigated and applied in different areas, and a lot of research and application results and findings have been published in thousands of journal articles and conference/workshops' proceedings, books, and special issues of remote sensing journals. An overview of literature in research and application studies of the TIR image data are given in the following sections in different areas including hydrology, biomass burning, urban thermal environments, and other disciples and areas. For a more detailed review of TIRRS applied in different disciples and areas, readers may read Chapters 7–10 in this book.

1.3.1 Hydrology

Remote sensing can provide a means of observing hydrological state variables over large areas, including retrieving LST, estimating surface soil moisture and evapotranspiration, and mapping ice/snow areas. For TIRRS applied hydrology, the application areas are estimating surface soil moisture, evapotranspiration, and mapping ice/snow areas.

1.3.1.1 Estimating Soil Moisture from TIR Remote Sensing Data

Based on the optical and thermal infrared remote sensing, many approaches have been developed by establishing relationships between soil moisture (SM) and soil reflectivity or surface temperature/vegetation coverage and soil thermal properties. SM estimation methods relying on optical and thermal remote sensing are collated and summarized as follows. (1) Optical remote sensing spectral/vegetation index-based methods; (2) triangle-trapezoid and thermal inertia methods; (3) machine learning (ML)-based methods in SM estimation. Spectral/vegetation-index-based methods are because some vegetation indices (VIs) are extremely sensitive to water stress, so they have been widely used to detect drought conditions, such as the normalized difference vegetation index (NDVI) and enhanced vegetation index (EVI).

Triangle-trapezoid-based methods are based on an interpretation of the image (pixel) distribution in surface radiant temperature (LST) to fractional vegetation cover (from NDVI) space. A triangle emerges because the range of surface radiant temperatures decrease as the vegetation cover increases; a vertex emerges showing the narrow range of surface radiant temperature over the dense vegetation (Carlson, 2007). Vegetation subject to that water stress usually experiences an increase in its radiant temperature compared to vegetation that is well watered. However, Moran et al. (1994) reported that the triangular space, in which the driest and wettest fully vegetated surfaces have the same LST, does not account for the impact of water stress on canopy transpiration and replaces the triangle with a trapezoid.

The application of thermal inertial-based methods is not straightforward as the knowledge of the physical properties of the target is required. So, the apparent thermal inertia (ATI) was proposed to make the determination of TI easier using the diurnal gradient of surface thermal responses (physical temperature) and thus can be derived directly from multi-spectral remotely sensed imagery (Xue and Cracknell, 1995). Cai et al. (2007) used MODIS images over the northern Chinese Plain and obtained a mean difference between measured and estimated soil moisture values of 4.3% over bare and sparsely vegetated fields. To infer reliable SM estimates derived from ATI, the temperature difference should be larger than 10°C.

Machine leaning (ML)-based methods offer some alternative approaches. Due to the high complexity of physical models, the popularity of ML for the remote-sensing based estimation of biophysical parameters has grown significantly over the last decade. The flexibility of ML approaches is further highlighted by their potential to be used in various hydro-meteorological applications, from the prediction of SM to precipitation, temperature, or wind. To address the complex nature of soil moisture, statistical learning theory tools such as Artificial Neural Networks (ANNs), Support Vector Machines (SVMs), and Relevance Vector Machines (RVMs) have been tested in recent years for their ability to spatially estimate distributed SM at the single-farm level. Examples of ML algorithms in spatial applications used in water resources and soil moisture applications can be found in Gill et al. (2007), Andris et al. (2013), Bachour et al. (2014), and Hassan-Esfahani et al. (2015).

1.3.1.2 Estimating Evapotranspiration from TIR Remote Sensing Data

As one of the most significant components, land surface evapotranspiration (ET—or latent heat flux, LE, in W/m^2) in the hydrological processes accounts for approximately 60% of the mean precipitation globally) and controls the water and energy transfer in the interface of land surface and atmosphere (Su, 2002; Oki and Kanae, 2006). Since the advent of Earth observation from space in 1957, there have been a large array of models developed to utilize satellite data for mapping ET across large spatial and temporal scales. Both empirical and physical models, which differ greatly in the model structure, input, and output, have been developed with varying degrees of complexity. Different combinations of the remote sensing spectral information from visible and NIR bands to mid and TIR bands can produce surface characteristic parameters that are indispensable to ET models. These parameters include NDVI, Leaf Area Index (LAI), fractional vegetation coverage (Fv), surface albedo, surface emissivity, and radiometric surface temperature (Mauser and Schädlich, 1998).

General methodologies of evapotranspiration retrieval can be roughly categorized into the (semi)empirical method and the analytical method. The (semi)empirical method is developed by establishing empirical regression functions of remote sensing observations and minimum ground-based measurements, while the analytical physics-based method mainly considers the physical processes at the scale of interest and generally requires a greater variety of direct and indirect measurements.

Physics-based methods for the one-source model include (1) Surface Energy Balance Index (SEBI), (2) Surface Energy Balance System (SEBS), (3) Simplified-Surface Energy Balance Index (S-SEBI), (4) Surface Energy Balance Algorithm for Land (SEBAL), and (5) Mapping Evapotranspiration at High Resolution and with Internalized Calibration (METRIC) models. This type of models makes use of LST observed by thermal sensors to estimate the sensible heat flux and to derive ET as a residual of the surface energy balance or to estimate the evaporative fraction and to derive ET from the available energy. They also include the Penman–Monteith (PM) evapotranspiration (ET) equation and the modified Priestley–Taylor method. The former is used to predict the rate of total evaporation and transpiration from the Earth's surface using common remote sensing data and meteorological data. Although the PM equation can in theory provide accurate ET estimates, it also has difficulty preparing the input variables for the parameterization of aerodynamic resistance and stomatal resistance, especially when applied globally. To overcome the problems encountered in the PM equation, Fisher et al. (2008) adapted the Priestley–Taylor (PT) equation to soil water stress conditions by exerting plant physiological limitations and soil drought constraint. This modified PT equation has been used by Fisher et al. (2008) to produce global LE at a monthly temporal scale and a 1° spatial scale during the years 1986–1993 using datasets from ISLSCP-II and AVHRR.

1.3.1.3 Mapping Ice/Snow Areas With TIR Remote Sensing Imagery

Ice/snow cover can be detected and monitored with a variety of remote sensing technologies. The greatest number of applications have been found in the VNIR region and TIR range of the electromagnetic spectrum. Many studies with optical and TIR remote sensing derived spectral/vegetation indices and LST conducted projects on mapping ice/snow areas via vegetation index thresholding, supervised classification, object-based image analysis, and LST and derived apparent thermal inertia approaches.

Lee and Choi (2022) proposed a daily snow cover mapping algorithm based on Geostationary-Korean Multi-Purpose Satellite-2A multitemporal images; a threshold with consideration for the characteristics of a geostationary satellite is applied to the normalized difference snow index (NDSI); and a daily snow cover map is generated utilizing a stacking process. Recently, a few studies reported the mapping of the Indian Himalayan glaciers using HSR data using object-based image analysis (OBIA), which include Mitkari et al. (2017), Jawak et al. (2018, 2019). The brief review suggests that researchers tried to evaluate the potential of OBIA for glacier cover/glacier facies mapping using the spectral properties. By considering mapping of the actual glacier area, in which the most significant parameter differentiating the debris covers is the surface temperature, Karimi et al. (2012) proposed an approach to distinguish the supraglacial and periglacial debris and outline the debris-covered glacier boundary by means of optical (Worldview-2 satellite imagery) and thermal (Landsat-TM sensor) remote sensing data. After collecting GCPs from different member of glacier in field operation, a Neural Network (NN) classification has been used to delineate the actual area of the glacier. Overall accuracy and Kappa coefficient with values of 92.9% and 0.84 respectively demonstrated the high efficiency of thermal–optical approach to delineate the actual area of this alpine glacier. In mapping rock glaciers and debris-covered glaciers, Brenning et al. (2012) explored the utility of remotely-sensed land surface temperature (LST) and derived apparent thermal inertia (ATI) for thermally characterizing and discriminating periglacial mountain environments, particularly rock glaciers and debris-covered glaciers. In their case study, they demonstrated this approach is useful by utilizing ASTER daytime and nighttime data in a study area in the Andes of Central Chile.

1.3.2 Biomass Burning

1.3.2.1 Active Fire Detection

Biomass burning impacts atmospheric chemistry, composition, climate, vegetation ecosystems, and, of course, human activities. Remote sensing can provide tools for observing biomass burning status spatiotemporally. Thermal remote sensing is used widely to map fire extents, examine fire regimes, characterize fire impacts, and estimate and characterize the chemical composition of fire emissions (Coheur et al., 2009; Wooster et al., 2013). During the last several decades, researchers have developed many active fire detection algorithms. There are general two types of active fire detection algorithms: fixed threshold algorithms and background contextual algorithms.

Fixed threshold algorithms are based on measuring the emitted energy for a pixel in either a single channel or over multiple channels and applying an empirical value (an absolute (or regional) maximum), and any pixel above a given threshold temperature is defined as containing a fire. Fixed threshold algorithms are simple to apply and use, but differences in fire characteristics among regions in the world lead to problems of false alarms and/or missed fires using this method. It limited in use to local and regional scales and only to the particular season under study. Thresholds that have been developed and optimized for one location—such as woodland or savannah, and time—such as summer or dry/wet season, when applied to other operating environments can lead to significant errors of omission or commission.

Contextual algorithms use adaptive thresholds for fire detection. Multispectral criteria are aimed at detecting the difference between a fire pixel (active fire) and the background temperature (environmental temperature in the proximity of the fire pixel) (Flasse and Ceccato, 1996; Giglio et al., 2003). These approaches generally have two pathways by which pixels containing actively burning fires can be identified. The first 'fixed threshold' pathway may use a single algorithm stage and is designed to detect pixels unambiguously containing large and/or intensely burning fires. The second 'contextual' pathway typically consists of two or more stages, whereby potential fire pixels (PFPs) are first identified using a fixed threshold algorithms based on relatively low thresholds (thus selecting many non-fire pixels as well as true fires), and then testing each PFP against the statistical properties of its immediately surrounding 'ambient background' pixels in order to confirm whether or not it is a 'true' fire pixel. The entire algorithm, incorporating the 'two pathways' procedure, is generally termed a 'contextual active fire detection algorithm.' Contextual algorithms have the advantage of making the detection process independent from the season and the zone analyzed since the thresholds are obtained by means of a statistical analysis of the environment. However, they have a serious drawback when they are applied to images in which the clouds have not been filtered since cloud edges cause false alarms. For the contextual algorithms, the selection of initial thresholds for identifying potential wildfire pixels is sensitive. Since the severity of wildfire varies considerably with fuel type, fuel amount, and weather conditions, the thresholds should also be contingent upon these variables. As a first approximation, one may use different thresholds over different land cover types in different seasons or set the threshold as a function of a vegetation index.

1.3.2.2 Burning Scar Mapping

Burned area (BA) estimates provide accurate spatial representations of wildfire extent and perimeters. Hotspots have been typically used as seeds or indicators of burn scars in various mapping approaches taking advantage of region growing algorithms. Active fire detection can also be used to locate training samples to define spectral characteristics for BAs (Roy et al., 2005). Active fire data have also directly been used as the basis of a global BA estimation method correlating active fire detection with BA mapping results (Giglio et al., 2006). In terms of approaches to exploiting thermal band data in BA mapping algorithms, probably the most successful has been the inclusion of active fire detections into BA data processing chains. Specifically, the locations of detected active fire pixels can very usefully act as 'seed locations' for optical waveband change detection methods aimed at identifying newly BAs from optical wavelength data. For example, the Hotspot And NDVI Differencing Synergy (HANDS) algorithm, used to map BAs across the Canadian boreal forest and North America is using a combination of hotspot with NDVI differencing to map BAs.

1.3.2.3 Fire Radiation Power Retrieval

Once detection of a fire pixel has been unambiguously confirmed, the fire activity contained within it can be characterized in terms of its fire radiation power (FRP), a measure of the rate of radiant energy emission from the fire integrated over all wavelengths and directions. Wooster et al. (2003, 2005), Ichoku and Kaufman (2005), and Freeborn et al. (2008) have provided a full description of how this measurement can be derived from satellite imagery and how it relates to rates of combustion and smoke emission.

There are two approaches that are generally used for measuring FRP, (1) the bi-spectral method coupled to the Stefan-Boltzmann Law, which employs the differential fire signal in two or more co-registered spectral channels (Zhukov et al., 2006) and (2) single band approaches that use the fire signal in the MIR channel only (Wooster et al., 2003). Bi-spectral or multi-spectral methods appear most appropriate for application to higher spatial resolution data (Wooster et al., 2003; Zhukov et al., 2005, 2006; Dennison et al., 2006), and their use with coarser spatial resolution imagery can be affected by difficulties in separating the fire signal from the background signal in channels and by interchannel spatial misregistration effects (Giglio and Kendall, 2001; Giglio and Justice, 2003). The single band approach or MIR radiance method of FRP estimation relies on quantifying the fire signal in the MIR only, where its strength is maximized (Wooster et al., 2003, 2005).

1.3.2.4 Smoking Plume Monitoring

Biomass burning emits a large amount of greenhouse gases and aerosol particles into the atmosphere. The commonly used methods to monitor smoking plumes may be classified into three categories: RGB false color composite image, multi-threshold, and machine-learning (ML) methods. The following is a brief introduction to the three categories of methods.

RGB false color composite image is the most common identifying smoke method that is to assign different colors to different channels or channel combinations (Randriambelo et al., 1998). The resulting false-color images can provide visual separation of smoke from other objects.

The multi-channel threshold uses more spectral information, multi-thresholds concerning the linear relationship among multiple bands to classify smoke features. Despite balancing computational efficiency and accuracy, these threshold methods require prior knowledge of specific bands and the thresholds among regions for smoke detection. Therefore, these methods are difficult to apply to other satellite datasets. Researchers have begun to explore simplified and adaptable approaches to detect smoke plumes.

Machine learning (ML) algorithms are designed to fit the complicated linear or non-linear relationships between input and output and find the optimal function to detect smoke. The ML algorithms have the capability to learn patterns whose complexity makes them difficult to analyze using other conventional approaches. In order to get satisfactory smoke plumes detection results, Li et al. (2013) proposed a smoke detection algorithm that uses a multi-threshold method for extracting training sample sets to train back-propagation neural network (BPNN) classification for smoke plumes detection. Their results indicated that the overall accuracy and Kappa coefficient 97.63% and 96.29%, respectively. The results of all cases are relatively satisfactory when compared to true-color composition images.

1.3.3 Urban Thermal Environments

Urbanization and global climate change have changed urban thermal environments significantly. Thermal remote sensing can provide tools of observing urban thermal environmental properties spatiotemporally, such as retrieved land surface temperature (LST) and urban biophysical parameters in evaluating urban heat islands (UHIs). Urban optical and TIR remote sensing data can be used to explore the interaction between urban environmental process and human activities. Various methods have been employed to analyze the mechanism of UHI, such as the comparison of statistical index, correlation, and regression methods. Most studies focused on biophysical and meteorological factors, such as vegetation coverage (Weng et al., 2004), impervious surface area (Weng et al., 2006), land cover characteristics (Hart and Sailor, 2009; Weng et al., 2007), and ecological setting (Imhoff et al., 2010). Social-economic factors have also been emphasized, such as pavement area, building area (Weng et al., 2008), building and population density (Chen et al., 2006), building energy use (Zhou et al., 2012), etc.

1.3.3.1 Factors Affecting UHI Intensity

LST is a major measure to evaluate UHI intensity. Satellite-derived surface temperature data (LST) have been utilized for urban climate and thermal environment analyses. Land surface emissivity (LSE) is a key parameter to retrieve LST from TIR satellite sensors. Socioeconomic factors including population, GDP, and night-light satellite data may be useful to evaluate UHI intensity. Climate/topographic factors (e.g., precipitation, temperature, wind speed, slope, aspect, elevation, and solar irradiance) are highly variable in the city environments to affect the UHI intensity. Vegetation abundance in the city areas describes the amount and nature of vegetation cover, and density and modulates the proportions of vegetation vs. non-vegetated ground (e.g., bare soil) visible to a sensor (Weng, 2009). The differences in radiative temperature between the vegetation canopy and the ground affect the measurement of LST (Sandholt et al., 2002). For any surface material, certain internal properties (soil properties), such as heat capacity, thermal conductivity, and inertia play important roles in controlling the temperature of a body at equilibrium with its surroundings. These thermal properties vary with soil type and its moisture content. Urban geometry is also another leading factor to UHI with canyon geometry as a result of narrow streets and tall buildings, which intercept air movement. The geometry also contributes to this influence through wind flow. The changes in land use/cover (LULC) pattern in relation to urbanization also impact the intensity and spatial pattern of the UHI effect (Chen et al., 2006). This has been affirmed with different land uses influencing UHI differently as a result of the surface properties (Tran et al., 2006). All those factors can be directly or indirectly extracted or retrieved from various TIR sensors' data.

1.3.3.2 Methods and Techniques of Evaluating UHI Intensity

UHI intensity calculation including empirical direct methods where TIR remotely sensed data are introduced directly in semi-empirical models to estimate LST is the simplified relationship between TIR remotely sensed and meteorological data (Courault et al., 2005). This method allows for the characterization of UHI intensity both at the local scale—using ground measurements—and over large areas, using satellite data, by calculating LST differences (Li et al., 2013). Many studies have documented the use of LST data to examine meso-scale temperature differences between urban and rural areas in cities worldwide to compare the difference between UHI areas and non-urban areas. The LST of UHI is generally higher than the surrounding rural areas and has a strong correlation with land cover. UHI intensity analysis is the most common method to compute the magnitude and extent of UHI by evaluating the LST difference between urban and surrounding non-urban areas (Gallo and Owen, 1998). These analyses are often supported with land surface information, such as land cover and impervious surface area (ISA). Many investigations have compared UHI to ecological indices, vegetation fraction, and percent ISA, finding strong correlations with mean LST by using urban ecological indices approaches. Statistical and machine learning (ML) models have also been proposed to measure UHI (Rajasekar and Weng, 2009). Among these studies, a Gaussian surface model has been utilized the most because it can provide not only the intensity but also the spatial extent and the central location of the UHI (Shi et al., 2021). The kernel convolution method has also been proposed to study UHI effects because of its high efficiency in characterizing the temperature values over space in a continuous surface (Rajasekar and Weng, 2009). The ML method has become popular in UHI but its use has remained restricted to predicting the natural world. ML techniques may not be the solution to all the problems remotely sensed data might have. However, these techniques indeed provide a powerful set of tools that deserve serious attention to deal with some relevant UHI remotely sensed data problems.

1.3.4 Other Areas

1.3.4.1 Mapping Geothermal Alteration and Mineralization Sites

Geothermal systems exist in regions of anomalously high crustal heat flow that may be associated with the occurrence of young igneous bodies or hot rocks located deeper in the crust. This high geothermal heat is normally transferred to the surface by the convection of ground waters that forms hydrothermal systems. Thermal infrared (TIR) remote sensing provides data with synoptic coverage for investigating the surface manifestations of geothermal systems as applied to both geothermal energy exploration and also for more fundamental research and monitoring. TIR remote sensing provides a method for rapid mapping and quantifying surface geothermal features in support of exploration and assessment of new resources. The use of TIR data to map and quantify surface geothermal features also supports long term monitoring of magmatic-related and/or protected systems. General techniques include a RGB false-color-composite technique that is extensively used for combining multispectral images through color coding, band ratioing techniques to reduce the effects of topography and enhancing the spectral differences between bands, and image transforms that are often commonly used to reduce the dimensionality of the input dataset and processing time focus processing on the information of interest within the input file and removing noise.

Identifying and classifying techniques/methods of geothermal alteration and mines that may include machine learning (ML) and deep learning (DL) algorithms as well as fuzzy logic modeling methods are briefly introduced, reviewed, and discussed. DL methods are a branch of ML methods that seeks to model high-level abstractions in data by applying multiple processing layers with complex structures. DL methods such as recurrent neural networks (RNNs), convolutional neural networks (CNNs), autoencoders, deep belief networks, and restricted Boltzmann machines have been successfully applied to transform different fields. Fuzzy logic modeling has been successfully applied for mineral prospectivity mapping in metallogenic provinces. Various factors such as host rock, structures, geophysical anomalies, remote sensing studies, and documented fieldworks can be considered information layers (fuzzy members) for mineral exploration.

1.3.4.2 Thermal Remote Sensing of Active Volcanoes

Observations from airborne and space-borne in the TIR region of the spectrum are useful when examining volcanic activity. This is because volcanic activity consists of the transfer of heat from within the Earth to the surface and observations of this heat frequently associated directly with the activity of a volcano at a specific time. The active volcano observed and expressed in the thermal signature might be a fresh lava flow or an active lava lake or may be more subtle, such as a degassing surface or warmed crater-lake. Volcanic hotspot detection algorithms and methods/techniques of mapping and classifying volcanos as well as challenges and future directions are briefly introduced here.

Volcanic hotspot detection algorithms depend on whether a hot spot detection technique operates by assessing a pixel's radiance (or temperature) in a spectral, spatial, or temporal space; hot spot detection algorithms can be divided into three categories: (1) fixed threshold, (2) contextual, and (3) temporal as follows.

(1) Spectral character (fixed threshold algorithms). That is, does the pixel have spectral properties, i.e., a sufficiently high brightness temperature, ΔT and/or NTI, where NTI is the normalized difference (ΔT) thermal index of T_{MIR} and T_{TIR} that are temperatures at MIR and TIR bands, respectively? This class of algorithms tends to apply to single or multiple but essentially fixed thresholds to determine whether a pixel is hot or not.
(2) Spatial context (contextual algorithms). That is, is the pixel brightness significantly different from that of its surrounding pixels? This class of algorithms assesses whether a pixel is thermally anomalous in a spatial sense (area).

(3) Temporal context (temporal algorithms). In other words, is the pixel brightness significantly different from that of its preceding history? Thus, this class of algorithms assesses whether a pixel is thermally anomalous in a temporal sense.

Mapping and classification of volcanos is needed to understand volcano physical type and property. Among these mapping techniques/methods, machine learning (ML) algorithms including deep learning (DL) algorithms have been widely used to automatically process remote sensing data in volcanic applications. Contemporary image recognition techniques based on convolutional neural network (CNN), as a widely used DL algorithm to classify images exploring spatially discriminative shape/texture features, are especially efficient at automatically classifying heat sources semantically on a global scale (Ma et al., 2019). The CNN-based classification could potentially be an alternative to visual inspection by human experts. The ML methods provide excellent results in terms of accuracy, precision, and sensitivity.

1.3.4.3 Thermal Remote Sensing of Plant Ecosystems

Thermal remote sensing can be used for mapping terrestrial ecosystem structure, including plant species and leaf area index and monitoring terrestrial ecosystem function, including monitoring water stress and ecosystem thermodynamics, etc.

The biological factors affecting leaf/canopy energy balance and ultimately leaf/canopy temperature, including transpiration, leaf angle, pubescence, leaf size, leaf shape and canopy position, number of leaves per stem, and canopy structure: canopy density and height vary within and across plant species and ambient conditions (Smith and Carter, 1988). The ecosystem structure variables may include plant species and composition, leaf area index (LAI), canopy density and height, etc. Studies concluded that due to the unique composition and structure of epidermal plant cells of different plant species (e.g., cuticular waxes), the reflectance peaks locate different positions between the wavelengths 8–14 μm (Salisbury and Milton, 1988). The results of Neinavaz et al. (2016a) showed that LAI might be successfully retrieved at the high value of approximately 5.5 when TIR hyperspectral data were used. The results also clearly showed that thermal image analysis can be applied as a nondestructive, rapid technique to characterize the temperature of crop canopy and then to estimate the LAI of wheat grown under different moisture stress conditions.

TIR data have been widely used for deriving canopy temperature as an indicator of water stress, canopy conductance, and transpiration for irrigation scheduling. The terrestrial ecosystem functions that may be characterized with TIR data may related to water stress, net primary product, and ecosystem thermodynamics, etc. Several temperature-based indices have been developed to normalize leaf temperatures to current environmental conditions for a more quantitative estimation of water stress. The normalized relative canopy temperature is associated closely and significantly with canopy water content and relative water content under arid and semi-arid conditions using hyperspectral VNIR/SWIR data and thermal imaging data.

Thermal remote sensing can provide environmental measuring means with capabilities for measuring ecosystem development and integrity. Recent advances in applying principles of nonequilibrium thermodynamics to ecology provide fundamental insights into energy partitioning in ecosystems (Quattrochi and Luvall, 2009). More developed terrestrial ecosystems will be more effective at dissipating the solar energy.

1.4 PERSPECTIVE OF TIR REMOTE SENSING

Given the great potential applications of TIR remote sensing as an overview earlier, it is necessary to develop new generations of—or continue using existing—TIR remote sensing sensors and missions, particularly space-borne TIR sensors and/or missions. At present, existing polar orbit moderate resolution and low (coarse)-resolution satellite TIR sensors/missions can continue to be used for research and applications in the future. The moderate resolution sensors/missions may include those onboard the Landsat series and Terra satellite with spatial resolution better than a few 100 m. For example, the TM onboard Landsats 4/5, the ETM+ onboard Landsat 7, and the Thermal Infrared Sensor (TIRS) onboard Landsat 8 measure land surface thermal energy in one or two thermal bands, while the ASTER sensor carried by Terra satellite senses land surface TIR energy in five multispectral thermal bands. The coarse resolution (~1,000 m) polar orbit satellite TIR sensors/missions usually have a typical high temporal resolution (0.5–2 days), including the NASA-MODIS sensor onboard the Terra/Aqua satellites and the NOAA-AVHRR. Other polar orbits coarse resolution TIR sensors also include the European Space Agency's Along-Track Scanning Radiometer (ATSR), Advanced ATSR (AATSR), and the Sea and Land Surface Temperature Radiometer (SLSTR) onboard Sentinel-3(A/B) and the Visible and Infrared Radiometer (VIRR) onboard Fengyun-3 (B/C). In addition to those polar orbit moderate and coarse resolution TIR sensors and missions, currently operating GOSE satellite TIR sensors and missions also can continue to be used. Geostationary orbits TIR sensors can provide TIR image data within the same region (dish area) of the Earth at high temporal resolution (e.g., 15 min) (Mao et al., 2021). They mainly include Imager onboard GOES, SEVIRI onboard MSG, and AGRI onboard Fengyun-4A, etc.

At present, the future of satellite TIR remote sensing is potentially promising, albeit somewhat uncertain. The new generations of satellite TIR sensors/missions are at a planned or concept stage. They mainly include ones operated by USA NASA/NOAA and the European Space Agency (ESA). In the USA, NASA owns the Hyperspectral Infrared Imager (HyspIRI) and the ECOsystem Spaceborne Thermal Radiometer Experiment on Space Station (ECOSTRESS) onboard the International Space Station (ISS). USA NASA and NOAA join to develop the Visible Infrared Imager Radiometer Suite (VIIRS) onboard the Joint Polar-orbiting Satellite System 3 (JPSS-3) satellite. In the Europe, ESA owns the Land Surface Temperature Monitoring (LSTM) onboard Sentinel-8, and Spectrometer and Broadband Imager sensors onboard High-resolution Temperature and Spectral Emissivity Mapping (HiTeSEM). Table 1.2 summarizes the spatial and spectral resolutions, TIR spectral wavelength covered by the TIR sensors, and number of TIR bands of the six future satellite TIR sensors/missions.

Guided by National Imperatives for the Next Decade and Beyond (NRC, 2007) for NASA for planning future Earth science remote sensing satellite instruments, the

TABLE 1.2
Future (Planned or Concept) Spaceborne TIR Sensors/Missions*

(Satellite)/ sensor or mission	(HyspIRI)/ HyspIRI	(JPSS-3)/ VIIRS	(Sentinel-8)/ LSTM	(HiTeSEM)/ Spectrometer	(HiTeSEM)/ Broadband Imager	(ISS)/ ECOSTRESS
Country	USA	USA	ESA	ESA	ESA	USA
GSD (m)	60	375, 750	30–50	60	20	40–60
Wavelength coverage (μm)	3.0–13.0	3.7–12.0	8.6–12.0	7.2–12.5	8.00–10.25, 10.25–12.50	8.0–12.5
Number of bands (TIR)	8	7	5	30–72	2	5
Table 1.2 Future (Planned or Concept) Spaceborne TIR Sensors/Missions*			0.018	0.06	-	0.9
Reference	JPL, 2018	(McIntire et al., 2022)	(Koetz et al., 2018)	(Udelhoven et al., 2017)	(Udelhoven et al., 2017)	(Stavros et al., 2017)

* Some abbreviation of satellites/sensors/missions can be seen in the text.VIIRS: Visible Infrared Imager Radiometer Suite
ESA: European Space Agency
Sentinel-8/LSTM (Land Surface Temperature Monitoring)
ISS: International Space Station
HiTeSEM: High-resolution Temperature and Spectral Emissivity Mapping
ECOsystem Spaceborne Thermal Radiometer Experiment on Space Station: ECOSTRESS)

HyspIRI mission has its heritage in the hyperspectral sensor Hyperion onboard NASA EO-1 satellite launched in 2000 and in ASTER carried on NASA EOS Terra morning satellite. The TIR part (covering 3–13 μm spectral range) of the HyspIRI possesses eight multispectral TIR bands with a spatial resolution of 60 m. Coupling with continuous spectral measurements in the visible to short-wavelength infrared (VSWIR: 0.38 to 2.51 μm) portion, two key science contributions of HyspIRI measurements are: (1) the global imaging spectroscope measurements of the terrestrial biosphere for delivering ecosystem composition and function and reducing the uncertainty in climate-carbon interactions and terrestrial biosphere feedback and (2) the global 8-band TIR multispectral measurements to provide improved TIR spectral information of volcano, fire-related emissions, and the evapotranspiration of terrestrial vegetation. Another satellite sensor: the fourth VIIRS (Visible Infrared Imager Radiometer Suite) onboard JPSS-3 (Joint Polar-orbiting Satellite System 3) platform is operated by a joint of NSAS and NOAA agencies. The JPSS-3 satellite is scheduled to launch scheduled in 2026. The VIIRS sensor on JPSS-3, an improved version of the first VIIRS on JPSS-1, has 7 multispectral TIR bands covering TIR spectral range 3.7–12.0 μm and with two spatial resolutions, 375 m and 750 m. The VIIRS sensor system will provide high quality data of the Earth from low Earth orbit covering the parts of the spectrum visible to TIR and for monitoring global environmental conditions, and its collecting and disseminating data may be used for studies and applications on weather, atmosphere, oceans, land, etc. In addition to those two planned space-borne TIR sensors/mission, one more TIR sensor, ECOSTRESS on ISS, operated by NASA, has to be mentioned. The ECOSTRESS covers a TIR spectral range of 8.0–12.5 μm with five multispectral TIR bands and at spatial resolutions of 40–60 m. This satellite TIR sensor, together with other satellite sensors/missions operated by NASA, will advance our ability to monitor and model terrestrial ecosystems between latitudes ~50° North and South and to measure the temperature of plants to retrieve plant physiological parameters. Specifically, the ECOSTRESS can measure key variables: LST and land surface emissivity (LSE), surface roughness, and vegetation canopy functional traits including evapotranspiration (Stavros et al., 2017).

There are three future TIR sensors/systems developed and operated by or related to the European Space Agency (ESA), including the LSTM onboard Sentinel-8 satellite and two sensors, Spectrometer and Broadband Imager, onboard the HiTeSEM satellite mission spaceborne platform. The high spatiotemporal resolution LSTM mission covers TIR spectral range 8.6–12.0 μm with five multispectral TIR bands at spatial resolutions 30–50 m. The primary objective to develop the LSTM requires to support monitoring evapotranspiration (ET) rate at European field scale by capturing the variability of land surface temperature (LST) (and hence ET) enabling more robust estimates of field-scale water productivity, while the secondary objective can be to support mapping and monitoring the soil composition and its dynamics through emissivity estimates, and the additional services include benefits from TIR observations such as coastal zone management and urban heat islands, etc. Both sensors/systems on the HiTeSEM platform cover the TIR spectral range 7.2–12.5 mm. The imaging spectrometer will collect 30–72 high spectral resolution TIR bands with a spectral resolution of 60 nm and a spatial resolution of 60 m, while the broadband imager will collect the TIR data in two broadbands but at a high spatial resolution of 20 m. The scientific goal of the HiTeSEM satellite mission with two sensors is to measure some key variables from the biosphere, hydrosphere, pedosphere, and geosphere associated with two global problems of significant societal relevance: food security and human health (Udelhoven et al., 2017). The key variables measured by the HiTeSEM two sensors may include land and sea surface radiation temperature and emissivity, surface moisture, thermal inertia, vegetation evapotranspiration, soil minerals and other properties, bio-physical/chemical variables, heat fluxes, etc.

In spite of the uncertainty in the future of satellite platforms carrying TIR sensors and systems there are several TIR airborne sensors (e.g., DAIS, MIVIS, TIMS, and ATLAS introduced in Quattrochi and Luvall, 1999) that can potentially be used to bridge this gap in the plan and concept of satellite thermal sensors/missions. Usually airborne systems, including UAV based sensors, have historically been used as a basis and complementary and calibration high resolution data for investigating and developing space based TIR sensors/systems. This is because, unlike satellite systems, the airborne systems are capable of collecting TIR data at varying spatial resolutions by flying at a flexible altitude.

Since the advent of TIRRS, many technical difficulties have been overcome in areas such as TIR sensor development and calibration, TIR data processing and mining (e.g., retrieving LST and LSE (land surface emissivity)), and temperature and emissivity separation (TES). However, there are still several main challenges we are facing today, which require improvement of solutions or to move this TIRRS technology toward more frequent operational use. Besides developing and calibrating the new TIR sensors/systems as discussed earlier, three issues or challenges regarding TIR data processing we need to solve or face in the future are addressed as follows. (1) Over the last four decades, most modern remote sensing sensors/systems can acquire TIR data at a low spatial resolution but high temporal resolution compared to optical remote sensing sensors (e.g., MODIS, AVHRR, and GOES Imager TIR data). Due to the issue of the trade-off between spatial and temporal resolutions in modern TIR remote sensing technologies, it is extremely difficult to acquire both high temporal and spatial resolution TIR imagery (Zhan et al., 2013). However, per different application purposes, both high spatial and temporal resolution TIR data are needed. Given that TIR data play a valuable role in controlling Earth's surface physical and biological processes, it is desirable to have TIR data at both high spatial and high temporal resolutions for many application purposes, such as mapping surface evapotranspiration in agricultural and arid regions (Anderson et al., 2012; Olivera-Guerra et al., 2017) and studying urban thermal environments (Bonafoni and Tosi, 2017). To overcome the dilemma (i.e., low spatial resolution but high temporal resolution) of most available satellite TIR remote sensing sensors, many studies have investigated downscaling LST (DLST) processes (from a relatively lower/coarser spatial resolution to a relatively finer resolution) to generate high spatial (also high temporal) LST data to meet many application purposes. Therefore, it is necessary to develop novel methods and techniques focuses on developing or refining existing advanced models and methods, such as deep learning and local and temporal weighted autoregressive regression. Those advanced and novel methods and techniques have demonstrated their excellent performance and robustness. However, more effort is still needed. (2) To process TIR data, especially satellite TIR data, besides addressing the low sensor signal to noise ratio and atmospheric correction, temperature and emissivity separation (TES) is one of the most challenging tasks to be solved (Gerhards et al., 2019). To retrieve both accurate surface temperatures and emissivity spectra from hyperspectral TIR data, two fundamental problems need to be solved (Gerhards et al., 2019). First, accurate atmospheric correction is needed. This is because (1) the spectral radiance measured by a sensor consists of not only the radiance emitted by the target of interest itself but also thermal radiation emitted by surroundings and reflected from the surface of the object (due to atmospheric down-welling radiance); (2) the spectral radiance measured from the target of interest is further affected by the atmospheric absorption, emission, and scattering (due to atmospheric upwelling radiance and transmittance). Second, a solution to the so-called TES-problem (i.e., an ill-posed problem) is required. The measured spectral radiance by a TIR sensor is, per Planck's Law, a function of the absolute temperature and the spectral emissivity of the observed object of interest. Thus, if the radiant temperature is measured in n bands, there will be $n + 1$ unknowns (n emissivities plus 1 surface temperature). Therefore, to derive accurate surface temperatures, the multiple spectral emissivities must be known and vice versa. Presently, although hyperspectral TIR data compared to multispectral data allow a good fit to Planck's radiance curve for a specific temperature, and as a variety of current algorithms differ in the underlying assumptions and the data basis, more efforts are needed to solve the TES ill-posed problem. (3) Similar to (1) but emphasizing data fusion procedures with different TIR sensors' data, to improve the effectiveness of TIR data for different applications, attention needs to be on operating a procedure for fusing different TIR sensors (e.g., MODIS with low spatial but high temporal and Landsat TIR data with a relatively high spatial resolution) to synergize high temporal resolution from one TIR sensor (e.g., MODIS) and high spatial resolution from another sensor (e.g., Landsat 8 TIRS) to produce high spatiotemporal Landsat-like time series TIR data products. Although the fusion procedures have been proved to be better than that using single sensors' data to generate high spatial resolution TIR data, they need to be improved. For example, the weighted combination of kernel-driven and fusion-based methods (termed CKFM) proved to have an excellent performance to enhance the spatiotemporal resolutions of time series LSTs (Xia et al., 2019), it can still be modified with a temperature cycle model (e.g., diurnal/annual temperature cycle models) to further improve the high spatiotemporal resolution time series LST product. Such fusion procedures may also be considered to fuse optical and/or microwave data and TIR data in the same sensor or different sensors to improve the accuracy of downscaling TIR data.

1.5 SUMMARY

TIR remote sensing (TIRRS) technology can provide TIR information from materials on the Earth's surface and in the atmosphere in an image and offers a way, out of human vision, of detecting and mapping many spatial phenomena on the Earth's surfaces and in the atmosphere by using radiant temperature information acquired by TIR imaging systems and sensors. In the beginning of this chapter, basic concepts of TIR energy/radiation and TIRRS are first introduced and differences between optical RS, TIRRS, and microwave RS are discussed. In Section 2, a brief history and development of TIRRS are looked back at chronologically following the three rough development stages: Stage 1 (early), initially discovering TIR radiation and measurement (1738–1900); Stage 2 (middle), TIRRS technology preliminarily used in the military (World Wars I and II); Stage 3 (modern), TIRRS technology developed and used for both military and civilian purposes (after 1960). After reviewing the TIRRS history, an overview of applications and investigations of TIRRS technology to different disciplines and areas is given in greater length in this chapter, which includes hydrology (ice/snow, water bodies) and soil moisture and other physical properties, biomass burning, urban thermal environments, volcanoes, geothermal phenomena detection and mapping, vegetation and ecosystems, thermal energy inventory, etc. The overview has placed an emphasis on the current status of applications and investigations of TIR imaging data to these disciplines and fields. And finally, a perspective of TIRRS is delivered and discussed based on USA NASA's and NOAA's and ESA's space-borne TIR imaging sensors/missions and programs that are being implemented and developed in the near future. Some main issues and challenges associated with TIRRS data processing for applications we need to solve are also discussed before ending this chapter.

2 Fundamentals of Thermal Infrared Remote Sensing

To well understand principles of TIR remote sensing techniques, it is necessary to know and understand basic concepts and fundamentals of TIR radiation. In this chapter, the thermal physical properties of materials are first introduced, including thermal conductivity, thermal capacity, and thermal inertia, also discussing thermal latent heat, thermal expansion, and thermal diffusivity. Then thermal radiation laws are discussed, including Planck's radiation law, the Stefan-Boltzmann law, Wien's Displacement law, and Kirchhoff's radiation laws, also reviewing emissivity and the inverse-square law. In the final section, thermal radiation of real-world materials (e.g., soil, rock, vegetation, and water) is reviewed, including kinetic energy/temperature, radiant temperature/flux, and brightness temperature; energy transferring ways; TIR atmospheric windows; irradiance, radiant exitance, albedo, radiance, and reflectance; and more theoretic basis of directional and hemispheric radiometric temperatures and emissivities.

2.1 THERMAL PROPERTIES OF MATERIALS

In the construction industry, understanding thermal properties of materials is important when people design heating and cooling devices (e.g., designing a precision instrument). On the Earth's surface and in the atmosphere, the material, such as water, rock, soil, vegetation, water vapor, or even human tissue, all have the ability to conduct heat directly from one onto another surface (thermal conductivity) and to store heat (thermal capacity). Some materials respond to temperature changes more rapidly or slowly than others (thermal inertia) (Jensen, 2007). Some materials, such as copper and silver, conduct their heat very well while other materials, such as wood or rubber, are poor heat conductors. Therefore, it is very useful to discuss thermal properties of material, as they have an impact on our ability to remotely sense thermal information from various types of materials. In the following, several thermal properties of materials are reviewed and discussed, and they may include thermal conductivity, thermal capacity, and thermal inertia, also thermal latent heat, thermal expansion, and thermal diffusivity.

2.1.1 Thermal Conductivity

Thermal conductivity (K) is a measure of the rate at which heat passes through a material. It is defined as a number of calories that will pass through 1 gram of material in 1 sec when two opposite faces are maintained at 1°C difference in temperature (cal g^{-1} sec^{-1} $°C^{-1}$).

The K is one of the important thermal properties of materials that helps us understand how easily a calorie of heat energy flows from one material to another. It may help identify the ability of materials to cause heat conduction in these materials.

DOI: 10.1201/9781003276173-2

There are different materials with different Ks, seen in our daily life. For example, metal has the highest K; alloy is less than metal; the lowest K is non-metal materials, and that is why we use aluminum in the kitchen. Therefore, the heat passes through metals much faster than through rocks; water has a higher K value than many other materials, but it takes a longer time for water to transfer heat over a given distance than many other materials. Many rocks and soils are extremely poor conductors of heat (Table 2.1). The K of a material also is variable due to soil moisture and particle size (Jensen, 2007). Table 2.1 presents thermal conductivities (Ks) of a list of selected materials. Per the table, different materials have significantly different Ks.

2.1.2 Thermal Capacity

Heat or thermal capacity (C) is a measure of the ability of a material to absorb heat energy. It is the quantity of heat required to raise the temperature of one gram of the material by

TABLE 2.1
Thermal Properties of Some Materials (in Indoor Temperature)*

Materials	Thermal conductivity (cal g^{-1} sec^{-1} °C^{-1})	Thermal capacity (cal g^{-1} °C^{-1})	Thermal density (g cm^{-3})	Thermal inertial (cal cm^{-2} $sec^{-1/2}$ °C^{-1})
General materials				
glass	0.002100	2.600000	0.160000	0.029000
wood	0.005000	0.500000	0.327000	0.009000
water, ice	0.005000	0.917000	1.007887	0.067979
water, liquid	0.001441	1.000000	1.000000	0.037959
water, gas	0.000055	0.000598	0.469885	0.000124
air	0.000060	0.001290	0.239962	0.000136
copper	0.950000	8.940000	0.092017	0.884026
aluminum	0.540010	2.698000	0.220124	0.566312
aluminum, liquid	0.220000	2.360000	0.259082	0.366764
Geologic materials				
basalt	0.005000	2.800000	0.200000	0.053000
dolomite	0.012000	2.600000	0.180000	0.075000
granite	0.007500	2.600000	0.160000	0.056000
gravel, sand	0.006000	2.100000	0.200000	0.050000
limestone	0.004800	2.500000	0.170000	0.045000
obsidian	0.003000	2.400000	0.170000	0.035000
sandstone	0.012000	2.500000	0.190000	0.075000
shale	0.004200	2.300000	0.170000	0.041000
slate	0.005000	2.800000	0.170000	0.049000
soil, sand	0.001400	1.800000	0.240000	0.024000
soil, moist clay	0.003000	1.700000	0.350000	0.042000

* Compiled from Janza, 1975; (Thermtest Instruments, 2022).

1°C (cal g^{-1} °C^{-1}) (Jensen, 2007). The C determines how well a material stores heat. For example, water has the highest heat capacity of any common substance (1.00). It absorbs heat very well relative to all the other materials (e.g., forest, but forest's C > grass's > land's). In other words, for a given amount of energy, warming up water is slower than many other materials. This also reflects that the temperature of a water body (e.g., a pond) usually varies very little between night and day. Conversely, rocks and metals do not store heat well and exhibit significantly different temperatures between the night and day. See more materials with different Cs in Table 2.1.

2.1.3 Thermal Inertia

Thermal inertia (P) is a measurement of the thermal response of a material to temperature changes and is defined in calories per cm^2 per second square root per 1°C (cal cm^{-2} $sec^{-1/2}$ °C^{-1}) (Jensen, 2007). It is of the resistance of a material to temperature change. The P can be computed using Eq. (2.1) as follows:

$$P = \sqrt{KCD} \tag{2.1}$$

where K is thermal conductivity; D is density (g cm^{-3}); and C is thermal capacity. Density (D) is a very key biophysical variable in this equation because thermal inertia generally increases linearly with increasing material density (Jensen, 2007). Therefore, the P increases with an increase in material conductivity, capacity, and density (Table 2.1). Generally, materials with high thermal inertia have more uniform surface temperatures throughout the day and night than materials of low thermal inertia. This is because materials with high P possess a strong inertial resistance to temperature fluctuations at a surface boundary, and they present less temperature variation per heating/cooling cycle (e.g., day and night) than those with lower thermal inertia (Garg, 2022).

To obtain P, K, C, and D must be measured by the in situ approach because remote sensing methods cannot directly measure them. However, it is possible to remotely sense and compute an apparent thermal inertia measurement per pixel in the following way. Maximum and minimum radiant temperature collected by remote sensing can be recorded from daytime and nighttime images. The change in radiant temperature, ΔT, for a specific pixel in a scene of images is determined by subtracting the nighttime image from the daytime image. Given the fact that ΔT is low for materials with high thermal inertia and vice versa, it is easy to determine a property called apparent thermal inertia (ATI) by the following equation:

$$ATI = \frac{1-\alpha}{\Delta T} \tag{2.2}$$

where α is the albedo (reflectance) measured in the visible band during the daytime for the pixel of interest (Jensen, 2007). Note that the ATI image must be interpreted carefully since ΔT may be affected by factors other than thermal inertia (Garg, 2022).

The apparent thermal inertia can be obtained with the measured temperature change, ΔT. Basically, a high ΔT value is usually related to terrain materials with a low thermal inertia value, while a low ΔT is usually associated with terrain materials with a high thermal inertia value. Using this in practice, geologists, who use remote sensing images, identify areas/rocks that have varying apparent thermal inertia characteristics to distinguish boundaries between bedrock and alluvial material, discriminate among rock units with similar spectral properties, and label zones of hydrothermal alteration.

Thermal capacity, conductivity and inertia, discussed earlier, are the three basic thermal properties of materials. To understand thermal properties well, there are still several thermal properties that need to be known, including specific latent heat, expansion, and diffusivity.

2.1.4 Specific Latent Heat

Specific latent heat is the energy per unit mass absorbed or evolved when a substance changes its phase. There are two types of latent heat and they are: (1) latent heat of fusion (the heat given out when a liquid changes into a solid), and (2) latent heat of vaporization (the heat absorbed when a liquid turns into a gas). Here we need to note that for a given substance, the liquid is in a higher energy state than the solid and the vapor is in a higher energy state than the liquid. For example, the specific latent heat of vaporization can be described as the energy that is required to change a unit mass of water from liquid to water vapor in a constant pressure and constant temperature, approximating 2.45 MJ kg^{-1} at 20°C (Allen et al., 1998). In addition, different from the specific latent heat, in practice, we also use specific heat to quantify the thermal property of a material, and it tells us how quickly a material gets heated. Therefore, the specific heat is defined as a quantity of heat that we need to change 1 unit degree (kelvin) per unit mass.

2.1.5 Thermal Expansion

Thermal expansion can be understood as a thermal property of a material, when heat is passed through the material, whose shape changes. Generally, a material expands when heated. There is a change in its dimension and this change comes in all directions. If we see it in the linear direction, then we will call it the linear expansion; if you look in an area, then there is an area expansion; if you talk in volume, then it will say volume expansion. For example, railway tracks often expand and, as a result, get misshapen due to extreme heat. This is an effect of thermal expansion on a railway track. The thermal stress associated with thermal expansion is as a material due to either thermal expansion or contraction. It can be potentially destructive in nature as it can make the material explode. For example, cracks can be seen on roads where the heat is extreme. The crack is a result of thermal stress. Thermal expansion may be quantified with a thermal expansion coefficient, which is the percentage of dimensional change one can expect per unit increase in temperature.

2.1.6 Thermal Diffusivity

Thermal diffusivity (k) is the measure related to thermal inertia. In a material with high thermal diffusivity, its heat moves rapidly through it because the material conducts heat quickly relative to its volumetric heat capacity. Thermal diffusivity (k) of a substance can be quantified by a rate at which heat is transferred within the substance and it may be expressed as follows (Garg, 2022):

$$k = \frac{K}{DC} \tag{2.3}$$

where K is the thermal conductivity; D is the thermal density; and C is the thermal capacity. The thermal diffusivity governs the rate at which heat is conducted from surface to depth in the daytime and from depth to surface in the nighttime. For example, water passes high specific heat, and thus minor changes in moisture content of soils lead to significant effects on thermal diffusivity of the soils.

2.2 THERMAL RADIATION LAWS

To understand and evaluate the thermal radiation principle, the concept of blackbody needs to be introduced first. A blackbody possesses these characteristics: absorbing all incident radiation that falls on it; emitting energy at the maximum possible rate per unit area at each wavelength for any given temperature; emitting radiation depending on wavelength and temperature but not on direction (diffuse radiation). As a perfect emitter and absorber, the blackbody is often used to compare it with real-world materials (selective body or radiator) and surfaces as a standard (Hu and Delichatsios, 2019). However, no objects in nature are true blackbodies; we may think of the Sun as approximating a 6,000 K blackbody and the Earth as a 300 K blackbody (Jensen, 2007). In this section, the properties of thermal radiation are described by the following thermal radiation laws to understand the thermal radiation principles, including Planck's, the Stefan-Boltzmann, Wien's Displacement, and Kirchhoff's radiation laws. And emissivity and the inverse-square law are also discussed.

2.2.1 PLANCK'S RADIATION LAW

As discussed earlier, the intensity of blackbody radiation depends on wavelength and its temperature. Planck's radiation law describes the spectral exitance or spectral radiation intensity (i.e., Planck's radiation distribution) of a blackbody with a temperature by the following function:

$$B(\lambda, T) = \frac{C_1}{\lambda^5 \left[exp\left(\frac{C_2}{\lambda T} \right) - 1 \right]} \tag{2.4}$$

where $B(\lambda, T)$ is the amount of energy of a blackbody and it gives off as radiation of different wavelengths and temperatures, w m^{-2} sr^{-1} μm^{-1}; C_1 and C_2 are the two physical constants equal to $C_1 = 1.19104 \times 10^8$ W m^{-2} sr^{-1} μm-1 and $C_2 = 1.43877 \times 10^4$ μm K, respectively.

Since a blackbody is a diffuse emitter, its spectral emissive power can be expressed by:

$$E(\lambda,T) = \pi B(\lambda,T) = \frac{\pi C_1}{\lambda^5 \left[exp\left(\frac{C_2}{\lambda T} \right) - 1 \right]} \quad (2.5)$$

Equation 2.5 is known as the Planck distribution having the following features: (1) radiation varies continuously with wavelength; (2) radiation increases with increasing temperature when wavelength is fixed; (3) with increasing temperature, radiation appears more at shorter wavelength (Hu and Delichatsios, 2019). See Figure 2.1 for the Planck's radiation distribution.

2.2.2 Stefan-Boltzmann Law

The total spectral radiant exitance ($M_b(T)$) measured in Watts m^{-2} leaving a blackbody is proportional to the fourth power of its temperature (T). This is known as the Stefan-Boltzmann law, which integrates the emitted radiance across the whole spectrum by an isotropic blackbody of a unit area (Tang and Li, 2013) and is expressed as:

$$M_b(T) = \pi \int_0^\infty B(\lambda,T) d\lambda = \sigma T^4 \quad (2.6)$$

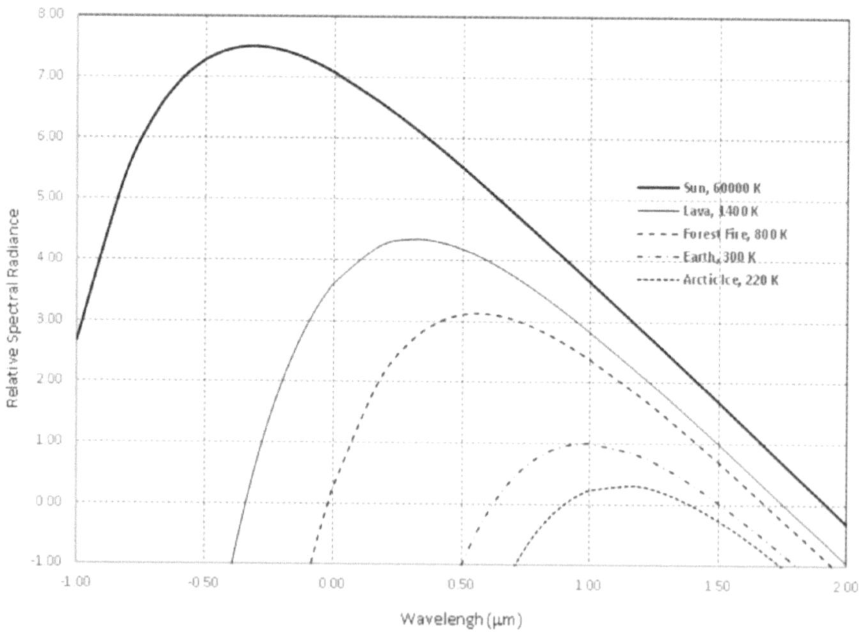

FIGURE 2.1 The Planck's law distribution.

where σ is the Stefan-Boltzmann constant equaling 5.6697 × 10^{-8} W m^{-2} K^{-4}, and T is temperature in degrees Kelvin. The law enables calculation of the total radiation energy of all directions and wavelengths of a blackbody with a temperature. The total radiant exitance is the integration of all the area under the blackbody radiation curve (Figure 2.2). Per the figure, note how the Sun produces more spectral radiant exitance ($M_b(T)$) at 6,000 K than that produced by the Earth at

300 K. As the temperature increases, the total amount of radiant energy measured in Watts per m^2 (the area under the curve) increases and the radiance peak value shifts toward shorter wavelengths. The wavelength corresponding to the radiant energy peak is called the dominant wavelength for a blackbody at a specific temperature, which can be determined by Wien's displacement law (Eq. 2.7) in the following section.

2.2.3 Wien's Displacement Law

In addition to computing the total amount of energy exiting a theoretical blackbody with Stefan-Boltzmann law, we can determine a relationship between the true temperature of a blackbody (T) in degrees Kelvin and its peak spectral exitance or dominant wavelength (λ_{max}) by Wien's displacement law:

$$\lambda_{max} = \frac{k}{T} \tag{2.7}$$

where k is a constant equaling 2898 μm K. We can determine the dominant wavelength of any object by substituting its temperature into Eq. 2.7 per Figures 2.2 and 2.3, the

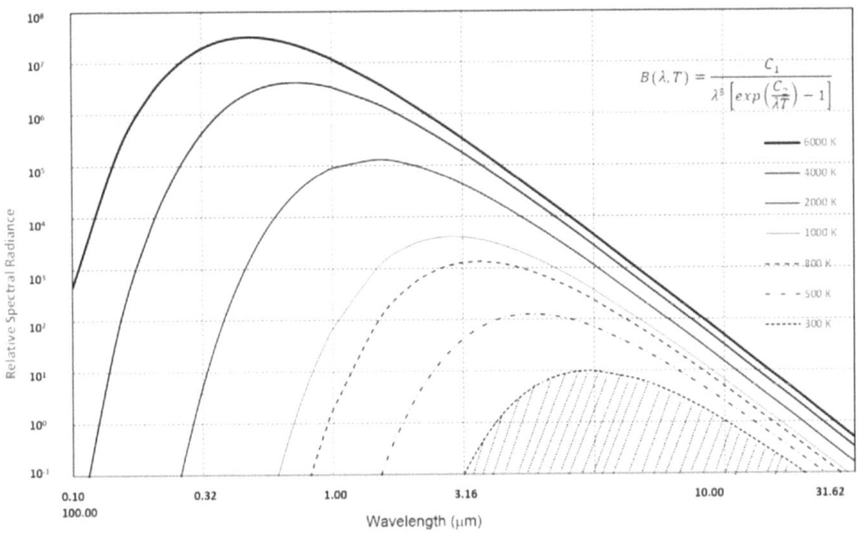

FIGURE 2.2 Stefan-Boltzmann law distribution and the filled area in slash line indicating the area of Earth at 300 K.

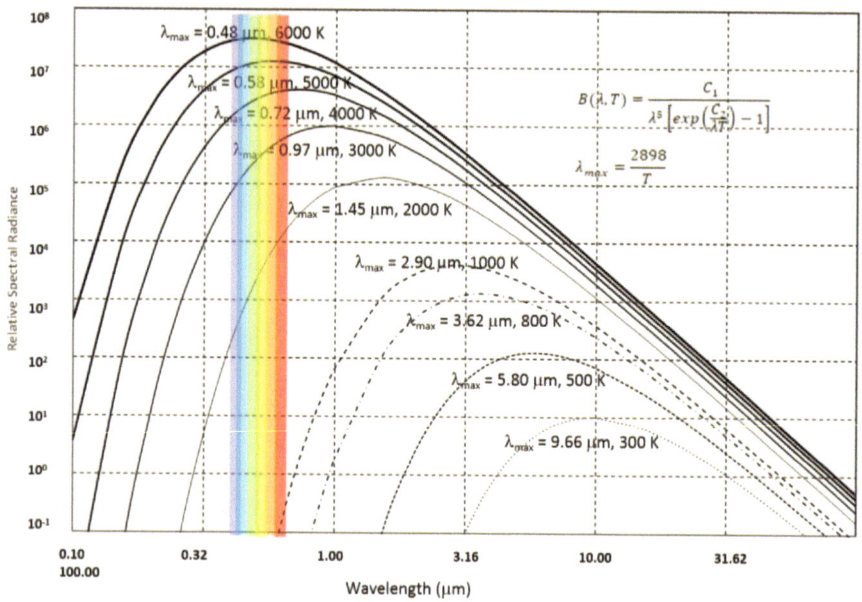

FIGURE 2.3 Wien's displacement law.

dominant wavelength of the 6,000 K for Sun is 0.48 μm, while the dominant wavelength for the Earth's surface (300 K) is 9.67 μm. Therefore, the dominant wavelength, λ_{max}, shifts toward the short wavelength portion of the spectrum as the temperature of the object increases.

In practice, the dominant wavelength determined by Wien's displacement law can provide valuable information regarding the part of the thermal infrared spectrum in which we might want to sense objects at different temperatures. For example, if we are looking for 800 K forest fires that have a dominant wavelength of approximately 3.62 μm, then the most appropriate TIR remote sensing sensors might include 3–5 μm TIR band(s) (e.g., AVHRR sensor with a channel 3 (3.9 μm)). Conversely, if we are interested in sensing soil, water, and rock ambient temperatures on the Earth's surface (300 K) with a dominant wavelength of 9.67 μm, then TIR detectors/bands operating in the 8–14 μm TIR spectral region might be most appropriate (e.g., Landsat 8 TIRS sensor acquiring two TIR bands data, covering 10.60–12.51 μm).

2.2.4 Spectral Emissivity (ελ)

In fact, the world is not composed of radiating blackbodies. Instead, it is composed of selectively radiating bodies, such as rock, soil, vegetation, and water, which emit a certain proportion of the energy emitted from a blackbody at the same temperature. Emissivity (ε) is a measure of the ability of a selective body to emit thermal radiation. The spectral emissivity (ε_λ), using the Planck's law, $B(\lambda,T)$, is defined as the ratio between the actual spectral radiance, $M_r(\lambda,T)$, emitted by a real world selective radiating body at the temperature and the spectral radiance, ($M_b(\lambda,T)$, emitted by a blackbody at the same temperature (Becker and Li, 1995):

$$\varepsilon_\lambda = \frac{M_r(\lambda,T)}{M_b(\lambda,T)} \tag{2.8}$$

where, $M_b(\lambda,T)$ and $M_r(\lambda,T)$ are the total radiances emitted from a blackbody and a real-world body, respectively, determined by Eqs. 2.6 and 2.10 (shown later).

Spectral emissivity (ε_λ) has the following properties:

- describing the actual absorption and emission properties of real objects or gray bodies.
- being wavelength dependent (ε_λ).
- being a measure of a material's ability to absorb and radiate energy.
- having a value from 0 to 1.

Figure 2.4 depicts the conceptual emissivity of a blackbody, a graybody, and a hypothetical selective radiator/body over the wavelength λ. Notice, in Figure 2.4a, how the spectral emissivity of the hypothetical selective radiator fluctuates across the wavelength λ, but the spectral emissivity of a blackbody retains a value of 1.0 and a constant of value between 0 and 1 for a graybody when three bodies have a same temperature. The spectral radiant emittance of each body with the same temperature is shown in Figure 2.4b. Note how the different emissivities give rise to dramatic changes in the spectral radiant emittance distribution of the selective radiator, where it is 1.0, it outputs the same amount of radiant energy as the blackbody. Where the selective radiator's emissivity is lower than that of the graybody, it emits lower spectral radiant emittance than that emitted by the graybody.

The spectral radiant emittance for three real world radiating bodies is shown in Figure 2.5. Note that the spectral radiant emittance for all the three real world radiating bodies is lower than that by the blackbody at the same temperature and that the radiant energy emitted by a real world radiating body (e.g., water, Figure 5a) is approximately the same as a blackbody for almost the entire spectral range, while for

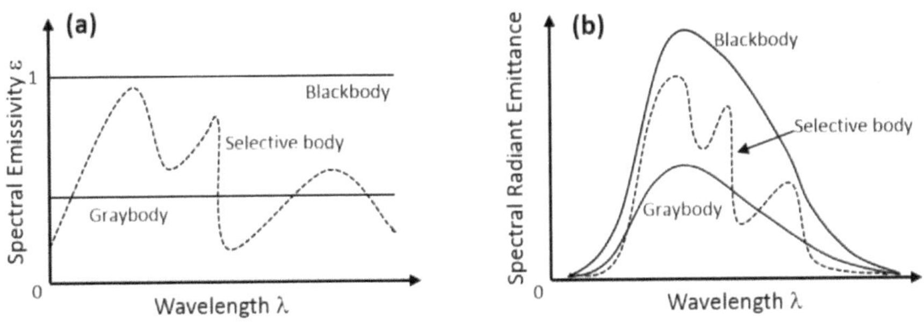

FIGURE 2.4 Spectral emissivity (a), spectral radiant emittance (b) for blackbody, graybody, and a hypothetical selective body/radiator.

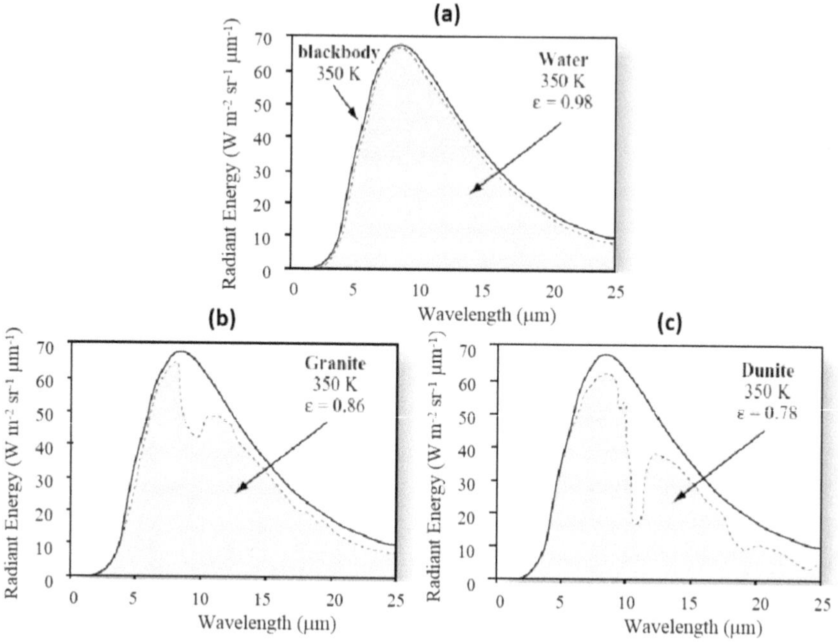

FIGURE 2.5 Radiant energy emitted from some selective bodies: a) water, b) granite, and c) dunite, all heated to 350 K with a blackbody at the same temperature as a comparison (modified from Jensen, 2007).

some real world radiating bodies (e.g., dunite, Figure 5c), the spectral radiant emittance curves between the blackbody and other real world materials depart in much of the spectral range. This is because some real materials have a close to 1.0 (but > 0) of emissivity (e.g., water) for almost the entire spectral range, but other real materials have much lower value than 1.0 (but > 0, e.g., dunite) for much of the spectral range. Table 2.2 shows emissivities of a list of selected real materials. From the table, some materials, like distilled water, have emissivities close to one (0.99) over the spectral range 8–14 μm, while others such as polished aluminum (0.08) and stainless steel (0.16) have very low emissivities.

In practice, two objects have the same true kinetic temperature but may have different apparent temperatures recorded by a TIR remote sensing sensor, which is simply because their emissivities are different. For example, two rocks could have the same true kinetic temperature but have different apparent temperatures because of different emissivities. The emissivity of an object may be influenced by a number of factors, including (Jensen, 2007):

- color—darker-colored objects are better absorbers and emitters (i.e., they have a higher emissivity) than lighter colored objects, which tend to reflect more of the incident energy.
- surface roughness—the greater the surface roughness of an object relative to the size of the incident wavelength, the greater the surface area of the object and potential for absorption and reemission of energy.

TABLE 2.2
Emissivities of a List of Selected Real Materials from 8–14 μm*

Material	Emissivity (ε)	Material	Emissivity (ε)
water, distilled	0.99	wood, planted oak	0.9
water	0.92–0.98	wood	0.87
water with petroleum film	0.972	glass	0.77–0.81
ice, rough	0.97–0.98	stainless steel	0.16
wet snow	0.98–0.99	aluminum, foil	0.05
dry snow	0.85–0.90	aluminum, polished	0.08
concrete	0.71–0.9	aluminum, painted	0.55
asphalt	0.95	polished metals	0.06–0.21
asphaltic concrete	0.94–9.97	oxidized steel	0.7
tar/stone	0.97	Copper, polished	0.04
loamy soil, dry	0.92	Copper, oxidized	0.87
loamy soil, wet	0.95	silver, polished	0.02
soil, sandy	0.9	silver, oxidized	0.04
dry mineral soil	0.92–0.96	granite	0.83–0.87
wet soil	0.95–0.98	dunite	0.78
brick, red and rough	0.93	basaltic rock	0.92–0.96
Quartz sand, large grains	0.91	limestone	0.92
sand	0.93	marble, polished	0.89–0.92
vegetation, closed canopy	0.98	paint	0.9–0.96
vegetation, open canopy	0.96	human skin	0.98
dry vegetation	0.88–0.94	white paper	0.9
grass	0.97		
deciduous forest	0.97–0.98		
coniferous forest	0.97–0.99		

* Complied from (Sabins, 1997; Garg, 2022; Wikipedia, 2022).

- moisture content—the more moisture an object contains, the greater its ability to absorb energy and become a good emitter, and thus wet soil particles have a high emissivity similar to water.
- compaction—the degree of soil compaction can affect emissivity, and a loose soil has a higher emissivity than that of a compacted soil.
- field of view—the emissivity of a single leaf measured with a very-high-resolution thermal radiometer will have a different emissivity than an entire tree crown viewed using a coarser spatial resolution radiometer.
- wavelength—the emissivity of an object is generally considered to be wavelength dependent. For example, when the emissivity of an object is constant throughout the 8–14 μm region, its emissivity in the 3–5 μm region may be different.
- viewing angle—the emissivity of an object can vary with sensor viewing angle, and a vertical/nadir direction will lead to a larger emissivity.

We must take into account real materials (objects)' emissivity when we use TIR remote sensing to measure their true kinetic temperature. This is done by applying Kirchhoff's radiation law, which is discussed in the following section.

2.2.5 Kirchhoff's Radiation Law

In the infrared portion of the electromagnetic spectrum, in general, an object that absorbs radiation may also emit radiation at the same wavelength. Kirchhoff's law states that the spectral absorptance (α_λ) of an object generally equals its spectral emissivity (ε_λ) at the same temperature across a TIR spectral range, i.e., $\alpha_\lambda = \varepsilon_\lambda$. The Kirchhoff's radiation law is often phrased as: 'good absorbers are good emitters and good reflectors are poor emitters.' Most real-world materials are opaque to thermal radiation, meaning a little radiant flux exits from the other side of terrain element. Therefore, transmittance, $\tau_\lambda = 0$. Substituting emissivity for absorptance and removing transmittance (τ_λ) from the radiation budget equation, $1 = \rho_\lambda + \alpha_\lambda + \tau_\lambda$, yields:

$$1 = \rho_\lambda + \varepsilon_\lambda \tag{2.9}$$

where ρ_λ is the reflectance of an object.

Because the terrain (object) does not lose any incident energy due to transmittance, all of the energy leaving the object must be accounted for by the relationship between reflectance (ρ_λ) and emissivity (ε_λ). For example, water absorbs almost all incident energy and reflects very little. Therefore, water is a very good emitter and has a high emissivity close to 1. Conversely, a sheet of metal roof reflects most of the incident energy and absorbs very little, yielding an emissivity much less than 1. Therefore, metal objects, such as cars, aircraft, and metal roofs, almost always look very cold (dark) on night thermal infrared imagery (Jensen, 2007). Based on the Kirchhoff's law principle, we can obtain the emissivities for TIR bands via spectral radiances measured by the TIR remote sensing sensor if we also know spectral reflectances of the bands (e.g., measured in laboratory). For example, AVHRR channel 3 (3.9 μm) has a total spectral radiance for a specific pixel in the image, consisting of reflected and emitted radiance (i.e., $1 = \rho_\lambda + \varepsilon_\lambda$), so, if one knows the reflectance, its emissivity can be obtained.

The goal of developing TIR remote sensing is to sense the apparent radiant temperature (T_{rad}), then derive the true kinetic temperature of the object (T_{kin}) on the Earth's surface and in the atmosphere. Unfortunately, it has the effect of the object's emissivity. If one knows the emissivity, the Stefan-Boltzmann law may be rewritten as:

$$M_r(T) = \varepsilon \sigma T^4 \tag{2.10}$$

where, $M_r(T)$ is the total spectral radiant flux of a real-world material (or a selective body), and σ is the Stefan-Boltzmann constant. It takes into account the temperature of an object and its emissivity to create a more accurate estimate of the radiant flux emitted by the object.

2.2.6 Inverse-Square Law

The inverse-square law states that the intensity of radiation emitted from a point source, S, varies as the inverse square of the distance, d, between source and detectors, D1 and D2, which means that halving the distance of a remote sensing detector,

D, from a point source, S, quadruples the infrared energy received by that detector. Therefore, we can obtain a more intense, stronger TIR signal if we can get the remote sensing detector as close to the ground (TIR source) as practical. According to Jensen (2007) and illustrated by Figure 2.6, for example, suppose a blackbody point source, S, with a certain temperature, T, and two remote detectors (D1 and D2) of equal sensitive area, say, 1 pixel area (e.g., 1 cm^2) and Detector D1 is located at a distance of d cm from S, and detector D2 is at a distance of 2d cm from S. By the Stefan-Boltzmann law (Eq. 2.6), the total infrared energies radiated by S into hemispheres with radii (d and 2d) are $M_{b(d)}(T)$ and $M_{b(2d)}(T)$ Watts/cm^2, respectively, of the radiating surface. Thus, $M_{b(d)}(T)$ is the total infrared energy radiated onto a hemispheric surface of radius d on which sensor D1 is located, which means onto a surface area of πd^2 cm^2. But $M_{b(2d)}(T)$ is also the total infrared energy radiated onto a hemispheric surface of radius 2d on which sensor D2 is located, specifically, which means onto a surface area of $\pi(2d)^2 = 4\pi d^2$ cm^2. Consequently, the infrared energy detected by D1 is $M_b(T)$' (Eq. 2.11) and by D2 is $M_b(T)$'/4 (Eq. 2.12) as follows:

$$M_{b(d)}(T) = \frac{M_b(T)}{\pi d^2} = M_b(T)' \tag{2.11}$$

$$M_{b(2d)}(T) = \frac{M_b(T)}{4\pi d^2} = \frac{M_b(T)'}{4} \tag{2.12}$$

2.3 THERMAL RADIATION OF REAL MATERIALS

Real materials do not behave like a blackbody that has the ability of absorbing the maximum radiant energy that falls on it and radiating maximum radiant energy from it. They emit only a fraction of the radiation emitted by a blackbody at the same temperature. This is an effect of emissivity (ε) of a real material (also called selective

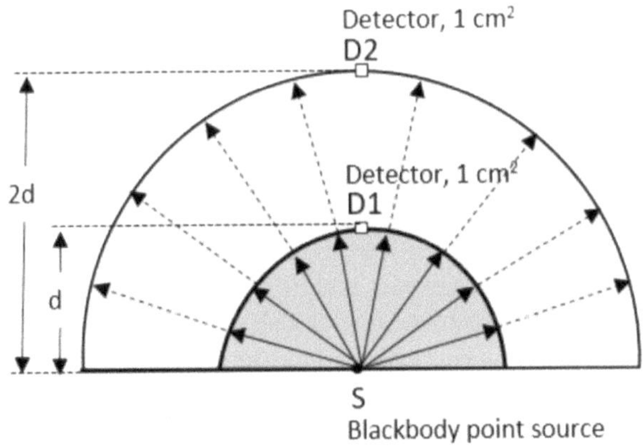

FIGURE 2.6 The inverse-square law indicating that the intensity of thermal radiation emitted from a blackbody point source, S, varies as the inverse square of the distance, d, between the source and remote detector receivers, D1 and D2 (modified from Jensen, 2007).

body). The emissivity can have values between 0 and 1 and is a measure of the ability of a material both to radiate and to absorb radiant energy. Within the TIR portion of the electromagnetic spectrum, if the emissivity of a material is known, its absolute temperature (i.e., kinetic temperature, T_{kin}) can be derived from the radiation it emits, and the radiation is called radiant energy, often measured by a TIR remote sensing sensor. If the emissivity is not considered, only the brightness temperature (i.e., radiant temperature, T_{rad}) of the material can be determined by modern TIR remote sensing technology. In this section, several basic concepts and principles related to thermal radiation of real-world materials are introduced in order to well understand principles of TIR remote sensing. They include kinetic energy/temperature (T_{kin}), radiant temperature/flux (T_{rad}), and brightness temperature; energy transferring ways; TIR atmospheric windows; irradiance, radiant exitance, albedo, radiance, and reflectance; and directional and hemispheric radiometric temperatures and emissivities.

2.3.1 Kinetic Energy/Temperature, Radiant Temperature/Flux, and Brightness Temperature

The energy of particles of matter in random motion is called kinetic heat (also referred to as internal, real, or true heat). All real-world materials having a temperature above absolute zero (0 K; −273.16 °C; and −459.69 °F) exhibit this random motion. When these particles collide, they change their energy state and emit electromagnetic radiation. The amount of heat can be measured in calories (the amount of heat required to raise the temperature of 1 g of water by 1°C). We can measure the true kinetic temperature (T_{kin}) or concentration of its heat using a thermometer to directly touch an object. For example, we can use a thermometer to measure your body temperature when you are sick, and we can also measure the true kinetic internal temperature of soil or water by physically touching them with a thermometer. In fact, an object's internal kinetic heat has the ability to converting its heat into radiant energy (also often called external or apparent kinetic heat), which allows us to utilize remote sensing technology to measure the radiant energy emitted by the object.

The electromagnetic radiation emitted by—or reflected from—an object is called radiant flux while the concentration of the amount of the radiant flux exiting an object is its radiant temperature (T_{rad}). For most real-world materials (except those composed of glass and metal) there is usually a high positive correlation between the true kinetic temperature of the object (T_{kin}) and the radiant temperature emitted from the object (T_{rad}) (Jensen, 2007). Therefore, we can utilize radiometers onboard an airborne or spaceborne platform to measure a real-world material's radiant temperature, which hopefully correlates well with the material's true kinetic temperature. This is the basis of TIR remote sensing and also a purpose of our developing TIR remote sensing technology. However, the relationship between the T_{rad} and T_{kin} is not perfect, creating a problem of the remote measurement of the T_{rad} always being slightly less than the true T_{kin} of the material. This is due to an effect of a thermal property called emissivity. In the last section, we discussed the emissivities of different real-world materials (Table 2.2). Therefore, we must consider the effect of a material's emissivity when using remote sensing measured T_{rad} of the material to estimate its T_{kin}.

Brightness temperature is a descriptive measure of radiation in terms of the temperature of a hypothetical blackbody emitting an identical amount of radiation at the same wavelength. The brightness temperature (T_b) is a directional temperature obtained by equating the measured radiance with the integral over wavelength of the Planck's blackbody law multiplied by the sensor response (Norman and Becker, 1995), which, in practice, is obtained by applying the inverse of the Planck function to the measured radiation as follows (Tang and Li, 2013):

$$T_b = B(\lambda, T)^{-1} \left[\frac{\int_{\lambda_1}^{\lambda_2} f_i(\lambda) R_\lambda d\lambda}{\int_{\lambda_1}^{\lambda_2} f_i(\lambda) d\lambda} \right] \tag{2.13}$$

where $B(\lambda, T)^{-1}$ is the inverse of Planck's function (Eq. 2.4); λ is the wavelength; λ_1 and λ_2 are the lower and upper wavelengths sensed by a sensor in band i; R_λ is the measured radiance at wavelength λ; and $f_i(\lambda)$ is the spectral response function of the sensor in band i and at wavelength λ.

2.3.2 Energy Transferring Ways

Energy or heat is often transferred from one object to another or from one place to another. There are three basic ways in which energy can be transferred, including conduction, convection, and radiation (Jensen, 2007). We are most familiar with conduction as a way to transfer energy. Energy may be conducted directly from one object to another, such as when a pan is in direct physical contact with a hot burner. In convection, the kinetic energy of bodies is transferred from one place to another by physically moving the bodies, such as the Sun bathing the Earth's surface with radiant energy causing the air near the ground to increase in temperature. The less dense air (wormer air) rises, creating convectional currents in the atmosphere. The transfer of energy by electromagnetic radiation is mostly of interest to remote sensing technologies because it is the only method of energy transfer that does not need to be physically touching or moving one object (place) to another, and it can take place in a vacuum, such as the region between the Sun and the Earth.

2.3.3 TIR Atmospheric Windows

We call those TIR regions TIR atmospheric windows, which allow a portion of the infrared energy to be transmitted from the terrain to the sensors/detectors efficiently. Figure 2.7 shows that the atmospheric windows in the TIR spectral range (3–14 μm), highlighted in red. Per the figure, it is clear that there are two general TIR windows in white areas: 3–5 μm and 8–14 μm where TIR remote sensing can be performed. Conversely, the black areas in Figure 2.7 denote regions of the electromagnetic spectrum where the atmosphere absorbs most of the infrared energy present, in which modern TIR remote sensing cannot be used because almost no energy can be passed through. Water vapor (H_2O), carbon dioxide (CO_2), and ozone (O_3) are responsible for most of the absorption. For example, atmospheric H_2O absorbs most

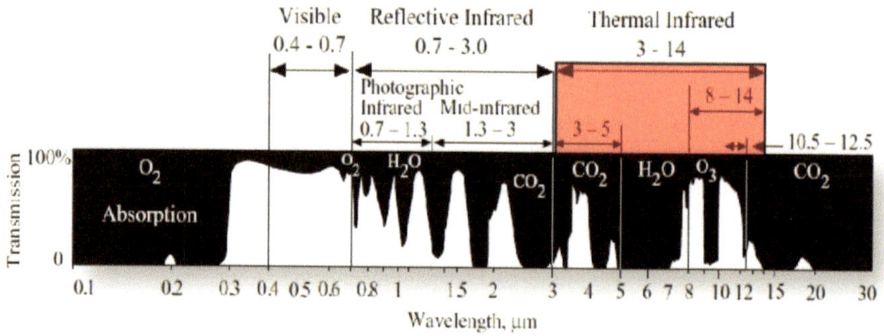

FIGURE 2.7 Atmospheric windows in the electromagnetic spectrum emphasizing TIR windows (commonly used: 3–5 mm and 8–14 mm), highlighted in a red range (modified from Jensen, 2007).

of the energy emitted from ground surface materials in the region from 5–7 μm, making it almost useless for TIR remote sensing.

In TIR remote sensing, the most used TIR range is the 8–14 μm window. However, the Earth's O_3 layer absorbs much of the thermal energy exiting the terrain in an absorption band from approximately 9.2–10.2 μm. Therefore, satellite TIR remote sensing systems often avoid the absorption band and only record data in the region from 8–9.2 μm and 10.2–12.6 μm. For example, Landsat 8 has two TIR bands: 10.60–11.19 μm and 11.50–12.51 μm. The Terra ASTER band 12 is 8.925–9.275 μm and band 13 is 10.25–10.95 μm. The region from 9.276–10.24 μm is not sensed due to atmospheric O_3 absorption.

2.3.4 Irradiance, Radiant Exitance, Albedo, Radiance, and Reflectance

To better understand the thermal radiation of real-world materials, Figure 2.8 helps us understand the several directional radiation terms: irradiance, radiant exitance, albedo, radiance, and spectral reflectance over the TIR region.

Irradiance (E) and Radiant exitance (M) ($W\ m^{-2}\ \mu m^{-1}$) Irradiance (E) is defined as the total incident radiant power or flux per unit area on the surface and usually is expressed in watts per square meter (Colwell et al., 1983):

$$E = \frac{dI}{dA} \tag{2.14}$$

And thus, the irradiance on the surface may be estimated by integrating incident radiance I over the whole 2π hemisphere (Tang and Li, 2013), namely:

$$E = \int_0^{2\pi}\int_0^{\pi/2} I(\theta,\varnothing) cos\theta sin\theta d\theta d\varnothing \tag{2.15}$$

Note that irradiance is not a function of direction but that radiance is. Compared to the irradiance that is the density of radiant flus incident on the surface, the radiant exitance (M) is defined as the density of radiant flux leaving a surface in all directions in the hemisphere over the surface. Its unit is the same as the irradiance.

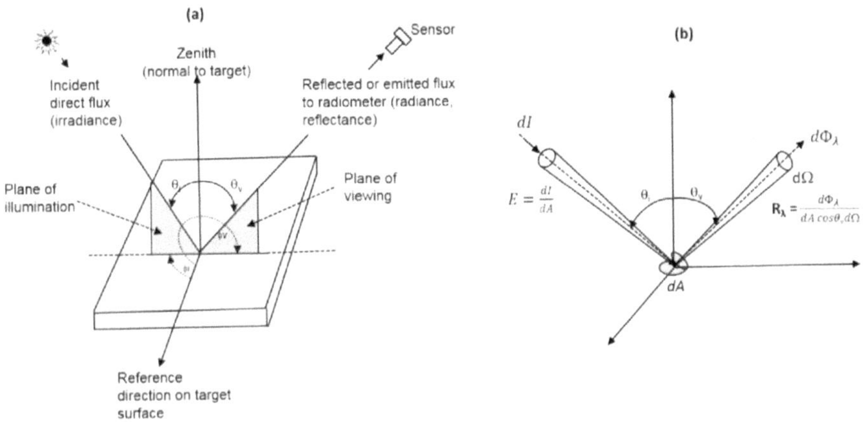

FIGURE 2.8 Illustrations of angles for concepts of irradiance, albedo, radiance, and reflectance (a) and detailed calculation for irradiance and radiance (b).

Albedo (α) Albedo (α) states the diffuse reflectivity or the reflecting power of a surface. It is defined the ratio of radiant exitance (*M*) from the surface to incident irradiance (*E*) over it (Schaepman-Strub et al., 2006), namely:

$$\alpha = \frac{M}{E} \tag{2.14}$$

Albedo is a dimensionless fraction (like spectral reflectance mentioned later) and it is measured at a scale from 0, for no reflecting power of a perfect black (body) surface, to 1, for a perfect reflection of a white (body) surface. The proportion reflected from a surface is determined not only by properties of the surface itself but also by the spectral and angular distribution of solar radiation reaching the Earth's surface. These factors affecting the albedo may vary with atmospheric composition, geographic location, and time.

Radiance (R_λ) *and reflectance* (ρ_λ) The radiance or spectral radiance (R_λ) ($Wm^{-2}\ sr^{-1}\ \mu m^{-1}$) is a measure of the quantity of radiation leaving a specific projected source area (A) on the ground, in a specific direction (θ), and within a specific solid angle (Ω). It is defined as radiant flux (Φ_λ) per meter squared per steradian in certain wavelengths leaving the projected source area (A) (Colwell et al., 1983):

$$R_\lambda = \frac{d\Phi_\lambda}{dAcos\theta d\Omega} \tag{2.15}$$

The radiance or spectral radiance (R_λ) is a physical quantity commonly used in remote sensing

The reflectance or hemispherical reflectance (ρ_λ) is defined as the dimensionless ratio of the radiant flux (Φ_λ) reflected from a surface to the radiant flux (E_λ) incident on it,

$$\rho_\lambda = \frac{d\Phi_\lambda}{dE_\lambda} \tag{2.16}$$

This reflectance quantity describes the general spectral reflectance characteristics of various phenomena, and usually in remote sensing, Eq. 2.16 is multiplied by 100% to become percent spectral reflectance.

2.3.5 Directional and Hemispheric Radiometric Surface Temperatures

According to Norman and Becker (1995), a directional radiometric surface temperature, $T_{R,\lambda}(\theta, \phi)$, may be derived from a radiative energy balance of a surface and it provides the best approximation to the kinetic temperature of the surface based on a measure of radiance. Given a narrow field-of-view sensor that is pointed toward a homogeneous, isothermal surface from a particular angle, the measured radiance is a combination of emission from the surface with reflection of incident, hemispherical flux (incident thermal irradiance) from the sky and surrounding objects into the direction of the thermal sensor. Therefore, the spectral radiance, $R_{R,\lambda}(\theta, \phi)$ sensed by a directional sensor can be estimated by the sum of the emitted blackbody radiation modified by the emissivity and the reflected radiation within an infinitesimal wavelength band around wavelength λ:

$$R_{R,\lambda}(\theta,\phi) = \varepsilon_\lambda(\theta,\phi) R_{B,\lambda}(\theta,\phi) + \int_0^{2\pi}\int_0^{\pi/2} \rho_{b,\lambda}(\alpha,\beta,\theta,\phi) \cos(\alpha) L_\lambda(\alpha,\beta)\sin(\alpha)\, d\alpha\, d\beta \tag{2.17}$$

where $R_{R,\lambda}(\theta, \phi)$ is the spectral radiance from the surface in the zenith angle θ and azimuth angle ϕ in the hemisphere of reflection and emission; $\rho_{b,\lambda}(\alpha, \beta, \theta, \phi)$ is the thermal spectral bidirectional reflectance distribution function for the surface; $\varepsilon_\lambda(\theta, \phi)$ is a directional emissivity; $R_{B,\lambda}(\theta, \phi)$ is the blackbody spectral radiance emitted from the surface based on the surface directional radiometric temperature ($T_{R,\lambda}(\theta, \phi)$; $L_\lambda(\alpha, \beta)$) is the incident radiance from the sky; and α and β are zenith and azimuth angles related to the incident hemisphere of view.

If the incident sky radiance is isotropic (i.e., $L_\lambda = L_\lambda(\alpha, \beta) = \text{constant}$), then the integral in Eq. 2.17 can be written as

$$L_\lambda \int_0^{2\pi}\int_0^{\pi/2} \rho_{b,\lambda}(\alpha,\beta,\theta,\phi)\cos(\alpha)\sin(\alpha)\, d\alpha\, d\beta = \rho_\lambda(\theta,\phi) L_\lambda \tag{2.18}$$

where $\rho_\lambda(\theta, \phi)$ is the hemispherical-directional spectral reflectance. Therefore, the relation between the directional emissivity ($\varepsilon_\lambda(\theta, \phi)$) and the hemispherical-directional reflectance ($\rho_\lambda(\theta, \phi)$) is given for opaque bodies at thermal equilibrium as

$$\varepsilon_\lambda(\theta,\phi) = 1 - \rho_\lambda(\theta,\phi) \tag{2.19}$$

at a specific wavelength λ.

A hemispherical radiometric temperature (T_H) for homogeneous media may be used to estimate the emitted radiance over a broad wavelength band i and full hemisphere of view, defined by

$$\varepsilon_i \sigma T_H^4 = \int_0^{2\pi} \int_0^{\pi/2} \int_{\lambda_1}^{\lambda_2} \varepsilon_\lambda(\theta,\phi) R_{B,\lambda}(\theta,\phi) \cos(\alpha) \sin(\alpha) \, d\lambda \, d\theta \, d\phi \quad (2.20)$$

2.3.6 Directional and Hemispheric Emissivities

Again according to Norman and Becker (1995), for isothermal, homogeneous emitters, the directional emissivity can be defined as the ratio of the actual emitted radiance to the radiance emitted from a blackbody at the same kinetic temperature. Surfaces possessing emissivity variations over wavelength ($\varepsilon_\lambda(\theta, \phi)$) are characterized by an average directional emissivity ($\varepsilon_i(\theta, \phi)$) for a given finite wavelength band i, which is given by

$$\varepsilon_i(\theta,\phi) = \frac{\displaystyle\int_{\lambda_1}^{\lambda_2} \frac{\varepsilon_\lambda(\theta,\phi) C_1}{\pi\lambda^5 \left[exp\left(\dfrac{C_2}{\lambda T_{R,i}(\theta,\phi)} \right) - 1 \right]} d\lambda}{\displaystyle\int_{\lambda_1}^{\lambda_2} \frac{C_1}{\pi\lambda^5 \left[exp\left(\dfrac{C_2}{\lambda T_{R,i}(\theta,\phi)} \right) - 1 \right]} d\lambda} \quad (2.21)$$

The hemispherical emissivity (ε_i) for a wavelength band i is defined as the directional emissivity as follows:

$$\varepsilon_i = \frac{1}{\pi} \int_0^{2\pi} \int_0^{\pi/2} \varepsilon_i(\theta,\phi) \cos(\theta) \sin(\alpha\theta) \, d\theta \, d\phi \quad (2.22)$$

For an isothermal surface for homogeneous media, the spectral hemispherical emissivity varies only slightly with wavelength, so,

$$\varepsilon_i = 1 - \rho_i \quad (2.23)$$

where ρ_i is the bihemispherical thermal reflectance for band i.

2.4 SUMMARY

In this chapter, thermal properties of material, the properties of thermal radiation laws, and thermal radiation of real-world materials are introduced and discussed. To better understand principles of TIR remote sensing techniques, several thermal properties of materials are first reviewed, and they may include (1) thermal conductivity, (2) thermal capacity, and (3) thermal inertia, also (4) thermal latent heat, (5) thermal expansion, and (6) thermal diffusivity. It is very useful and necessary to review those thermal properties of material, as they have an impact on our ability to remotely sense thermal information from various types of materials. To understand and evaluate the thermal radiation principles, the properties of thermal radiation are described by several thermal radiation laws and then discussed, including (1) Planck's radiation law, (2) the Stefan-Boltzmann law, (3) Wien's Displacement law, and (4) Kirchhoff's radiation law, and (5) emissivity and (6) inverse-square law are also reviewed. To better understand principles of TIR remote sensing, in the third section, several basic concepts and principles associated with thermal radiation of real-world materials are finally introduced. They include (1) kinetic energy/temperature (T_{kin}), radiant temperature/flux (T_{rad}), and brightness temperature; (2) energy transferring ways; (3) TIR atmospheric windows; (4) irradiance, radiant exitance, albedo, radiance and reflectance; and (5) more theoretical basis of directional and hemispheric radiometric temperatures and emissivities.

3 Thermal Infrared Remote Sensing Sensors, Systems, and Missions

The successful application of thermal infrared (TIR) remote sensing in diverse fields, spanning land, ocean, and atmospheric monitoring, necessitates a thorough understanding of the underlying technology. This chapter aims to provide readers with essential knowledge about the working principles, technical characteristics, and current status of operational airborne and spaceborne TIR sensors, systems, and missions. By illuminating these aspects, we equip readers with the tools to effectively utilize and interpret the rich datasets produced by these advanced instruments, facilitating informed decision-making and scientific inquiry across various disciplines.

3.1 WORKING PRINCIPLES OF TIR REMOTE SENSING IMAGING

TIR image sensors/systems are radiation detectors (in photo-electronics or charge couple device (CCD)) that expose a change in temperature from TIR sources (TIR targets) due to absorption of infrared radiation and convert this temperature variation into an electric output signal. The temperature of the pixel in a TIR image is directly proportional to the power of the absorbed TIR radiation (Budzier and Gerlach, 2011). Figure 3.1 outlines a signal process of TIR sensor/system, including TIR radiation received by a lens system from TIR sources through two basic scanning systems (across-track and along-track), then the TIR received radiation exposed to radiation devises/detectors after passing a prism and filter, and finally TIR image created and displayed via a conversion of electronic signal to thermography (i.e., digital TIR image). Thermal infrared sensors are also called photon or quantum sensors that record charge carriers generated by photons of the TIR radiation because of different photoelectric effects (Budzier and Gerlach, 2011). To enable one to effectively detect the low-energy (temperature) infrared radiation, photon sensors have to be cooled substantially below ambient temperature, so, TIR sensors can be work at ambient temperature. In order to measure and record the TIR radiation by detectors, although there are many different operating sensors/systems, depending upon the mechanism of scanning and data acquisition, there are two basic operating principles or methods of scanning that are used for collecting single-band, multi-/hyperspectral-spectral TIR image data: across-track thermal scanners and pushbroom linear and area-array CCD detectors (Figure 3.2) (Jensen, 2007).

DOI: 10.1201/9781003276173-3

3.1.1 Across-Track Scanning by Rotating a Scanning Mirror

Figure 3.2a shows the working principle of an across-track scanner by rotating a scanning mirror. Across-track scanners can scan the Earth in a series of lines. A rotating scan mirror in a scanner sweeps from one edge of the swath to the other. Since the scanning lines are oriented perpendicular to the direction of motion of the sensor platform (airborne or spaceborne), the scanning mode is called across-track scanning. As the platform moves forward over the Earth, the successive scanning lines build up a two-dimensional image to cover the Earth's surface. The incoming energy reflected or emitted radiation from Earth's surface is separated into several spectral components by a prism and filter, which are detected independently by detectors. The detectors measure the energy for each spectral band as an electrical signal, and then it is converted to digital data and recorded for spectral (e.g., thermal) image. In the scanned image, each pixel corresponds to each ground instantaneous field of view (IFOV). The ground pixel size and thus the spatial resolution of the image is determined by the IFOV of the sensor and the altitude of the platform. That is, per the Figure 3.2a, the spatial resolution is defined and calculated by a diameter of the circular ground area viewed by the sensor, which is a function of the product of the IFOV of the scanner measured in milliradians and the altitude of the platform above ground level measured in meters (Jensen, 2007). Usually, airborne scanners typically sweep large angles (FOV in Figure 3.2a between 90° and 120°), while satellite scanners sweep fairly small angles (FOV:10–20°) to cover a broad region due to a higher altitude of the platform. By this way of the cross-track scanning, imaging is carried out pixel by pixel and line by line, as the sensor-craft keeps moving forward.

Because the distance from the sensor to the target increases toward the edges of the swath (i.e., moves away from nadir on either side), the circle becomes an

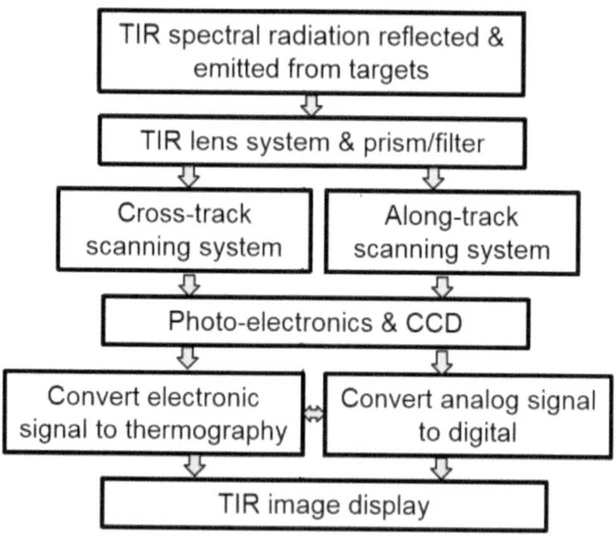

FIGURE 3.1 A block diagram of signal process of TIR sensor/system.

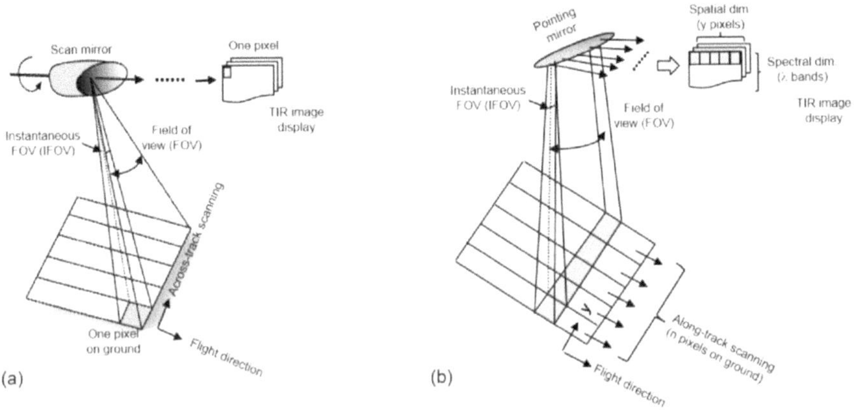

FIGURE 3.2 The basic types of TIR imaging principles. (a) Across-track scanning (by rotating a scan mirror); (b) along-track scanning (by pushbroom linear area).

ellipsoid, and thus the pixel size on the ground also becomes larger and introduces geometric distortions to the images, which is called one-dimensional relief displacement (Jensen, 2007). The displacement is introduced in both directions away from nadir for each scanning line with the across-track mirror. Also, the length of time the IFOV 'sees' a ground resolution cell as the rotating mirror scans (called the dwell time) is generally quite short and directly influences the design of the spatial, spectral, and radiometric resolution of the sensor.

There are many multispectral TIR imaging sensors/systems that adopt the across-track scanning principle to collect optical and TIR imagery. Typically, they include Landsat Thematic Mapper (TM, including TIR band), Landsat 7 Enhanced Thematic Mapper Plus (ETM+, including TIR band), NOAA Advanced Very High-Resolution Radiometer (AVHRR, typical channels 3–4), NOAA Geostationary Operational Environmental Satellite (GOES Imager bands 2–5), and NPOESS Preparatory Project (NPP sensors).

3.1.2 Along-Track Scanning by Using a Pushbroom Linear Array or Area-Array

Along-track scanners also use the forward motion of the platform to record successive scan lines and build up a two-dimensional image (Figure 3.2b). Instead of a scanning mirror, the along-track scanning uses a linear array of detectors located at the focal plane of the image formed by lens systems, which are 'pushed' along in the flight track direction. These systems are also referred to as pushbroom scanners, as the motion of the detector array is analogous to the bristles of a broom being pushed along a floor. Therefore, based on the along-track scanning principle, the pushbroom imaging sensor can use a linear array or area-array CCD-area array of detectors to collect multi-/hyper-spectral TIR imagery. In a linear array mode (Figure 3.2b), the number of pixels equals that of ground cells for a given swath (a line). The motion of platform (airborne and satellite) realizes the scan in the along-track direction, and thus the line frequency equals the pixel dwell time. Each individual detector in a linear array measures the energy for a single ground resolution cell and thus the altitude and IFOV of the detectors determine the spatial resolution of the imaging system. For a multispectral scanner, a separate linear array is required to measure each spectral band or channel. For each scan line, the energy detected by each detector of each linear array is filtered spectrally, sampled electronically, and recorded digitally. For a hyperspatial scanner, imaging detectors in an area-array is required to measure each hyperspectral spectral band (spectral dimension, λ) for each pixel (spatial dimension, y) to form a $\lambda \times y$ area array of image (i.e., an image with λ bands with y pixels along one line). The rest of the pushbroom scanning mechanism remains the same as in the linear array mode.

Along-track scanning with linear array or area array has several advantages over across-track mirror scanning. (1) Due to the spatial scanning completed by solid scanning of device, the array of detectors combined with the pushbroom motion allows each detector to 'view' and measure the energy from each ground resolution cell for a longer time (dwell time). This allows more energy to be detected and improves the radiometric resolution and thus increases the imaging SNR and improves the sensitivity and spatial resolution of the imaging system. (2) Because detectors are usually solid-state microelectronic devices, such scanners are lighter, smaller, and less complex than the across-track counterparts because of fewer moving parts. However, their major disadvantage is that cross-calibrating thousands of detectors to achieve uniform sensitivity across the array is necessary, complicated, and time consuming.

There are many TIR imaging systems, especially hyperspectral TIR imaging sensors/systems, which adopt the along-track scanning principle. For example, NASA Terra/Aqua Moderate Resolution Imaging Spectrometer (MODIS, typical TIR channels 31 and 32), NASA Preparatory HyspIRI mission (TIR multispectral sensor), AISA-OWL airborne hyperspectral TIR sensor, Airborne Thematic Thermal InfraRed and Electro-Optical Imaging System (ATTIREOIS), and Airborne Hyperspectral Thermal Imager (TASI-600).

3.2 AIRBORNE TIR REMOTE SENSING SENSORS/SYSTEMS

During the last four decades, many airborne TIR imaging sensors/systems were developed and tested. Table 3.1 presents a summary of existing airborne multi-/hyper-spectral TIR sensors and systems including basic technical parameters, characteristics, date of development, and manufacturers. A survey of a total of 18 selected important airborne multi-/hyper-spectral TIR imaging sensors and systems in their alphabetic order is summarized in this section by referring to Kramer (2002), Pu (2017), Gerhards et al. (2019), Liu et al. (2020), and Neinavaz et al. (2021).

3.2.1 Airborne Hyperspectral Imager (AHI)

The Airborne Hyperspectral Imager (AHI) was designed and developed by The University of Hawaii, which consists of a long-wave infrared pushbroom hyperspectral

TABLE 3.1
A Summary of Existing Airborne Multi-/Hyper-Spectral TIR Imaging Sensors and Systems (Typical Wavelength 7–15 µm)*

Sensor/system	Number of bands	Spectral range (µm)	Spatial resolution	Band width (FWHM, nm)	IFOV (mrad)	FOV (⁰)	Operation	Country/Agency
AHI	32 or 256	7.5–11.5		~125 or ~15	2		Since 1998	USA/The University of Hawaii
AHS	10	7.95–13.14	2.0 m @ 1 km AGL		2.5	90		USA/SensyTech Inc.
AISA-OWL	Up to 96	7.5–12.3	1.1–1.5 m @ 1 km AGL	100	1.1	24 or 32.3	Since 1993	Finland/SPECIM, Spectral Imaging, Ltd
ATHIS	155	8.0–12.5		38	1	40	Since 2019	China/Shanghai Institute of Technical Physics
ATLAS	6	8.2–12.2	2.0 m @ 1 km AGL	400–1,000	2	72	Since 1998	USA/NASA
ATTIREOIS	1	10.0–12.5	0.27 m	2,000			Since 2011	USA/NASA/Flight Landata Inc (FLD)
CIS	2	10.5–12.5		1,000	1.2 x 1.2		Since 1993	China/Shanghai Institute of Technical Physics
DAIS-7915	6	8.7–12.7	10 m @ 3 km AGL	600	3.3	78	Since 1994	USA/GER Corporation
DAIS-16115	12	8.0–12.0	10 m @ 3 km AGL	333	3	78	Since 1994	USA/GER Corporation
DAIS-21115	6	8.0–12.3	20 m @ 1 km AGL				Since 1994	USA/GER Corporation
G-LiHT	1	8.0–14.0	1 m @ 335 m AGL			30	Since 2011	USA/NASA
HyMAP	32	8.0–12.0	3–10 m @ 1.5–5 km AGL	100–200	2.5, 2.0	61.3	Since 1996	Australia/Australian Integrated Spectronics Company
Hyper-Cam-LW	Up to 550	7.4–12.5	0.3 m @ 1 km AGL	Up to 4 (cm^{-1})	0.35	14 × 11	Since 2009	Canada/Telops
HyTES	256	7.5–12.0	1.8 m @ 1 km AGL	18	1.7	50	Since 2013	USA/NASA/JPL
MAIS	7	8.2–12.2		400–800	3	90	1987–1990	China/Shanghai Institute of Technical Physics of CAS
MIVIS	10	8.2–12.7		400–500	2	71	Since 1993	USA/Sensys Technologies Inc.
MAKO	128	7.8–13.4	1 m @ 1830 m AGL	44	2	90	Since 2010	USA/Aerospace Corporation

(*Continued*)

TABLE 3.1 (*Continued*)
A Summary of Existing Airborne Multi-/Hyper-Spectral TIR Imaging Sensors and Systems (Typical Wavelength 7–15 μm)*

Sensor/system	Number of bands	Spectral range (μm)	Spatial resolution	Band width (FWHM, nm)	IFOV (mrad)	FOV (⁰)	Operation	Country/Agency
MUSIC	90	6.0–14.5		60–1400	0.5		Since 1989	USA/Lockheed Palo Alto Research Laboratory, CA
SEBASS	128	7.5–13.5	1.0 m @ 1 km AGL	46	1.1	7.3	Since 1995	USA/Aerospace Corporation
TASI-600	32	8.0–11.5	0.85 m @ 1 km AGL	110	1.2	40	Since 2006	Canada/ITRES Research
TIMS	6	8.2–12.2	1.4 m @ 1 km AGL	400–1000	2.5	80	Since 1984	USA/NASA/JPL

* references: Kramer (2002), Pu (2017), Gerhards et al. (2019), Liu et al. (2020), and Neinavaz et al. (2021).AHI: Airborne Hyperspectral Imager
AHS: Airborne Hyperspectral Scanner
AISA: Airborne Imaging Spectrometer for different Applications
ATHIS: Airborne Thermal infrared Hyperspectral Imaging System
ATLAS: Airborne Thermal/Visible Land Application Sensor
ATTIREOIS: Airborne Thematic Thermal InfraRed and Electro-Optical Imaging System
CIS: Chinese Imaging Spectrometer
DAIS: Digital Airborne Imaging Spectrometers
G-LiHT: Goddard's LiDAR, Hyperspectral and Thermal airborne imager
HyMap: Hyperspectral Mapper
Hyper-Cam: Hyperspectral Cameras
HyTES: Hyperspectral Thermal Emission Spectrometer
MAIS: Modular Airborne Imaging Spectrometer
MIVIS: Multispectral Infrared and Visible Imaging Spectrometer
MAKO: airborne thermal infrared imaging spectrometer
MUSIC: MUltiSpectral Infrared Camera
SEBASS: Spatially-Enhanced Broadband Array Spectrograph System
TASI: Airborne Hyperspectral Thermal Imager
TIMS: Thermal Infrared Multispectral Scanner

imager and a bore sighted three-color visible high resolution CCD line-scan camera. The system is a helicopter-borne TIR hyperspectral imager with real-time on-board radiometric calibration and mine detection. AHI is an imaging spectrometer covering the spectral range 7.5–11.5 μm and collects narrow-band images in up to 256 spectral bands. The spectrometer of the AHI system does not use low-temperature optical refrigeration, and thus it has 32 bands and the spectral resolution is approximately 125 nm at the initial design stage. AHI noise equivalent delta temperature is < 0.1 K at 300 K at all wavelengths. The IR sensor consists of four subsystems: telescope, spectrograph, background suppressor, and focal plane array and associated electronics. The initial design for the AHI included a 12-GB hot-swappable RAID array. Data was sampled off of the 256 × 256 focal plane to 32-band resolution. A recent upgraded version to the instruments' data system has allowed full resolution data to be streamed off of the sensor directly to a single IDE disk drive to allow the instrument's full 256 band spectral resolution to be captured (Winter et al., 2004). AHI was funded by the Defense Advanced Research Projects Agency to investigate the utility of hyperspectral imaging for detection of buried land mines in a project called the Hyperspectral Mine Detection Program. The project was very successful, and the AHI system is currently used by the US Army for counter-mine and other research purposes.

3.2.2 Airborne Hyperspectral Scanner (AHS)

The airborne hyperspectral scanner (AHS), developed by SensyTech Inc., currently ArgonST, USA, is operated by the Spanish Institute of Aeronautics and it was placed onboard the aircraft CASA 212–200 Paternina. The AHS sensor is based on the integration of many advanced technologies developed by SenSyTech under R&D contracts over a few years before 2006. The AHS incorporates advanced components to ensure high performance while maintaining the ruggedness to provide operational reliability in a survey aircraft. The main AHS technical parameters and the arrangement of the spectral bands are 80 bands in four ports (VIS, NIR, SWIR, MIR, and LWIR), while bands 71–80 cover the thermal spectral range (LWIR). The AHS has a 90° (± 45°) FOV, 2.5mrad IFOV, and 2–6 m GFOV at 140 kt cruise speed.

3.2.3 Airborne Imaging Spectrometer for Different Applications, (AISA-OWL)

AISA is a commercially available hyperspectral imaging spectrometer of Spectral Imaging Ltd. (Specim) of Finland. The objective of developing the AISA is to provide users with a wide variety of application purposes with different models. The AISA family consists of several models including AISA-OWL, the TIR model. Other models cover VNIR, SWIR, and MIR spectral ranges. Specim's thermal airborne hyperspectral sensor AISA-OWL covers the contiguous spectral range from 7.6 to 12.3 μm in 96 bands at spectral resolution of ~100 nm and IFOV = 1.1 mrad, and it has the sensitivity to detect and classify even gases. A cooled HgCdTe array detector is adopted in AISA-OWL. The AISA-OWL is a fast pushbroom hyperspectral system designed to provide the remote sensing market with a high performance LWIR hyperspectral imager in the smallest achievable size. Thanks to its design and sensitivity, AISA-OWL acquires all its spectral bands for its whole swath-width of 384 pixels simultaneously, and sustains the 100 frames per second storage speed of the whole data cube continuously. As a result, the high sensor or target dynamics are possible without degradation of image quality. It operated at the beginning of 1993. AISA is a programmable pushbroom imaging spectrometer based on CCD technology (Kramer, 2002). Many application areas of AISA include hydrology, geology, forestry, and agriculture (Pu, 2017). The AISA can be flown on a light aircraft at 1 km altitude.

3.2.4 Airborne Thermal Infrared Hyperspectral Imaging System (ATHIS)

The airborne thermal infrared hyperspectral imaging system (ATHIS) is deployed by China/Shanghai Institute of Technical Physics on an airborne platform and its IFOV corresponds to the TIR hyperspectral information of a line on the ground. The ATHIS adopts the technical scheme of spectrometer refrigeration. To improve the data-acquisition efficiency, the ATHIS consists of three identical TIR hyperspectral cameras to achieve 40° observation FOV by post mosaic and the FOV of a single camera is approximately 14°. To reduce the volume of ATHIS, three reflective triplet spectrometers of the same design are placed in a cold box with a low temperature of 100 K. The weight of ATHIS is approximately 35 kg. Two high-power Stirling refrigerators are used to achieve the goal of low-temperature refrigeration. The three telescopes are of independent structure, which are placed outside the cold box and are separately connected with their spectrometer components inside the cold box (Liu et al., 2020). To ensure the operation of ATHIS under different conditions, the electronic control cabinet is also designed for power supply and detector signal acquisition. An independent high-precision IMU (PosPac 610 Applanix) is rigidly connected to the platform, which can be used for geo-referencing of the collected data (Liu et al., 2020).

3.2.5 Airborne Thermal/Visible Land Application Sensor (ATLAS)

The Airborne Terrestrial Applications Sensor (ATLAS) operates in the visual and infrared bands. The ATLAS is one of NASA aircraft-based sensors, including the Daedalus TM Simulator aboard the ER-2 aircraft and the ATLAS aboard the Leaijet. The ATLAS can detect 15 multispectral channels of the radiation through the visible, near infrared, and thermal spectrums. Among them, bands 10–15 cover TIR range (8.2–12.2 μm) with a spectral resolution 400–1000 nm and FOV = 72°. The ATLAS mission request was designed to best suit agricultural research. In 1998, the USDA-ARS U.S. Water Conservation Laboratory and NASA Stennis Space Center investigated the use of ATLAS spectral imagery for farm management applications (Moran, 1998). The ATLAS Mission of San Juan, Puerto Rico was conducted during February of 2004 to investigate the impact of the urban growth and landscape in the climate of this tropical city. This ATLAS sensor has been used in other field campaigns to investigate the UHI in Atlanta, Salt Lake City, Baton Rouge, and Sacramento, all in the continental mass of the United States of North America.

3.2.6 Airborne Thematic Thermal InfraRed and Electro-Optical Imaging System (ATTIREOIS)

Funded by NASA-GSFC, Flight Landata Inc (FLD) developed an advanced Airborne Thematic Thermal InfraRed and Electro-Optical Imaging System (ATTIREOIS). ATTIREOIS features two sets of advanced Focal Plane Arrays: a broadband Thermal InfraRed Sensor (TIRS) and a four band Multispectral Electro-Optical Sensor to approximate Landsat ETM+ bands 1–4, and 6, and LDCM bands 2–5, and 10+11. The sensor component itself of the airborne TIRS is basically the same as that initially developed for LDCM by NASA-GSFC. More specifically, TIRS for ATTIREOIS uses the same performance-proven Quantum Well Infrared Photodetector (QWIP) focal plane array device as the TIRS for LDCM, which is developed at NASA GSFC. Those QWIP imagers have a broadband spectral response in the 10.0–12.5 μm range and are operational at 43K cooled by high performance cryocoolers. By integrating three QWIPs on the same focal plane and pushbroom-operating the sensor on a stabilized platform, the design objectives of wide swath width (1850 pixels), fast data throughput, risk mitigation, and quick implementation of the LWIR imaging have been achieved (Sun and Shu, 2011). For NASA applications, ATTIREOIS may be used to under flight Landsat/LDCM satellites to support the satellite measurement calibration and validation observations. In addition, ATTIREOIS will fulfill commercial applications, for which one of FLD's business is interested in starting a private Google Earth Globe that layers timely updated ATTIREOIS high resolution LWIR/EO regional data over the moderate resolution Landsat/LDCM global data.

3.2.7 Digital Airborne Imaging Spectrometers (DAIS Series TIR Sensors)

DAIS-7915 is a 79-channel/15-bit quantization electronical-mechanical scanner developed and built by the Geographical Environmental Research Corporation (GER) of Milbrook, NY. A DAIS-7915 sensor covers the spectral range from the visible to the thermal infrared wavelengths at variable spatial resolution from 3 to 20 m depending on the carrier aircraft flight altitude (1.5–4.5 km AGL). For the TIR spectral range, there are six multispectral TIR bands with the bandwidth 600 nm. DAIS is accessible through DLR (German Aerospace Center) to serve European Airborne Remote Sensing Capabilities (EARSEC), as well as for other contractual applications worldwide (Kramer, 2002). Image data are measured with a radiometric resolution of 15 bits/pixels and coregistered bands. The DAIS scan mechanism is of Kennedy type where a cubic polygon mirror scans the terrain below through the opened window hatch in the bottom of the aircraft. The scan mirror rotates anti-clockwise with respect to the aircraft heading to provide a ground element cross track scanning motion while the forward motion of the aircraft provides a line-by-line scan. DLR's DAIS-7915 is operated from a Dornier DO-228 aircraft and has a FOV = 26°, which is covered with 512 pixels per scanline. The IFOV currently used is 3.3 mrad, giving a typical pixel size of 10 m from a flight altitude of 3 km AGL. The DAIS-7915 has flown since 1994. The DAIS-16115 has 161 channels (among which are 12 multispectral TIR bands) but still at 15-bit quantization. All spectrometer channels are spatially registered (image cube) and its potential applications are the same as DAIS-7915. Compared to DAIS-7915, DAIS-21115 has a very different band setting for VNIR and SWIR spectral ranges, but for TIR range, very similar multispectral TIR band setting (i.e., six TIR bands covering TIR spectral range). The DAIS imaging spectrometers can be used in environmental monitoring of land and marine ecosystems, vegetation stress research, agriculture and forestry resources mapping, geological mapping, mineral exploration, and provision of data for geographic information systems (Pu, 2017).

3.2.8 Goddard's LiDAR, Hyperspectral and Thermal (G-LiHT) Airborne Imager

Goddard's LiDAR, Hyperspectral and Thermal (G-LiHT) airborne imager is a unique system that allows simultaneous measurements of vegetation structure, foliar spectra, and surface temperatures at very high spatial resolution (~1 m) on a wide range of airborne platforms. G-LiHT was designed as a relatively inexpensive, robust, and portable research tool for evaluating the potential benefits of data fusion for studies of terrestrial ecosystems. With the G-LiHT sensor system, hyperspectral imaging spectroscopy provides quantitative information on vegetation cover, species composition, and biophysical and chemical properties that can be derived from measurements of reflected sunlight in the visible through SWIR wavelengths; LiDAR provides quantitative, 3D information on terrain and vegetation cover, height and distribution of canopy elements; land surface temperature provides data needed to estimate evapotranspiration and other surface energy fluxes and can be an indicator of soil or vegetation moisture status. The G-LiHT's thermal remote sensing capability originates from the Gobi-384 thermal imaging camera (see its measurement characteristics in Table 3.2). The instrument may measure TIR radiation over a broad spectral range (8 to 14 μm) using an uncooled amorphous silica microbolometer detector with 384 across track pixels and a 30° FOV. The comprehensive measurements map plant species composition, plant functional types, biodiversity, biomass and carbon stocks, and plant growth. In addition, G-LiHT data enhance our ability

TABLE 3.2
Measurement Characteristics of G-LiHT Thermal Camera (Cook et al., 2013)

Spectral range	8–14 mm
Swath width/FOV	173 m/30°
Imaging array size	384 × 288
Sensitivity (NETD)	>50 mK at 30°C
Acquisition rate	25 Hz
Radiometric resolution	16 bit
Spatial resolution	1 m at 335 m and 110 kt

NETD: Noise Equivalent Temperature Difference

to validate data from existing satellite missions and support NASA Earth Science research. Low-cost deployment of the G-LiHT system opens a wide range of applications for targeted, airborne remote sensing, including diurnal and seasonal processes in terrestrial ecosystems at ~1 m spatial resolution.

3.2.9 Hyperspectral Mapper (HyMap)

The HyMap Imaging Spectrometer was built by Australian Integrated Spectronics Ltd. and operated by HyVista Corp. It is a 'whiskbroom style' instrument, similar to Figure 3.2a, an across-track scanning mode. The HyMap systems have set the benchmark for commercially available hyperspectral sensors in terms of SNR, image quality, stability, adaptability, and ease of use. The instrument offers an extremely high SNR of more than 500:1, which is more than most existing imaging spectrometers including AVIRIS can realize. The HyMap provides hyperspectral spectral ranges covering across the solar wavelengths (0.4–2.5 μm) with 126 bands and 32 bands in the TIR (8–12 μm). The spectral bandwidths of HyMap hyperspectral data are 100–200 nm in the TIR spectral range (see more technical parameters in Table 3.3). The design of the HyMap series of airborne hyperspectral sensors features an opto-mechanically scanned fore-optics combined with modular, high efficiency spectrographs and optimized detector arrays. The HyMap sensor is mounted on a stabilized platform. Position and orientation during the flight are measured continuously by means of Differential-GPS and INS (Bucher and Lehmann, 2000). The HyMap records an image by using a rotating scan mirror, which allows the image to build line by line as the aircraft flies forward. The HyMap offers rapid and efficient wide-area imaging, especially for mineral exploration and environmental monitoring applications and it has been operated in Europe, Australia, the USA, and Africa in 24 specific campaigns.

3.2.10 Hyperspectral Cameras (Hyper-Cam xLW)

Hyper-Cam xLW is a Fourier-transform TIR hyperspectral imager developed by Telops of Canada. Hyper-Cam has two models, hyper-cam MW (3–5 μm) and

TABLE 3.3
HyMap Sensor Technical Characteristics (Kramer, 2002)

Imager type	Whiskbroom scanner
Detector array	Line array
Wavelength range	450–2,500 nm, 8,000–12,000 nm
Spectral resolution (FWHM)	15–16 nm (0.4–2.5 μm), 100–200 nm (8.0–12.0 μm)
SNR	> 500:1
FOV	60° (512 pixels across track)
IFOV	2.5 mrad (along track), 2.0 mrad (across track)
Spatial resolution	3–10 mw
Aircraft	Cessna 402B, WB-57–JSC
Altitudes	1,500–5,000 m AGL

hyper-cam xLW (7.7–11.8 μm) (Liu et al., 2020). The Hyper-Cam is an advanced passive infrared hyperspectral imaging system that combines high spatial and spectral resolution. It provides real-time radiometrically calibrated data for gas and mineral detection and identification. The spectral range of the latest version, Hyper-Cam Mini xLW, is 7.35–13.5 μm. The14 × 11° FOV with subwindowing permits to fit the desired FOV. Hyper-Cam Mini xLW is an advanced compact infrared hyperspectral imaging system that combines high spatial, spectral, and temporal resolution for remote detection, identification, and quantification. Its spatial resolution is 0.35 mrad and its spectral resolution is up to 4 cm–1. The Hyper-Cam Airborne Mini is a revolutionary hyperspectral imaging system that is designed to fit small aircrafts and other compact vehicles. This lightweight imaging sensor is a versatile tool for hyperspectral IR surveys and a unique instrument for real-time gas detection, identification, and quantification. The data collected by Hyper-Cam include all the relevant information for orthorectification and stitching. Visible images are simultaneously recorded along with the infrared data cubes using a boresight CCD camera on the airborne platform.

3.2.11 Hyperspectral Thermal Emission Spectrometer (HyTES)

The Hyperspectral Thermal Emission Spectrometer (HyTES) was developed by NSAS/Jet Propulsion Laboratory (JPL) in 2013 as a part of the risk reduction activities associated with the Hyperspectral Infrared Imager (HyspIRI). HyTES, an airborne TIR hyperspectral imager, adopts a similar scheme and represents the highest level of the development of the TIR hyperspectral imager. Its technical difficulty even exceeds most of the space hyperspectral imagers in the same time period. The science objectives of HyTES are the same as those of HyspIRI. Therefore, the HyTES provides the HyspIRI group data at much higher spatial and spectral resolutions to help determine the optimum band positions for the HyspIRI-TIR instrument (Liu et al., 2020). The HyTES pushbroom design has 512 spatial pixels over a 50° field of view and 256 contiguous spectral bands between 7.5 and 12 μm in the TIR wavelength region. HyTES includes many key enabling state-of-the-art technologies including a high-performance concave diffraction grating, a quantum well infrared photodetector focal plane array, and a compact Dyson-based optical design. The system uses two mechanical cryocoolers to maintain instrument temperature. The first cooler holds the focal plane array at 40 K and the second cooler holds the remainder of the cryovacuum system at 100 K. The HyTES may be used to help address the TIR questions including five main science themes: volcanoes, wildfires, water use and availability, urbanization, and land surface composition and change.

3.2.12 Airborne Thermal Infrared Imaging Spectrometer (MAKO)

The Airborne thermal infrared imaging spectrometer (MAKO) was designed and built by USA/Aerospace Corporation, USA in 2007 and began flight testing in 2010. The core of the MAKO instrument is the cryogenic Dyson spectrometer. The spectrometer disperses the 7.8–13.4 μm TIR spectral range into 128 spectral bands defined by the rows of the array. Its slit length also spans 128 spatial pixels, and each spatial pixel is dispersed along an array column (i.e., 128 bands). The external telescope has an angular magnification of 3.66X, reducing the 2 mrad intrinsic IFOV of the cryostat optics to 0. 547 mrad, or 1 meter GSD at 1.8 km AGL. The telescope also relays the system cold stop to an entrance pupil close to the pointing mirror—so the size of the pointing mirror and blackbody calibration. Sources are minimized. Recently, the maximum cross-track scan width has been increased, at which the sensor can view the ground without vignetting by the aircraft port through which it looks, from ±45° to ±56.4°. This gives 3,600 across-track pixels per whisk and results in an area coverage rate of 33 km^2/minute at a nadir ground sampling distance of 2 m at AGL = 3660 m and typical 51.4 m/s ground speed of the Twin Otter aircraft. An additional improvement is the integration of a new commercial 3D stabilization mount, which is significantly better at compensating for aircraft motions and thereby maintains scan performance under quite turbulent flying conditions.

3.2.13 Modular Airborne Imaging Spectrometer (MAIS)

The MAIS was designed and operated by the Shanghai Institute of Technical Physics (SITP) of the Chinese Academy of Sciences (CAS). The instrument is the first real imaging spectrometer in China. It provides 71 spectral bands; the first 64 bands having a bandwidth of 20–30 nm in the spectral range 0.44–2.50 μm; the second seven bands cover the thermal infrared range 8.2 –12.2 μm. The MAIS features a modular design as most hyperspectral imager. Its design is based on the linear array detectors added to an optical mechanical scanner frame. The whole system consists of a tilted 45° rotation mirror optical mechanical scanner unit, main optical system, VNIR spectrometer, SWIR spectrometer, TIR spectrometer, data acquisition system, motor drive, and power supply system. Therefore, separate calibration of each module is possible. The MAIS was first used in 1990 and thereafter in a number of remote sensing applications by SITP for geological and environmental survey. During the period September–October 1991, the MAIS was installed on a Citation S/II aircraft of the CAS and flown successfully in a joint Sino-Australian remote sensing campaign near Darwin and at several other test sites in Western Australia (Pu, 2017).

3.2.14 Multispectral Infrared and Visible Imaging Spectrometer (MIVIS)

The Daedalus AA5000 Multispectral Infrared and Visible Imaging Spectrometer (MIVIS) is a very promising whisk-broom system, operating at 12 bits resolution in 102 narrow spectral bands from the visible to the thermal infrared, among which there are ten TIR bands covering the 8.2–12.7 μm spectral range. It is a product made by Sensys Technologies Inc, USA. The MIVIS is a hyperspectral imaging system that is flown onboard of a CASA 212 aircraft. The instrument operates in the whisk-broom configuration, and the light, reflected by the frontal rotating mirror, is collected by the parabolic telescope and feeds four spectrometers, providing spectral coverage from the visible to the thermal infrared range. There are 755 cross-track pixels along a scanning line, which is imaged through 2.0 mrad IFOV and 71.059° FOV. MIVIS provides the highest digitalization accuracy if compared with similar remote sensing imaging systems, like the ten bits of the Airborne Visible/infrared Imaging Spectrometer (AVIRIS) and the only eight bits of the Thermal Infrared

Multispectral Scanner (TIMS). Besides the MIR spectral region between 2.0 and 2.5 μm being very useful for geological studies, an important feature of this sensor is the availability of ten narrow bands in the TIR range from 8.21 to 12.70 μm. Compared to the TIMS, which has only six spectral channels in the same spectral range, the MIVIS provides a better signal-to-noise ratio.

3.2.15 MUltiSpectral Infrared Camera (MUSIC)

The MUltiSpectral Infrared Camera (MUSIC) was developed by and built at the Lockheed Palo Alto Research Laboratory, California. The current version of the MUSIC airborne hyperspectral imager has been operated since 1989 and flew on NASA research aircraft at up to 20 km AGL. The MUSIC measures infrared image data simultaneously in the MIR (2.5–7.0 μm) and TIR (6.0–14.5 μm) spectral ranges (Kramer, 2002). The spatial resolution is associated with 0.5 mrad IFOV and each infrared range is covered with a total of 90 spectral bands. The sensor consists of two parallel optical telescopes, each with its own set of optical filters and detector arrays. Each detector array is with a 45 × 90 pixel matrix for a total 4,050 detectors in each array and all detectors collect data in parallel at up to 80 frames/s. The detector arrays are cooled with liquid helium. Due to its cooling, the internal background temperature is very low and detectors operate close to the BLIP limit as defined by the external signal through the cooled spectral filters (Kramer, 2002). The system also offers a quicklook capability. The objective of developing the MUSIC sensor is for chemical vapor sensing, plume diagnostics, and extraction of spectral signatures (Pu, 2017).

3.2.16 Spatially-Enhanced Broadband Array Spectrograph System (SEBASS)

The Spatially-Enhanced Broadband Array Spectrograph System (SEBASS) was developed by Aerospace Corporation, USA, in 1995. The instrument uses two spherical-faced prisms to operate simultaneously in the atmospheric transmission windows found between 2.0 and 5.2 μm (MIR) and between 7.8 and 13.4 μm (LWIR). The system records both MIR and TIR incoming energy by dispersing the spectral range of either the MIR or TIR atmospheric window band onto 128 ×128-pixel focal plane arrays. A set of low-temperature optical systems with all-optical-circuit refrigeration was designed for the instrument. The entire system is cooled to 10 K by liquid helium to reduce the influence of the background radiation of the light machine and improve the overall performance of the instrument. The SEBASS instrument uses an additional aplanatic lens, which improves the optical performance of the spectrograph over a two-dimensional field and provides demagnification of the image of the slit at the focal plane. Long-wavelength light is dispersed by a spherical-faced prism/conic mirror combination and is imaged onto a focal plane array through a zinc selenide lens. Although the SEBASS instrument is designed primarily for use from an aircraft platform, it was used in March 1996 for a tower-based collection. In general, the SEBASS is intended to explore the utility of hyperspectral infrared sensors for remotely identifying solids, liquids, gases, and chemical vapors in the 2 to 14 μm 'chemical fingerprint' spectral region.

3.2.17 Airborne Hyperspectral Thermal Imager (TASI-600)

Airborne Hyperspectral Thermal Imager (TASI-600) was made by Canadian ITRES Research Limited in 2006. It is an advanced airborne hyperspectral thermal pushbroom imager features with 32 spectral channels, a spectral range of 8 to 11.5 μm, a 40° FOV, and a 600 spatial pixels line image. The spectral smile and key distortion are less than 0.5 pixels. The TASI 600 is distinguished from other thermal sensors in two important ways: it is a 'mapping sensor,' so that every pixel in a thermal image is referenced with geographic coordinates and an elevation value; and it is 'hyperspectral'—instead of offering a single grayscale image covering the entire thermal spectrum, the TASI provides 32 spectral bands so images can be viewed in 'hypercolor.' The TASI-600 is fully compatible with the Itres real-time processing system, which is capable of real-time data backup, radiometric calibration, diagnostics alerts, image processing, and target detection. Instead of receiving only a relative measure of temperatures, TASI-600 users can determine emissivity values and absolute temperatures and identify materials, such as rock types, soil classes, metals, plastics, and other manmade materials.

Therefore, TASI-600 is well-suited to applications including mining exploration, geophysical mapping, and landmine and ordnance detection.

3.2.18 Thermal Infrared Multispectral Scanner (TIMS)

Thermal Infrared Multispectral Scanner (TIMS) is a six-channel scanner with high radiometric sensitivity that consists of a 19-cm diameter Newtonian reflector telescope mounted behind an object-plane 45° flat scanning mirror followed by a Czerny-Turner spectrometer. TIMS covers TIR spectral range 8.2–12.2 μm with six bands and it uses the IFOV of 2.5 mrad and a total FOV 80° to collect images. The sensitivity ranges between 0.1° and 0.3°C noise-equivalent temperature change at 300 K. This sensitivity is comparable to a noise-equivalent change in spectral emissivity of 0.002 to 0.006. This sensitivity and the internal reference sources allow an accurate estimate of the brightness temperature of water and the detection of stress in plants (through changes in temperature). The instrument may be used whenever an accurate measure of the spectral radiance or brightness temperature of the Earth's surface or atmosphere is required in the TIR atmospheric windows within the TIR spectral region (8.2–12.2 μm).

3.3 SPACEBORNE TIR REMOTE SENSING SENSORS/MISSIONS

In this section, a survey of ten selected important and current/operational spaceborne multi-/hyper-spectral TIR imaging sensors and missions is summarized by referring to Yamaguchi et al. (1998), Jensen (2007), Huang et al. (2013), Pu (2017), and Yao et al. (2020). Table 3.4 presents a summary of existing spaceborne multi-/hyper-spectral TIR sensors, systems, and missions in terms of their series and group,

TABLE 3.4
A Summary of Existing Satellite Multi-/Hyper-Spectral TIR Imaging Sensors/Systems (Except HYspIRI; Typical Wavelength 7–15 μm)

Satellite	Sensor/system	Revisit period*	No. of TIR bands	TIR spectral range, wavelength (μm)	Spatial resolution (m)	Swath width (km)	Agency	Operation, launch year	Information source**
Polar orbits									
ERS-2/EnviSAt	ATSR/AATSR			10.8–12.0	1,000	5,121/500	ESA< UKSA	since 1995	https://earth.esa.int/eogateway/instruments/aatsr
CBERS-3,4,4b	IRSCAM	26 D		10.4–12.5	80	120	CRESDA, INPE	since 2014	https://space.oscar.wmo.int/satelliteprogrammes/view/cbers,
Envisat	AATSR	35 D	2	11.0, 12.0	1,000	500	ESA, UKSA	2002–2012	http://www.atsrsensors.org/aatsr.htm
ERS-1,2	ATSR-1,2	35 D (2 D?)	2	10.35–12.50	1,000	512	ESA, UKSA, CSIRO	Since 1991	http://www.atsrsensors.org/aboutATSR.htm
FY-3A,3B	MERSI	1 D			250	2,800	NRSCC, CAST, NSM a.o.	2008, 2010	https://navigator.eumetsat.int/product/EO:EUM:DAT:0100
Fengyun-3(A/B/C)	VIRR	5.5 D	2	10.3–12.5	1,100			Since 2008	https://fy4.nsmc.org.cn/data/en/instrument/VIRR.html
GCOM-C1, SGlI	SGLI		2	10.8–12.0	500		Japan (Jxax)	Since 2009	https://ieeexplore.ieee.org/stamp/stamp.jsp?tp=&arnumber=6721288
Himawari-8 & 9	AHI		7	7.30–13.28	2,000		Japan (JMA)	Since 1977	https://spaceflight101.com/spacecraft/himawari-8-and-9/
HJ-1B	InfraredCam	31 D			300	720	CRESDA, CAST, NRSCC	Since 2008	https://space.oscar.wmo.int/instruments

Landsat-4,–5	TM	16 D	1	10.4–12.5	120	185	NASA	1982–2011	https://earthexplorer.usgs.gov/
Landsat-7	ETM+	16 D	1	10.4–12.5	60	185	NASA	Since 1999	https://earthexplorer.usgs.gov/
Landsat-8	TIRS	16 D	2	10.60–12.51	100	185	NASA	Since 2013	https://earthexplorer.usgs.gov/
NASDA ADEOS-II	GLI		7	3.725–11.95	1,000	1,600	Japan	Since 2002	https://crisp.nus.edu.sg/~research/tutorial/gli.htm
NOAA	AVHRR	0.5 D	2	10.5–12.5	1,100	2,700	NOAA	Since 1978	https://www.avl.class.noaa.gov/release/data_available/avhrr/index.htm
Sentinel-3(A/B)	SLSTR	1 D	2	10.00–12.02	1,000			Since 2016	https://sentinel.esa.int/web/sentinel/home
Suomi NPP	VIIRS	<1 D	4	8.55–12.01	750(1,600)	3,000	NASA/NOAA	Since 2011	https://en.wikipedia.org/wiki/Suomi_NPP
Suomi NPP	CrIS	<1 D			1,400	2,200	NASA/NOAA	Since 2011	https://en.wikipedia.org/wiki/Suomi_NPP
Suomi NPP	CERES	<1 D			2,000	3,000	NASA/NOAA	Since 2011	https://en.wikipedia.org/wiki/Suomi_NPP
Terra	ASTER	4–16 D	5	8.125–11.650	90	60	NASA joint with Japan's Ministry of International Trade and Industry	Since 1999	https://earthexplorer.usgs.gov/
Terra/Aqua	MODIS	1–2 D	10	6.535–14.385	1,000	2,300	NASA	Since 1999	https://modis.gsfc.nasa.gov/
TET	TET-1	10 D			356	180	DLR	Since 2012	https://www.dlr.de/content/en/articles/missions-projects/firebird/tet-1.html

(Continued)

TABLE 3.4 (*Continued*)
A Summary of Existing Satellite Multi-/Hyper-Spectral TIR Imaging Sensors/Systems (Except HYspIRI; Typical Wavelength 7–15 μm)

Satellite	Sensor/ system	Revisit period*	No. of TIR bands	TIR spectral range, wavelength (μm)	Spatial resolution (m)	Swath width (km)	Agency	Operation, launch year	Information source**
Geostationary orbits									
Fengyun-4A	AGRI	15 M	4	8.0–13.8	4,000			Since 2016	https://fy4.nsmc.org.cn/nsmc/en/theme/FY4A_instrument.html
GOES	Imager	3 H	2	10.2–12.5	4,000	full earth disk		Since 1975	https://www.goes.noaa.gov/
GEOS 16 & 17	ABI		7	7.24–13.60	2,000	full earth disk			
INSA–2A,3A	VHRR	30 M	1	10.5–12.5	8,000	full earth disk	ISRO	Since 1992	https://www.mosdac.gov.in/insat-3a
Meteosat-3,4,5,6,7	MVIRI	30 M	1	10.5–12.5	5,000	full earth disk	EUMETSAT, ESA	Since 1988	
Meteosat-8~11, MSG	SEVIRI	15 M	5	8.3–14.4	3,000	full earth disk	ESA/EUMETSAT	Since 2002	https://www.eumetsat.int/seviri
MOS-1&1b	VTIR		2	10.5–12.5	2,700				https://space.skyrocket.de/doc_sdat/meteosat-1.htm
MTI	MTI		3	8.01–10.70	2,014		DoE, USAF STP	Since 2012	http://www.astronautix.com/m/mti.html
MTSAT-1,2,3	Imager		2	10.3–12.5	4,000	full earth disk	JMA	Since 1999	https://www.jma.go.jp/jma/jma-eng/satellite/introduction/MTSAT_series.html
SeaSat	VIRR		1	10.2–12.5	4,400		NAA (JPL)	Since 1978	https://en.wikipedia.org/wiki/Seasat#Instruments

*: D-day, H-hour, M-minute (limited in the table); ** accessed on November 14, 2021. For Abbreviations see the text.

existing platform/satellite and obit, operational spectral range and spatial/spectral resolution, swath width, and features of sensors and missions, etc.

3.3.1 Landsat TIR Sensor Series (TM, ETM+, TIRS Sensors, Landsat 4/5, 7, 8)

Since 1972, NASA has launched eight Landsat satellites (Landsat 1–8), among which four Landsat satellites carried four popularly used TIR sensors for research and application. They are the Thematic Mapper (TM) onboard Landsat 4/5, the Enhanced Thematic Mapper Plus (ETM+) onboard Landsat 7, and the Thermal Infrared Sensor (TIRS) onboard Landsat 8 that measure land surface thermal energy in one or two thermal bands (Table 3.5). Landsat TIR sensor series usually have a moderate spatial resolution (60–120 m) but a relatively low temporal resolution (~ two weeks), and they are designed and cover a typical TIR spectral range and are suitable for collecting TIR data from space for characterizing Earth's surface materials and features (e.g., vegetation, soil, rocks, water bodies) and some atmospheric properties. The TM and ETM+ instruments are whiskbroom radiometers, with an oscillating scan mirror in front of the telescope optics that sweeps back and forth across the field of view, roughly perpendicular to the direction of spacecraft travel. The Landsat Data Continuity Mission (LDCM, i.e., Landsat

8) was launched with two instruments that will be operated simultaneously: (1) the optical sensor, the operational land imager (OLI) that operates in the visible, near-infrared, and short-wave infrared, and (2) the thermal infrared sensor (TIRS) that images in typical TIR spectral range. The LDCM specifications are compatible and comparable with most of the heritage Landsat specifications, including ground resolution, swath width, radiometric and geometric accuracy, and bandwidths for both optical sensor and TIR sensor. Landsat 5 TM TIR sensor has worked for a longer time (>25 years) compared to other Landsat series and provided—and cumulated—a long history and high quality TIR data with a wider range of min/max temperature compared to e.g., ETM+ and TIR sensor. ETM+, compared to the TM sensor, has been added with a low-gain and high-gain mode. Although ETM+ thermal data has been available since 1999, in May 2003 the ETM+ sensor had a technical failure in a scan mirror, which led to 22% of the datasets being lost. However, the renaming 8% are unaffected and can be ordered. The new sensor TIRS onboard Landsat 8 acquiring data in two bands in the typical thermal domain can effectively perform the LDCM mission.

Per their TIR application aspect, most of the TIR data in original resolution of downscaled to higher resolutions (e.g., downscaled land surface temperature (DLST) at 30 m) are used for mapping and characterizing thermal environments with heterogenous landscapes, such as heterogenous urban landscapes. For example, Huang et al. (2013) and Yao et al. (2020) utilized DLST maps at high resolutions, downscaled from the Landsat TIR sensors to characterize and monitor urban heat islands (UHIs). Zawadzka et al. (2020) generated the very high resolution DLST maps for characterizing heterogenous urban landscapes.

TABLE 3.5
Characteristics of Landsat TIR Sensor Series

TIR sensor	Landsat #	Band wavelength (μm)	Spatial resolution (m)	Launch year	Optical # of MS bands (Pan); spatial resolution (m)
TM	4, 5	10.40–10.50	120	1982, 1984	TM: 6; 30
ETM+	7	10.40–12.50	60	1999	ETM+: 6 (1); 30 (15)
TIRS	8	10.60–11.19	100	2013	OLI (Operational Land
		11.50–12.51	100		Imager): 8 (1); 30 (15)

3.3.2 NOAA-AVHRR Sensor Series (AVHRR Sensor Series, NOAA-POSE Satellite Series)

The Advanced Very High-Resolution Radiometer (AVHRR) sensor series was developed and operated by NOAA agency and is onboard NOAA Polar-orbiting Operational Environmental Satellites (POES). This series of satellites commenced with TIROS-N (launched in 1978) and continued with NOAA-8 (launched in 1983) through the current NOAA-19 (launched in 2009) (Jensen, 2015). The NOAA Sun-synchronous POES satellites carry the AVHRR sensors. Different AVHRR sensors have different versions. NOAA-POSE satellites orbit at approximately 833 km above Earth at an inclination of 98.9° and continuously record data in a swath 2,700 km wide at 1.1 × 1.1 km spatial resolution at nadir. The AVHRR is a cross-track scanning system. The scanning rate of the AVHRR is 360 scans per minute. A total of 2,048 samples (pixels) are obtained per Earth scan, which spans an angle of ±55.4° off-nadir. The IFOV of each band is approximately 1.4 mrad leading to a resolution at the satellite subpoint of 1.1 × 1.1 km. The band characteristics of the various AVHRR sensors/satellites are summarized in Table 3.6.

The AVHRR sensors acquire images of the entire Earth two times daily, daytime and nighttime, per one NOAA-POSE satellite. Usually, two NOAA series satellites are operational at one time (one odd, one even) so that theoretically it is possible to acquire AVHRR data every six hours. Such a high frequency of coverage enhances the likelihood that cloud free observations can be obtained for specific temporal windows and makes it possible to monitor change in land-cover conditions over short periods, such as a growing season (Jensen, 2015). There are two different resolution AVHRR products: (1) full-resolution AVHRR data at 1.1 × 1.1 km are called local area coverage (LAC) data, and (2) they may be resampled to 1.1 × 4 km resolution are called global area coverage (GAC) data. AVHRR data are mainly for land-cover characterization and the mapping of daytime and nighttime clouds, snow, ice, and surface temperature (Table 3.6). Thermal data from AVHRR have been widely used for applications related to land surface and sea surface temperature (e.g., Pu et al., 2007). A large advantage of the AVHRR data is its availability of data covering the last four decades. It is therefore the only sensor series in space that allows for real long-term trend analyses. Moreover, the resolution (1.1 × 1.1 km) of the AVHRR data makes it possible to collect, store, and process global datasets. For these reasons, NASA and NOAA initiated the AVHRR Pathfinder Program to create universally available global long-term remotely sensed datasets that can be used to study global climate change. Therefore, AVHRR

TABLE 3.6
Characteristics of NOAA-AVHRR Sensor Series*

Band	AVHRR with NOAA-6,8,10,12, bandwidth (μm)	AVHRR/2 with NOAA-7,9,11,13,14, bandwidth (μm)	AVHRR/3 with NOAA-15 ~ 19, bandwidth (μm)	Applications
1, Red	0.58–0. 68	0.58–0. 68	0.58–0. 68	Daytime clouds, urban, snow/ice, vegetation
2, NIR	0.725–1.10	0.725–1.10	0.725–1.10	Daytime vegetation, land/water boundaries, snow/ice, flooding
3, SWIR	3.55–3.93	3.55–3.93	3A: 1.58–1.64	Daytime vegetation, snow/ice, dust monitoring
			3B: 3.55–3.93	Surface temperature (T), wildfire, snow/ice; nighttime only for 3B, cloud, sea surface T, volcanic eruption
4, TIR 1	10.50–11.50	10.30–11.30	10.30–11.30	Surface T, wildfire detection, cloud, volcanic eruption
5, TIR 2	None	11.50–12.50	11.50–12.50	Cloud and surface T, water vapor path radiation

* references: GISGeography, 2022; (NOAA-AVHRR, 2022).

data are particularly relevant to study climate change and environmental degradation because of the comparatively long records of data already accumulated (over 40 years).

3.3.3 EOS-MODIS Sensor (MODIS, EOS-Terra/Aqua Satellites)

The Moderate Resolution Imaging Spectrometer (MODIS) is onboard NASA's EOS Terra (AM) and Aqua (PM) satellites launched 1999 and 2002, respectively. MODIS can provide long-term Earth's observations to obtain knowledge about global dynamics and processes occurring on the surface of the Earth (land and ocean) and in the lower atmosphere (NASA-MODIS, 2022). MODIS flies in a 705-km Sun-synchronous orbit. MODIS views the entire surface of the Earth every one to two days. It has a FOV of ±55° off-nadir, which creates a swath width of 2,330 km. MODIS can acquire high radiometric resolution images (12-bit) of daylight-reflected solar radiation and daytime/nighttime thermal emission over all regions of the globe (Jensen, 2015). MODIS is a whiskbroom scanning imaging radiometer consisting of a across-track scan mirror, collecting optics, and a set of linear detector arrays with spectral interference filters located in four focal planes. MODIS collects data in 36 spectral bands: 26 bands covering visible SWIR spectral range from 0.4 –6 μm and ten bands covering TIR spectral range from 6–15 μm. MODIS' spatial resolution ranges from 250 × 250 m (bands 1–2) to 500 × 500 m (bands 3–7) and 1,000 × 1,000 m (bands 8–36). The band number, band width, and band primary utilities, emphasized on TIR spectral bands, are summarized in Table 3.7.

MODIS data are being processed to create numerous global datasets, including (NASA-MODIS, 2022) (1) land ecosystem variables (e.g., vegetation indices, LAI, fraction of photosynthetically active radiation, vegetation net primary production), (2) atmospheric variables (e.g., cloud fraction, cloud optical thickness, aerosol optical depth, etc.), and (3) ocean variables (e.g., sea-surface temperature and chlorophyll). MODIS data have been processed into different global coverage data products, including land surface temperature and emissivity and thermal anomalies and fire products (NASA-MODIS, 2022). More detailed MODIS data products can be found from NASA-MODIS (2022) where the information sites are listed as follows: (1) MODIS level 1 data, geolocation, cloud mask, and atmosphere products (https://ladsweb.modaps.eosdis.nasa.gov/); (2) the LP (Land Processing) DAAC (https://lpdaac.usgs.gov/), which operates as a partnership with the U.S. Geological Survey (USGS); (3) the NSIDC (National Snow & Ice Data Center) DAAC (Distributed Active Archive Center) MODIS data collection (https://nsidc.org/data/modis/index.

TABLE 3.7
Characteristics of Bands of the Terra Satellite Moderate Resolution Imaging Spectrometer (MODIS) Covering TIR Spectral Range (7–15 μm)*

Band	Bandwidth (mm)	Band utility
27	6.535–6.895	Atmospheric water vapor
28	7.175–7.475	
29	8.400–8.700	
30	9.580–9.880	Atmospheric ozone
31	10.780–11.280	Earth surface and atmospheric cloud temperature
32	11.770–12.270	
33	13.185–13.485	Atmospheric cloud-top altitude properties
34	13.485–13.785	
35	13.785–14.085	
36	14.085–14.385	

* Bands 1–2, VNIR, 250 m resolution, land-cover classification, chlorophyll monitoring and LAI mapping
Bands 3–7, Vis-MIR, 500 m resolution, Land, cloud, and aerosol properties
Bands 8–16, VNIR, 1,000 m resolution, Ocean color, phytoplankton, biogeochemistry
Bands 17–19, NIR, 1,000 m resolution, atmospheric water vapor
Bands 20–23, SWIR-TIR, 1,000 m resolution, Earth surface and atmospheric cloud temperature
Bands 24–25, TIR, 1,000 m resolution, atmospheric temperature
Band 26, MIR, 1,000 m resolution, Cirrus clouds properties

html); and (4) the NASA's OBPG (Ocean Biology Processing Group) Ocean Biology DAAC (https://oceancolor.gsfc.nasa.gov/).

3.3.4 Terra-ASTER Sensor (ASTER TIR Sensor, EOS-Terra Satellite)

The Advanced Spaceborne Thermal Emission and Reflection Radiometer (ASTER) is a cooperative effort between NASA and Japan's Ministry of Economy Trade and Industry (METI). The ASTER instrument provides the next generation in remote sensing imaging capabilities when compared to the older Landsat TM and Japanese Earth Resources Satellite (JERS-1) the Optical Sensor (OPS) (NASA-ASTER, 2022). ASTER captures high spatial resolution data in 14 bands—from the visible to the thermal infrared wavelengths—and provides stereo viewing capability for digital elevation model creation. The sensor system consists of three separate instrument subsystems. As the 'zoom lens' for Terra, ASTER data are used by other Terra and space-borne instruments for validation and calibration (NASA-ASTER, 2022).

The instrument has three separate optical subsystems: the visible and near-infrared (VNIR) radiometer, middle-infrared (MIR) radiometer, and thermal infrared (TIR) radiometer. Table 3.8 summarizes the characteristic of Terra satellite ASTER three subsystems. There are three spectral bands in the VNIR (NIR band with two modes, nadir and backward), six bands in the MIR, and five bands in the TIR regions, with 15, 30, and 90 m ground resolution, respectively. Because the data will have wide spectral coverage and relatively high spatial resolution, a variety of surface materials will be enabled to discriminate and relevant problems in some lower spatial resolution data resulting from mixed pixels will be reduced (Yamaguchi et al., 1998). The three VNIR bands have band passes similar to those of the Landsat TM and the OPS of the Japanese JERS-1. The VNIR will be especially useful for topographic interpretation because it has along-track stereo coverage in band 3, with nadir (3N) and backward (3B) views with 15-m spatial resolution. The VNIR bands will be useful in assessing vegetation and iron-oxide minerals in surface soils and rocks. The six MIR bands were selected mainly for the purpose of surface soil and mineral mapping. Band 4 is centered at the 1.65 μm spectral region, and bands 5–9 target the characteristic absorption features of phyllosilicate and carbonate minerals in the 2.1–2.4 μm spectral region. The ASTER MIR subsystem will permit more detailed surface soil and lithologic mapping than Landsat TM and JERS-1 OPS. The ASTER TIR subsystem has five bands in the TIR spectral region, as shown in Table 3.8. Emissivity patterns derived from the five TIR bands will be used to estimate silica content (Yamaguchi et al., 1998), which is important in characterizing silicate rocks. Having multispectral TIR data allows for a more accurate determination of the variable spectral emissivity of the land surface and, hence, a more accurate determination of the land surface temperature.

3.3.5 CBERS Series TIR Sensors

CBERS series TIIR sensors are related to Chine-Brazil Earth Resource Satellite program (CBERS, 2022). CBERS-1 has been operating from 1999 to 2003, while CBERS-2 was in orbit between 2003 and 2007. The CBERS-3, originally planned for 2010, then autumn 2012, is still launched in December 2013. However, a problem with the carrier rocket has led to the loss of the satellite, which carries an improved

TABLE 3.8
Characteristics of the Terra Satellite Advanced Spaceborne Thermal Emission and Reflection Radiometer (ASTER)

Subsystem	Band	Bandwidth (μm)	Spatial resolution (m)	Radiometric resolution (bit)
VNIR	1	0.52–0.60	15 × 15	8
	2	0.63–0.69		
	3N	0.76–0.86		
	3B	0.76–0.86		
MIR	4	1.600–1.700	30 × 30	8
	5	2.145–2.185		
	6	2.185–2.225		
	7	2.235–2.285		
	8	2.295–2.365		
	9	2.360–2.430		
TIR	10	8.125–8.475	90 × 90	12
	11	8.475–8.825		
	12	8.925–9.275		
	13	10.25–10.95		
	14	10.95–11.65		

infrared scanner that delivers thermal data (10.4–12.5 μm) with 80 m SR (spatial resolution) and a swath width of 120 km. Depth of acquisition is 8 m. The panchromatic band has a SR of 5 m and the multispectral 10 m and a swath width of 60 km. With CBERS-2, Brazil adopted an open data distribution policy, ensuring free access through the internet to its catalogue and to full resolution images. It offers online registration: any user can browse the catalogue, choose as many images as he wants, and download them for immediate use, with no cost or bureaucracy and working with a simple and fast catalogue system. The same policy was adopted for CBERS-2B and will be adopted for CBERS-3 when it is launched. CBERS-4 is on a sun-synchronous orbit from the end of 2014 (till 2017), altitude = 778 km; inclination = 98.55°; repeat cycle = 26 days, orbital period = 100.26 minutes; local solar time on descending node (LTDN) at 10:30 hours, although planned for 2015, replacing the failed mission of CBERS-3. The CBERS-4b are in the production phase and are not yet in orbit, although it is planned for 2016. Committed to shortening the gap in data caused by the launch failure, assembly of CBERS-4 was finished in early 2014 to allow several months of testing before the satellite was delivered to the launch site on October 17.

3.3.6 ATSR (AATSR) Aeries TIR Sensors

The Advanced Along-Track Scanning Radiometer (AATSR) was one of nine instruments on board the Envisat satellite. It was the third in a series of instruments designed primarily to measure sea surface temperature (SST) remotely, following on from ATSR-1 and ATSR-2 on board ERS-1 and ERS-2, respectively. As its name implies, this sensor embodies some improvements over its predecessors, ATSR-1 and ATSR-2, and continues the ATSR-1 and ATSR-2 mission datasets of precise sea surface temperature (SST), thereby ensuring the production of a ten year near-continuous dataset from the ERS-1 and ERS-2 missions at the levels of accuracy of 0.3 K or better for climate research (AATSR, 2022). AATSR was launched on board the Envisat satellite on March 1, 2002. Also, the AATSR embodied substantial redesign of the structure as well as a different closed-cycle cooler system and reference blackbodies, both of which had become available commercially since the start of the ATSR program in1991.

The primary scientific objective of the AATSR is to establish continuity of the ATSR-1 and ATSR-2 datasets of precise SST, thereby ensuring the production of a unique ten year near-continuous dataset at the levels of accuracy required (0.5 K or better) for climate research and for the community of operational as well as scientific users developed through the ERS-1 and ERS-2 missions (AATSR, 2022). The second objective is to perform quantitative measurements over land surfaces. The land and cloud measurement objectives will be met through the use of an additional visible focal plane assembly.

The instrument had thermal infrared channels that were used to measure upwelling radiance from the sea surface and the atmosphere centered at 3.7, 11 and 12 μm in the thermal infrared. The calibrated top of the atmosphere brightness temperatures (BT) derived from these channels were then used to calculate SST (11/12 μm channels during the day and 3.7/11/12 μm channels at night).

AATSR also included three visible/near infrared channels centered at 0.55, 0.67 and 0.87 μm. These were first introduced on ATSR-2 and extended the instrument's capabilities over land, particularly for the study of vegetation.

AATSR did not retrieve cloud and atmospheric parameters on a routine basis. However, the different AATSR channels were used to provide information on the basic location, extent, and structure of clouds. Two fields in the Level 2 product were also reserved for cloud top temperature (CTT) and cloud top height (CTH). Work with ATSR-2 data also demonstrated that the visible and thermal channels could be combined to estimate a number of specific properties within the cloud field.

AATSR visible and thermal channel data can be used for studies of the cryosphere. Used on their own or more often synergistically with other sensors, applications include snow cover monitoring, monitoring variations in annual sea ice growth and decay, and location of boundaries and features of the Greenland and Antarctic ice sheet, and feature mapping for radar altimetry over land and ice/snow is a topic of particular research interest

3.3.7 FY Series TIR Sensors

The FengYung-3 (FY-3) (Fengyun-3 satellite) series is the second generation of Chinese polar-orbiting meteorological satellites, following on from the FY-1 series. The series has been developed collaboratively by CMA (Chinese Meteorological Administration) and CNSA (China National Space Administration) and consists of five satellites, FY-3A, FY-3B, FY-3C, FY-3D, and FY-3E, launched in May 2008, November 2010, September 2013, November 2017, and July 2021, respectively. FY-3A was decommissioned in 2018, and an additional five satellites (FY-3F, FY-3G, FY-3H, FY-3I, and FY-3J) are scheduled to be launched between 2022 and 2025. The series objectives are to provide global measurements of 3D temperature and moisture soundings, cloud and precipitation parameters, meteorological and hydrological events, biosphere anomalies, and geophysical parameters in support of global change and climate monitoring.

Fy-3 (A/B/C) instrument sensors are listed in Table 3.9. Measurements include atmosphere; ocean; gravity and magnetic fields; land; snow and ice; atmospheric winds; vegetation; radiation budget; albedo and reflectance; atmospheric humidity fields; cloud type, amount, and cloud top temperature; atmospheric temperature fields; surface temperature (ocean); liquid water and precipitation rate; cloud particle properties and profile; surface temperature (land); ozone; ocean color/biology; trace gases (excluding ozone); snow cover, edge, and depth; ocean surface wind; gravity, magnetic, and geodynamic measurements; soil moisture; multi-purpose imagery (land); sea ice cover, edge, and thickness. Instrument types cover cloud profile and rain radars, atmospheric temperature and humidity sounders, scatter meters, atmospheric chemistry, earth radiation budget radiometers, space environment, imaging multi-spectral radiometers (Vis/IR), and imaging multi-spectral radiometers (passive microwave).

3.3.8 Suomi NPP Series TIR Sensors

Launched in October 2011, the Suomi National Polar-orbiting Partnership (Suomi NPP) is a weather satellite aimed at demonstrating sensor capabilities, as well as providing data continuity to previous Earth Observing System (EOS) missions initiated by the National Aeronautics and Space Administration (NASA). Suomi NPP is operated by NASA and the National Oceanic and Atmospheric Administration (NOAA), with the aim of monitoring the environment on Earth and the planet's climate. The NPOESS Preparatory Project (NPP) is intended to bridge the gap between old Earth Observing Systems (EOS) and new systems (JPSS) by flying new instruments, on a new satellite bus, using a new ground data network. Originally planned for launch five years earlier as a joint NASA/NOAA/DoD project, the project continued as a civilian weather forecasting replacement for the NOAA Polar Operational Environmental Satellites (POES) series and ensured continuity of climate measurements begun by the EOS of NASA. The spacecraft was launched on October 18, 2011 from Vandenberg Space Force Base. Suomi NPP is operated with the aim of monitoring the environment on Earth and the planet's climate.

TABLE 3.9
FY-3(ABC) MERSI Sensor Technical Characteristics (https://www.eoportal.org/satellite-missions/fy-3)

Agency	CMA (China)	CMA (China)	CMA (China)
Satellite:	FY-3A	FY-3B	FY-3C
Launch date	27/05/08	5/11/2010	23/09/13
Swath (km)	2,900	2,900	2,900
Resolution (m)	250/1,000	250/1,000	250/1,000
# of bands	20	20	20
Spectral coverage (mm)	402–2,155	402–2,155	402–2,155

Understanding, monitoring, and predicting the course of long-term climate change and short-term weather conditions remain tasks of profound importance. Economic competitiveness, human health and welfare, and global security all depend in part on our ability to understand and adapt to global environmental changes. Over the last dozen years, NASA has launched a series of satellites—known collectively as the EOS—that has provided critical insights into the dynamics of the entire Earth system: clouds, oceans, vegetation, ice, solid Earth, and atmosphere. Now NASA is helping to create a new generation of satellites to extend and improve upon the Earth system data records established by EOS. Suomi NPP, formerly known as the NPOESS Preparatory Project, will serve as a bridge between the EOS satellites and the forthcoming series of Joint Polar Satellite System (JPSS) satellites. Suomi NPP represents a critical first step in building this next-generation satellite system. The JPSS satellites, previously called the National Polar-orbiting Operational Environmental Satellite System (NPOESS), will be developed by NASA for the NOAA.

Key science objectives and capabilities of Suomi NPP include:

- climate change—contribute to long-term records of global environmental data critical for understanding the dynamics of climate change.
- health of the ozone layer—daily measurements of the atmospheric ozone layer that will determine whether the ozone layer is recovering as expected.
- natural disasters—monitor wildfires, volcanic eruptions, snowstorms, droughts, floods, hurricanes, and dust plumes.
- weather predictions—a sounding instrument will collect information about cloud cover, atmospheric temperatures, humidity, and other variables critical to accurate weather prediction.
- vegetation—map global land vegetation and quantify changes in plant productivity to understand the global carbon cycle and monitor agricultural processes to predict and respond to food shortages and famines.
- global ice cover—monitor changes to Earth's sea ice, land ice, and glaciers to track the pace of climate change.
- air pollution—monitor the spread of health-sapping pollutants such as soot, particulate matter, nitrogen dioxide, and sulfur dioxide.
- temperatures—maintain a global record of atmospheric, land surface and sea surface temperatures critical to understanding the long-term dynamics of climate change.
- Earth's energy budget—make measurements to determine how much energy is entering and exiting Earth's atmosphere.

The Suomi NPP carries three TIR sensors, which are Visible Infrared Imaging Radiometer Suite (VIIRS), Clouds and the Earth's Radiant Energy System (CERES), and Cross-track Infrared Sounder (CrIS). VIIRS is one of the key instruments onboard the Suomi NPP spacecraft, which was successfully launched on October 18, 2011. The VIIRS nadir door was opened on November 21, 2011, which enables a new generation of operational moderate resolution-imaging capabilities following the legacy of the AVHRR on NOAA and MODIS on Terra and Aqua satellites. The VIIRS empowers operational environmental monitoring and numerical weather forecasting, with 22 imaging and radiometric bands covering wavelengths from 0.41 to 12.5 μm, providing the sensor data records for more than 20 environmental data records including clouds, sea surface temperature, ocean color, polar wind, vegetation fraction, aerosol, fire, snow and ice, vegetation, and other applications. Results from the on-orbit verification in the postlaunch check-out and intensive calibration and validation have shown that VIIRS is performing very well. CERES will be used to study the Earth's radiation budget, monitoring the amount of energy emitted and reflected by the planet, and it measures both solar energy reflected by the Earth and heat emitted by our planet. This solar and thermal energy are key parts of the Earth's radiation budget. CERES continues a multi-year record of the amount of energy entering and exiting from the top of the atmosphere of Earth. It will provide scientists with needed long-term, stable datasets to make accurate projections of global climate change. CrIS has 1305 spectral channels and will produce high-resolution, three-dimensional temperature, pressure, and moisture profiles. It measures continuous channels in the infrared region and has the ability to measure temperature profiles with improved accuracy over its predecessors. These profiles will be used to enhance weather forecasting models and will facilitate both short- and long-term weather forecasting. Over longer timescales, they will help improve understanding of climate phenomena.

3.3.9 GLI (Global Imager, NASDA ADEOS-II Satellite)

GLI is an optical he National Space Development Agency of Japan (NASDA) core sensor on the Advanced Earth Observing Satellite 2 (ADEOS II) launched by NASDA, NASA, and CNES in December 2002. GLI operates in spectral range from 0.38 to 11.95 μm with a total of 36 spectral bands. The GLI imaging spectrometer is an opto-mechanical instrument featuring a cross-track mirror and an off-axis parabolic mirror as the collecting optics and focal planes in which the detectors are arrayed in the along-track direction with spectral interference (dichroic) filters. GLI can tilt the scan mirror ±20° from nadir in order to avoid sun glitter. GLI has five focal planes, two for VNIR, two for SWIR, and one for MWIR/TIR. Two VNIR focal planes have detector arrays for 13 and ten bands, respectively. Two SWIR focal planes have detector arrays for four and two bands, while the MWIR/TIR regions have one focal plane with a detector array for seven bands. One SWIR and the MWIR/TIR focal planes are cooled to 220 K and 80 K by a multistage Peltier element and Stirling cycle mechanical cooler, respectively. The VNIR detector material is Si, and the SWIR is InGaAs, while the MWIR/TIR material is CMT.

GLI is for studying and monitoring the carbon cycle in the ocean, principally as to biological processes. It observes the reflected solar radiation from the Earth's surface, including land, oceans, and clouds and/or infrared radiation with a multichannel system for measuring the biological content, such as chlorophyll, organic substance, and vegetation index as well as temperature, snow and ice, and cloud distribution. These data will be used for understanding the global circulation of carbon and climate changes. Multispectral observations from the near UV to the near IR reflected solar radiation from the Earth's surface including land, ocean, and clouds. Out of the 36 observation bands of GLI (up to about ten), extending 380 nm to 865 nm bands could be used for ocean color remote sensing with 1 km spatial resolution at nadir. Substantially, the GLI sensor's data can be used for determination of chlorophyll pigment, phycobilin, and dissolved organic matter in the ocean and classification of phytoplankton according to their pigment. The GLI data can also be utilized for measuring sea surface temperature, cloud distribution, land coverage, vegetation index, etc.

3.3.10 GOES Series TIR Sensors (Imager, ABI)

NOAA operates two series of remote sensing satellites: the Geostationary Operational Environmental Satellites (GOES) and the Polar-orbiting Operational Environmental Satellites (POES). Both are currently based on multispectral scanner technology. The U.S. National Weather Service uses data from these sensors to forecast the weather. GOES is operated by the NOAA National Environmental Satellite Data and Information Service (NESDIS). NOAA's most recent generation of geostationary satellites began with GOES-8, which was launched in April, 1994. GOES-9 was launched on May 23, 1995. GOES-10 West was launched April 25, 1997, and became operational July 7, 1998. GOES-12 East was launched July 23, 2001 and became operational on April 1, 2003. The GOES system consists of several observing subsystems:

- GOES Imager (provides multispectral image data).
- GOES Sounder (provides hourly 19-channel soundings).
- a data-collection system (DCS) that relays data from in situ sites at or near the Earth's surface to other locations.

The GOES spacecraft is a three-axis (x, y, z) stabilized design capable of continuously pointing the optical line of sight of the imaging and sounding radiometers toward the Earth. GOES are placed in geostationary orbits approximately 35,790 km (22,240 statute miles) above the equator. The satellites remain at a stationary point above the equator and rotate at the same speed and direction as Earth. This enables the sensors to stare at a portion of the Earth from the geosynchronous orbit and thus more frequently obtain images of clouds, monitor the Earth's surface temperature and water vapor characteristics, and sound the Earth's atmosphere for its vertical thermal and water vapor structures (Jensen, 2007).

GOES Imager: the Imager is a five-channel multispectral scanner. The bandwidths and spatial resolution are summarized in Table 3.10. The visible energy passes through the initial beam splitter and is focused onto eight silicon visible detector elements. Each of the eight visible detectors has an IFOV of approximately 1 × 1 km at the satellite's suborbital point on the Earth. Thermal infrared energy is routed to the specialized detectors in the radiative cooler. The thermal infrared energy is further separated into the 3.9-, 6.75-, 10.7-, and 12-μm channels. Each of the four infrared channels has a separate set of detectors: four-element indium-antimonide (InSb) detectors for band 2; two-element mercury-cadmium-telluride (Hg:Cd:Te) detectors for band 3; and four-element mercury cadmium-telluride (Hg:Cd:Te) detectors for both bands 4 and 5. The GOES channels have 10-bit radiometric precision. The

TABLE 3.10
Imager Sensor (GOE-8, 9, 10,12) Technical Characteristics (Jensen, 2007)

Band	Spectral resolution (mm)	Spatial resolution (km)	Utility
1	0.52–0.72	1 × 1	Clouds, pollution, haze detection, and identification of severe storms
2	3.78–4.03	4 × 4	Fog detection; discriminates between water, clouds, snow or ice clouds during daytime; detects fires and volcanoes, nighttime sea-surface temperature (SST)
3	6.47–7.02	8 × 8	Estimation of mid- and upper-level water vapor, detects advection, and tracks mid-level atmospheric motion
4	10.2–11.2	4 × 4	Cloud-drift winds, severe storms, cloud-top heights, heavy rainfall
5	11.5–12.5	4 × 4	Identification of low-level water vapor, SST, and dust and volcanic ash

primary utility of the visible band 1 (1 × 1 km) is in the daytime monitoring of thunderstorms, frontal systems, and tropical cyclones. Band 2 (4 × 4 km) responds to both emitted terrestrial radiation and reflected solar radiation. It is useful for identifying fog and discriminating between water and ice clouds and between snow and clouds and for identifying large or very intense fires. It can be used at night to track low-level clouds and monitor near-surface wind circulation. Band 3 (8 × 8 km) responds to mid- and upper-level water vapor and clouds. It is useful for identifying the jet stream, upper-level wind fields, and thunderstorms. Energy recorded by band 4 (4 × 4 km) is not absorbed to any significant degree by atmospheric gases. It is ideal for measuring cloud-top heights, identifying cloud-top features, assessing the severity of some thunderstorms, and tracking clouds and frontal systems at night. Thermal band 5 (4 × 4 km) is similar to band 4 except that this wavelength region has a unique sensitivity to low-level water vapor hemisphere once every three hrs in 'routine' scheduling mode.

The GOES Sounder uses 1 visible and 18 infrared sounding channels to record data in a north-to-south swath across an east-to-west path. The Sounder and Imager both provide full Earth imagery, sector imagery, and area scans of local regions.

The 19 bands yield the prime sounding products of vertical atmospheric temperature profiles, vertical moisture profiles, atmospheric layer mean temperature, layer mean moisture, total precipitable water, and the lifted index (a measure of stability). These products are used to augment data from the Imager to provide information on atmospheric temperature and moisture profiles, surface and cloud-top temperatures, and the distribution of atmospheric ozone (Jensen, 2007).

3.4 SUMMARY

In this chapter, the working principles, performance, and technical characteristics of current airborne and spaceborne multi-/hyper-spectral sensors, systems, and missions are briefly introduced. Based on the mechanism of scanning and data acquisition, two types of commonly used scanning principles of TIR imaging instruments, across-track scanning by rotating a scanning mirror and along-track scanning by using a pushbroom linear array or area-array, are first described. Then, in Section 3.2, a survey of selected important airborne multi-/hyper-spectral TIR sensors and systems is summarized in terms of basic technical parameters, characteristics, date of development, and application areas. In such a survey, a total of 18 selected airborne TIR sensors and systems are summarized and include (1) Airborne Hyperspectral Imager (AHI), (2) Airborne Hyperspectral Scanner (AHS), (3) Airborne Imaging Spectrometer for different Applications, (AISA-OWL), (4) Airborne Thermal infrared Hyperspectral Imaging System (ATHIS), (5) Airborne Thermal/Visible Land Application Sensor (ATLAS), (6) Airborne Thematic Thermal InfraRed and Electro-Optical Imaging System (ATTIREOIS), (7) Digital Airborne Imaging Spectrometers (DAIS series TIR sensors), (8) Goddard's LiDAR, Hyperspectral and Thermal (G-LiHT) airborne imager, (9) Hyperspectral Mapper (HyMap), (10) Hyperspectral Cameras (Hyper-Cam xLW), (11) Hyperspectral Thermal Emission Spectrometer (HyTES), (12) Airborne thermal infrared imaging spectrometer (MAKO), (13) Modular Airborne Imaging Spectrometer (MAIS), (14) Multispectral Infrared and Visible Imaging Spectrometer (MIVIS), (15) MUltiSpectral Infrared Camera (MUSIC), (16) Spatially-Enhanced Broadband Array Spectrograph System (SEBASS), (17) Airborne Hyperspectral Thermal Imager (TASI-600), and (18) Thermal Infrared Multispectral Scanner (TIMS). In Section 3.3, a survey of ten selected important spaceborne multi-/hyper-spectral sensors and missions is summarized in terms of their launch year, actual platform/satellite, operational spectral range and spatial/spectral resolution, and swath width, features of sensors, and missions. The ten current/operational spaceborne hyperspectral sensors and missions are summarized and they include (1) Landsat TIR Sensor Series (TM, ETM+, TIRS sensors, Landsat 4/5, 7, 8), (2) NOAA-AVHRR Sensor scrics (AVHRR sensor series, NOAA-POSE satellite series), (3) EOS-MODIS sensor (MODIS, EOS-Terra/Aqua satellites), (4) Terra-ASTER sensor (ASTER TIR sensor, EOS-Terra satellite), (5) CBERS series TIR sensors, (6) ATSR (AATSR) aeries TIR sensors, (7) FY series TIR sensors, (8) Suomi NPP series TIR sensors, (9) GLI sensor (Global Imager, NASDA ADEOS-II satellite), and (10) GOES series TIR sensors (Imager, ABI).

4 TIR Radiative Transfer in the Atmosphere and TIR Data Correction

The radiance recorded at a sensor (airborne and spaceborne) is not fully representative of Earth's surface features. It is affected by radiometric errors caused by the remote sensing systems themselves, as well as atmospheric and terrain surface effects. To accurately and reliably utilize remotely sensed TIR data, it is necessary to conduct imaging radiometric correction especially correcting various atmospheric effects to retrieve surface temperature before applying the imaging data for various research and application purposes. Therefore, in this chapter, the reasons why it is necessary to perform TIR imaging data correction (Section 4.1), TIR radiative transfer principle in the atmosphere (Section 4.2), and methods, principles, and algorithms of atmospheric correction for TIR images (Section 4.3) are concisely introduced and discussed.

4.1 INTRODUCTION

Usually, the radiance recorded at a sensor is not fully representative of Earth's surface features within the sensor's field of view. This is because some radiation errors are introduced in remotely sensed data by the sensor system itself and the radiance recorded in remote sensing images is usually altered by the atmosphere effects. Several of the radiance errors introduced by multi-/hyper-spectral sensors include line or column striping, spatial and spectral misregistration across the spatial axis and spectral axis in pushbroom imaging systems, etc. The solar radiation on the Sun-surface-sensor ray path is subject to molecular and aerosol scattering and absorption by gases (Gao et al., 2009). The atmospheric scattering (e.g., Rayleigh and aerosol) with the shorter wavelength region below 1.00 μm modifies the radiances of adjacent fields of different reflectance (adjacency effect). Therefore, dark areas surrounded by bright areas appear to have a higher reflectance than the intrinsic reflectance. In the TIR spectral region, water vapor and ozone are major absorbers see Figure 2.7 in Pu (2017). To remove and/or suppress the atmospheric effects on target spectra, there are many ways to atmospherically correct remotely sensed data. Relatively straightforward methods include the earlier empirical line method, flat field method, and image/scene-based method, whereas more complex methods include recently developed rigorous radiative transfer modeling approaches. For example, panchromatic, multispectral, and hyperspectral imagery from the solar spectral region and thermal region can be processed to obtain surface reflectance and temperature images by the ATCOR4.

It is also necessary to atmospherically correct the remote sensing data even over a single data image of quantitative analyses by combining field survey data

DOI: 10.1201/9781003276173-4

with spectral image data for applications such as extracting biophysical parameters from vegetation (e.g., biomass, leaf area index, chlorophyll, and crown closure) are required (Pu et al., 2003). Since the effects of atmospheric scattering and absorption on multi-/hyper-spectral remote sensing data are critical, the atmospheric correction is particularly important to optical and TIR imaging data for retrieval of surface reflectance and temperature and atmospheric constituents. If the data are not atmospherically corrected, the subtle spectral difference in reflectance (or emittance) among the important components may be lost (Jensen, 2007). Therefore, TIR radiative transfer in the atmosphere and representative methods of atmospherically correcting TIR imaging data are briefly introduced and discussed in this chapter as follows.

4.2 TIR RADIATIVE TRANSFER IN THE ATMOSPHERE

In the TIR spectral region, the TIR radiative transfer is not only influenced by atmospheric scatting and absorption, but it is also affected by the atmosphere itself emitting irradiation energy; under clear sky conditions the radiance ($Lsensor(\lambda)$) received at sensor is expressed as follows (Li et al., 2013; Zhang et al., 2020):

$$\begin{aligned} Lsensor(\lambda) = \{ & \varepsilon(\lambda)L(\lambda,Ts) + [(1-\varepsilon(\lambda)(Latm\downarrow(\lambda)+Lsun\downarrow(\lambda))] \\ & + \rho b(\lambda)Ecos\theta sun \cdot \tau(\lambda,\theta sun)\}\tau(\lambda) \\ & + Latm\uparrow(\lambda) + Lsun\uparrow(\lambda) \end{aligned} \tag{4.1}$$

where $\varepsilon(\lambda)$ represents the specral emissivity of surfave target; $L(\lambda,Ts)$ is thespectral radiance value at surface temperature T_s; subscripts $Latm\uparrow(\lambda)$ and $Latm\downarrow(\lambda)$ represent the upward and downward atmospheric path radiance, respectively; subscripts $Lsun\uparrow(\lambda)$ *and* $Lsun\downarrow(\lambda)$ represent the values of atmospheric scatting energy to that solar light upward and downward hemisphere radiance divided by π; $\rho b(\lambda)$ is the bi-directional reflectivity of the surface; $Ecos\theta sun$ is the solar irradiance at the TOA at the solar zenith angle θ, and τ is the atmospheric tramsmittance, assuming the spectral variation in emissivity is small, and using Kirchhoff's law, $\rho = 1-\varepsilon(\lambda)$ to express the hemispherical-directional reflectance as directional emissivity (Hulley, 2013). In the spectral TIR range 8–14 μm, TIR radiance part in the solar irradiance may be ignored (Zhang et al., 2020), therefore, Eq. (4.1) can simply be written as

$$Lsensor(\lambda) = \{\varepsilon(\lambda)L(\lambda,Ts) + [(1-\varepsilon(\lambda)(Latm\downarrow(\lambda))]\}\tau(\lambda) + Latm\uparrow(\lambda) \tag{4.2}$$

When satellite TIR sensor measures within TIR spectral band k, the band average equivalent radiance observation can be expressed as follows:

$$L_k = \frac{\int_{\lambda_1}^{\lambda_2} f_k(\lambda)L(\lambda)d}{\int_{\lambda_1}^{\lambda_2} f_k(\lambda)d} \tag{4.3}$$

where $f_k(\lambda)$ represents the spectral response function of satellite TIR sensor spectral band k; λ_1 and λ_2 are the start and end band wavelength, respectively; and the relationship between the equivalent radiance of satellite TIR sensor band k and its digital number is

$$L_k = G_k \cdot \left(DN_k - DN_{space}\right) \tag{4.4}$$

where, G_k is the calibration slope value of satellite TIR sensor spectral band k, and $DN_k and\ DN_{space}$ are the digital number of satellite TIR sensor spectral band k and its clod space observation digital number.

4.3 ATMOSPHERIC CORRECTION FOR TIR DATA

Accurate calibration of TIR date from either a satellite or an aircraft-based sensor is necessary because of atmospheric emission, attenuation, and effect (mainly including scattering and absorption). Atmospheric radiance must be accounted to obtain calibrated surface temperatures. Although the band passes for satellite and aircraft thermal channels used for mapping surface temperature fall within the atmospheric window for atmospheric long wave transmittance (8.0–12.0 μm), the maximum transmittance across this region is only about 80% (Figure 4.1). Figure 4.1 indicates thermal atmospheric transmittance in the TIR spectral range is only about 80% (Quattrochi and Luvall, 2009).

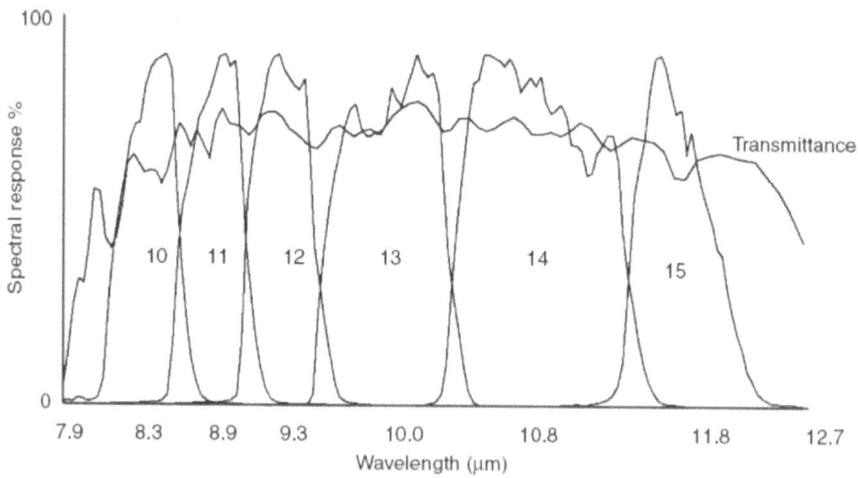

FIGURE 4.1 Graph of thermal atmospheric transmittance and Airborne Thermal and Land Applications Sensor (ATLAS) thermal spectral responses. Numbers in the graph represent ATLAS band numbers, and those below the graph indicate the band centers in micrometers.

Source: Quattrochi and Luvall, 2009.

4.3.1 General Atmospheric Correction

There are three typical approaches to atmospheric correction for TIR sensors. The first approach is the split-window technique (e.g., Wan and Dozier, 1996) and uses differential absorption characteristics of atmospheric water vapor between two bands in the TIR spectral region (11–12 μm). Variations of this method include the split window (SW) approach (Wan and Dozier, 1996; Coll and Caselles, 1997); sensors such as the Moderate Resolution Imaging Spectroradiometer (MODIS), Advanced Along-Track Scanning Radiometer (AATSR), and Advanced Very High Resolution Radiometer (AVHRR) all use a variation of the split-window method to retrieve LST over land by assuming the emissivity can be estimated from land cover classification maps by assigning fixed emissivities based on cover type (Snyder et al., 1998). In the split-window technique, which is in widespread use, the brightness temperatures T_{b1} and T_{b2} are measured in two different but closely spectral bands, for example one centered at 11 mm and one at 12 μm. The brightness temperature T_{b0} of the radiation leaving the surface is then calculated as a linear combination of the measured brightness temperatures

$$T_{b0} = a_0 + a_1 T_1 + A_2 T_2 \tag{4.5}$$

where the coefficients a_0, a_1, and a_2 are determined empirically. Because of the contribution from reflected sky radiation, the coefficients have different values during the day and during the night (Hulley, 2013). The technique is reasonably accurate (to within 0.5 K) if the emissivity of the surface is constant, so it is reliable for sea surface temperature measurements, but over land surfaces its usefulness is more limited. With an increased number of spectral bands, such as those available from MODIS, Eq. (4.5) is replaced by a more general linear combination of brightness temperatures, and the technique is generally more accurate (Hulley, 2013). As we have already noted, it is possible to estimate surface emissivity from thermal inertia measurements. Split-window based algorithms are generally tuned to have high LST accuracy over graybody surfaces where the emissivity is well known (e.g., water, vegetation), whereas the residual effects are significant from incomplete atmospheric correction degrading the accuracy of TES (temperature and emissivity separation) LST products over graybody surfaces (Gustafson et al., 2006). The residual effects can be due to a misclassification in the original cover type, a lack of fidelity in the cover type map, or a dynamic change in the cover type map for example from an increase in soil moisture. The accuracy of the TES algorithm is limited by residual errors from incomplete atmospheric correction, which results in a larger apparent emissivity spectral contrast. This intrinsic weakness of the TES algorithm has been systematically analyzed by several authors (Gillespie et al., 1998; Gustafson et al., 2006), and its effect is greatest over graybody surfaces that have a true spectral contrast that approaches zero.

The second is the single-channel method (or mono-window algorithm, MWA), which usually requires a good radiative transfer model and atmospheric profiles that must be provided by either satellite soundings or conventional radiosonde (Qin et al., 2001a). Thus, the single-band algorithm combined with global assimilated data is currently the most applicable algorithm to operational atmospheric correction for high spatial-resolution TIR multispectral data over land, but it is not always accurate (Hulley, 2013). For example, Qin et al. (2001b) correct and retrieve land surface temperature LST using Landsat TM band 6 (TIR band).

The third method uses full radiative transfer calculations to estimate atmospheric effects on a pixel-by-pixel and band-by-band basis. The atmospheric transmittance, path radiance, and downward sky irradiance are estimated using a radiative transfer model such as MODerate resolution atmospheric TRANsmission (MODTRAN) (Kneizys et al. 1996). Because the main correction is due to water vapor, which is highly variable both spatially and temporally, physical modeling is rather unsatisfactory unless a detailed characterization of the atmosphere is available. This can be obtained by local radiosonde observations or by airborne or spaceborne atmospheric sounding.

4.3.2 A Water Vapor Scaling (WVS) Method for Improving Atmospheric Correction of TIR Data

4.3.2.1 A Water Vapor Scaling (WVS) Model

In order to minimize these errors from incomplete atmospheric correction, a Water Vapor Scaling (WVS) method will be introduced here to improve the accuracy of the atmospheric parameters on a band-by-band basis for each observation using an Extended Multi-Channel/Water Vapor Dependent (EMC/WVD) algorithm (Tonooka, 2001; Hulley, 2013; Malakar and Hulley, 2016) as follows.

Using Eqs. (4.2) and (4.3), the surface radiance for band i can be written as a combination of two terms: the Earth-emitted radiance and the reflected downward irradiance from the sky and surroundings:

$$L_{s,i} = \varepsilon_0 B_i\left(T_s\right) + \left(1-\varepsilon_\lambda\right)L_\lambda^{\downarrow} = \frac{L_i(\theta) - L_i^{\uparrow}}{\tau_i(\theta)} \tag{4.6}$$

where the atmospheric parameters, $L_\lambda^{\downarrow}, \tau_i(\theta)$ and $L_i^{\uparrow}$, are the atmospheric downward irradiance, transmittance, and atmospheric path radiance and estimated with a radiative transfer model such as MODTRAN using input atmospheric fields of air temperature, relative humidity, geopotential height, etc.

The WVS method can improve the accuracy of atmospheric parameters output by MODTRAN in Eq. (4.6) on a pixel-by-pixel basis using the EMC/WVD equation (Tonooka, 2001; Hulley, 2013). The EMC/WVD equation estimates the at-surface brightness temperature, given the at-sensor brightness temperature, along with an estimate of the total water vapor amount, which is expressed as

$$T_{g,i} = a_0 + \sum_{k=1}^{N} a_{i,k} T_k, \quad a_{i,k} = p_{i,k} + q_{i,k} W + r_{i,k} W^2 \tag{4.7}$$

where i is the band number; n is the number of bands; W is the rough estimate of total precipitable water vapor (cm) from global assimilated data etc.; p, q, and r are the regression coefficients for each band; T_k is the measured brightness temperatures of band k (K), and $T_{g,i}$ is the surface brightness temperature for band i.

The coefficients in Eq. (4.7) of the EMC/WVD equation are determined using a global-based simulation model with profile data typically from numerical weather model data, such as the National Centers for Environmental Prediction (NCEP) Climate Data Assimilation System (CDAS) reanalysis project (Hulley, 2013).

The scaling factor, γ, used for improving a water profile, is based on the assumption that the transmissivity, τ_i, can be expressed by the Pierluissi double exponential band model formulation (Kneizys et al., 1996). The scaling factor is computed for each graybody pixel on a scene using $T_{g,i}$ computed from Eq. (4,7) and τ_i computed using two different γ values that are selected a priori (Hulley, 2013):

$$\gamma^{a_i} = \frac{\ln\left[\frac{\tau_i(\theta\gamma_2)^{\gamma_1 a_i}}{\tau_i(\theta\gamma_1)^{\gamma_2 a_i}} \cdot \left(\frac{B_i(T_{g,i}) - L_i^{\uparrow}(\theta,\gamma_1)/(1-\tau_i(\theta,\gamma_1))}{L_i - L_i^{\uparrow}(\theta,\gamma_1)/(1-\tau_i(\theta,\gamma_1))}\right)^{\gamma_1^{a_i}-\gamma_2^{a_i}}\right]}{\ln(\tau_i(\theta,\gamma_2)/(\tau_i(\theta,\gamma_1))} \tag{4.8}$$

where a_i is the band model parameter; $\tau_i(\theta\gamma_{1,2})$ is the transmittance calculated with water profile scaled by γ; $L_i^{\uparrow}(\theta,\gamma_{1,2})$ is the atmospheric path radiance calculated with water profile scaled by γ; and $B_i(T_{g,i})$ is the brightness temperature of ground surface temperature of graybody at band i.

Typical values for γ are $\gamma_1 = 1$, and $\gamma_2 = 0.7$. Tonooka (2005a) found that the γ calculated by Eq. (4.8) will not only reduce biases in the water vapor profile but will also simultaneously reduce errors in the air temperature profiles and/or elevation. In the WVS method, the γ values of nongray pixels are horizontally interpolated from those of surrounding gray pixels irregularly distributed. This procedure is similar to spatial horizontal interpolation in grid cells in numerical forecast (Tonooka, 2001).

4.3.2.2 Scaling Atmospheric Parameters

Once the MODTRAN run has completed and the γ image has been interpolated and smoothed, it can be computed for the gray pixels and be used to scale the atmospheric parameters that transmittance τ_i and path radiance $L_i^{\uparrow}$ are modified as follows (Berk et al., 2014):

$$\tau_i(\theta.\gamma) = \tau_i(\theta,\gamma_1)^{\frac{\gamma_i^{a_i}-\gamma 2^{a_i}}{\gamma_1^{a_i}-\gamma 2^{a_i}}} \cdot \tau_i(\theta,\gamma_2)^{\frac{\gamma_i^{a_i}-\gamma 2^{a_i}}{\gamma_1^{a_i}-\gamma 2^{a_i}}} \tag{4.9}$$

$$L_i^{\uparrow}(\theta.\gamma) = L_i^{\uparrow}(\theta.\gamma_1) \cdot \frac{1-\tau_i(\theta,\gamma)}{1-\tau_i(\theta,\gamma_1)} \tag{4.10}$$

Once the atmospheric transmittance and path radiance have been adjusted using the scaling factor, the surface radiance can be re-computed with the updated atmospheric parameters using Eq. (4.2).

In the WVS simulation model, the downward sky irradiance can be estimated using the atmospheric path radiance, transmittance, and view angle as parameters

(Tonooka, 2001). To simulate the downward sky irradiance, in a MODTRAN run, the sensor target is placed a few meters above the surface, with surface emission set to zero, reflectance set to 100%, and view angle set at prescribed angles, e.g., Gaussian angles (θ = 0°, 11.6°, 26.1°, 40.3°, 53.7°, and 65°) (Malakar and Hulley, 2016). In this way, the only radiance contribution is from the reflected downwelling sky irradiance at a given view angle. The total sky irradiance contribution is then calculated by summing up the contribution of all view angles over the entire hemisphere (Hulley, 2013):

$$L_i^{\downarrow} = \int_0^{2\pi} \int_0^{\frac{\pi}{2}} L_i^{\downarrow}(\theta) \cdot sin\theta \cdot cos\theta \cdot d\theta \cdot d\delta \tag{4.11}$$

where θ is the view angle and δ is the azimuth angle. However, to minimize computational time in the MODTRAN runs, the downward atmospheric sky irradiance can be estimated as a non-linear function of atmospheric path radiance at nadir view (Tonooka, 2005a):

$$L_i^{\downarrow}(\gamma) = a_i + b_i \cdot L_i^{\uparrow}(0,\gamma) + c_i \cdot L_i^{\uparrow}(0,\gamma)^2 \tag{4.12}$$

where, a_i, b_i, and c_i are regression coefficients, and $L_i^{\downarrow}(\gamma)$ is computed by:

$$L_i^{\downarrow}(0,\gamma) = L_i^{\uparrow}(0,\gamma_1) \cdot \frac{1-\tau_i(\theta,\gamma)^{cos\theta}}{1-\tau_i(\theta,\gamma)} \tag{4.13}$$

Tonooka (2005a) found RMSEs of less than 0.07 W/m2/sr/µm for ASTER bands 10–14 when using Eq. (4.12) as opposed to Eq. (4.11).

4.3.2.3 Determining EMC/WVD Coefficients

The EMC/WVD coefficients, p, q, and r from Eq. (4.7) can be determined using a global simulation model with input atmospheric data from either radiosonde or numerical weather model sources. (Malakar and Hulley, 2016). Radiosonde databases such as the TIGR, SeeBor, and CLAR contain uniformly distributed global atmospheric soundings acquired both for day and night in order to capture the full-scale natural atmospheric variability (Hulley, 2013). Geophysical profiles of air temperature, relative humidity, and geopotential height are used in combination with surface temperature and emissivity to simulate at-sensor brightness temperatures for the global set of profiles distributed uniformly over land (Hulley, 2013). For example, the air temperature profiles are changed by −2, 0, and +2 K, while the humidity profiles are scaled by factors of 0.8, 1.0, and 1.2 to capture more variability for that specific atmospheric profile shape. These types of perturbations will help simulate a full range of atmospheric conditions. Furthermore, the surface temperatures are modified by −5, 0, 5, and 10 K, and the surface emissivity provided consists of a set of ten spectra typically from graybodies selected from the ASTER spectral library; for example, water, vegetation, snow, ice, and some types of soils (Hulley, 2013). These emissivity spectra typically have values greater than 0.95. This ensures that the simulation results are not affected by uncertainties in surface emissivity. The at-sensor radiance is then calculated using MODTRAN for the full set of profiles and perturbations ($3 \times 3 \times 4 \times 10 = 360$). The surface elevation is taken from a global digital elevation model (DEM), and the view angle is assumed to be nadir. Furthermore, a noise-equivalent differential temperature (NEΔT) appropriate for the sensor is applied using a normalized random number generator (Hulley, 2013). Finally, using the simulated at-sensor brightness temperature T_k, at-surface T_g brightness temperatures, and an estimate of the total precipitable water vapor, the coefficients in Eq. (4.7) can be determined by using a linear least squares method. For example, WVS coefficients computed using MODTRAN4 for MODIS bands 29, 31, and 32 can be found in (Tonooka, 2005b) and for ASTER bands 10–14 in (Tonooka, 2005a).

4.3.2.4 Impacts of WVS on LST and Emissivity Retrievals

In this section, we exam the impact of using WVS to improve to the accuracy of LST and emissivity retrievals using the TES algorithm for ASTER and MODIS observations over the Salton Sea on July 15, 2000 for example (Hulley, 2013). The emissivity of water is high (~0.98) and spectrally flat, and so this provides a good test for assessing the accuracy of the emissivity retrieval, and consequently the LST retrieval. Any warping of the spectra is usually caused by the result of incomplete atmospheric correction. The LST and emissivity were retrieved in the same manner for both ASTER and MODIS data using the methods outlined in Hulley and Hook (2011). The results show that emissivity spectra from the standard atmospheric correction (STD) method are too low for both ASTER and MODIS, especially in the 8–10 μm region when compared to the lab spectra from the ASTER spectral library. Using the WVS coefficients results in a dramatic improvement in emissivity accuracy in both magnitude (up to 0.06 for ASTER band 11 and 0.09 for MODIS band 29) and spectral shape. Table 4.1 shows LST and emissivity differences between the WVS and STD approach for 1 MODIS pixel (1 km) and 10×10 ASTER pixels (90 m) over the same area of the Salton Sea. LST differences between the STD and WVS method are

TABLE 4.1
Land Surface Temperature (LST) and Emissivity Differences Between Using the Water Vapor Scaling (WVS) and Standard (STD) Atmospheric Correction Methods for MODIS (1 km pixel) and ASTER (10 × 10, 90 m pixels) Data Over the Salton Sea, CA, July 15, 2000 (Hulley, 2013)

	ASTER		MODIS	
	STD	WVS	STD	WVS
Ts (K)	309.65	306.76	310.22	305.06
MMD	0.0522	0.0458	0.0597	0.0083
emin	0.9120	0.9611	0.9100	0.9700

Note: MMD represents the minimum-maximum difference in emissivity for a water pixel spectrum, while emin is the minimum emissivity for the water spectrum as large as 5 K for MODIS and just under 3 K for ASTER. For water, the Minimum-Maximum Difference (MMD) in emissivity should approach near zero, and there is a significant decrease in MMD when using WVS as opposed to the STD approach. Results show improvements when using the Water Vapor Scaling (WVS) method as opposed to the STD method. Total precipitable water estimated from the MOD07 atmospheric product on this day was high (4 cm), indicating very humid conditions.

4.4 SUMMARY

In the beginning, the importance and necessity of TIR image radiometric correction and atmospheric correction are discussed for most multi-/hyper-spectral remote sensing data applications. This is because the remotely sensed data are not fully representative of Earth's surface features and suffer from radiometric errors caused by both remote sensors themselves and atmospheric effects, and sometimes terrain surface effects as well. To better understand and accurately correct various atmospheric effects, why we need to conduct atmospheric and radiometric corrections are first introduced before substantial discussion of atmospheric correction (AC) methods and algorithms. Then we introduce the radiative transfer in the atmosphere. In Section 4.3, atmospheric correction for TIR data is discussed, including (1) three general atmospheric correction types: the split-window technique with two or more TIR bands, the single-channel method (or mono-window algorithm), and the full radiative transfer calculations to estimate atmospheric effects on a pixel-by-pixel and band-by-band basis; (2) e a Water Vapor Scaling (WVS) method for improving atmospheric correction for TIR sensor data, including a Water Vapor Scaling (WVS) model, scaling atmospheric parameters, determining EMC/WVD coefficients, and impacts of WVS on land surface temperature and emissivity retrievals, etc.

5 Methods and Techniques of LST and Emissivity Retrieval and Estimation With TIR Data

In order to accurately and reliably retrieve land surface temperature (LST) and estimate land surface emissivity (LSE) from various sensed TIR data, it is necessary to review and introduce many existing methods and techniques of LST retrieval and LSE estimation. This chapter will explore the necessity of performing LST retrieval from TIR imaging data (Section 5.1), provide the basic theoretical background (Section 5.2), present the methods and techniques for LST retrieval and LSE estimation (Sections 5.3 and 5.4), and discuss the limitations and future directions of these approaches.

5.1 INTRODUCTION

As a direct driving force in the exchange of long-wave radiation and turbulent heat fluxes at the surface–atmosphere interface, land surface temperature (LST) is one of the most important parameters in the physical processes of surface energy and water balance at local and global scales and it is commonly used for meteorological, hydrological, agricultural, and surface energy balance studies. LST is defined as the temperature that is felt when there is an exchange of long-wave radiation and turbulent heat fluxes through the surface-atmosphere interface (Avdan and Jovanovska, 2016). Land surface emissivity (LSE) is defined as a ratio of the radiation energy emitted from a selective body to the radiation energy emitted from the blackbody at the same temperature. LST has been recognized as one of the high-priority parameters of the International Geosphere and Biosphere Program (IGBP) (Townshend et al., 1994).

Due to the strong heterogeneity of land surface characteristics such as vegetation, topography, and soil, LST changes rapidly in space time and an adequate characterization of LST distribution and its temporal evolution, therefore, requires measurements with detailed spatial and temporal information. Given the complexity of Earth's surface temperature over land, ground measurements cannot practically provide values over large areas. With the development of remote sensing from space, satellite data offer the only possibility for measuring LST over the entire globe with sufficiently temporal resolution and with complete spatially averaged rather than point data. LST is affected by various factors such as land surface type, surface moisture, illumination, and atmospheric conditions. Thus, it is difficult to measure in-situ LST that is representative of large areas. Remote sensing observations are reliable only when they offer the possibility of measuring LST at high spatial resolution and

DOI: 10.1201/9781003276173-5

temporal frequency for a large area. Satellite-based thermal infrared (TIR) data is directly linked to the LST through the radiative transfer equation. The retrieval of the LST from remotely sensed TIR data has attracted much attention, and its history dates back to the 1970s (McMillin, 1975). However, direct estimation of LST from the radiation emitted in the TIR spectral region is difficult to perform with that accuracy, since the radiances measured by the radiometers onboard satellites depend not only on surface parameters (temperature and emissivity) but also on atmospheric effects (Li and Becker, 1993; Ottlé and Stoll, 1993; Prata et al., 1995). Therefore, besides radiometric calibration and cloud screening, the determination of LSTs from space-based TIR measurements requires both emissivity and atmospheric corrections (Li and Becker, 1993).

Satellite TIR sensors receive unenergetic electromagnetic radiation (EMR), which can be quantified in the form of measurements of top of atmosphere (TOA) radiances. This includes the upwelling radiance emitted from the ground, the upwelling radiance from the atmosphere, and the downwelling radiance emitted by the atmosphere and reflected from the ground. During the day there is both emission and reflection of EMR, but during the night sensed EMR is restricted to only emission. The inverse of Planck's Law is used to derive blackbody/brightness temperatures from TOA radiances (Figure 5.1). TOA radiances are then converted to LST by correcting for three main effects; atmospheric attenuation, angular effects, and spectral emissivity values at the surface. In the 8–12 μm TIR spectral region, aerosol absorption and scattering are negligible and generally ignored (Prata et al., 1995); water vapor is principally responsible for atmospheric effects. Other gases, e.g., ionospheric O_3 and CO_2 also influence atmospheric transmission, but unlike water vapor, O_3 and CO_2 vary slowly. Hence, water vapor, which is poorly mixed and varies on short time-scales, is the most relevant gas. Consequently, frequent information about the state of the atmosphere, especially the temperature and water-vapor, are needed. Radiosonde data provide such information and yield results that, at the location of the radiosonde, are more accurate than any other method for the troposphere, but the soundings have to be synchronous and co-located with the satellite measurements. Therefore, information from vertical sounders, e.g., TIROS operational vertical sounder (TOVS) onboard NOAA polar-orbiters, are better suited for large-scale studies (Reutter et al., 1994). Spectral emissivity refers to the relative ability of a surface to emit radiation and can be highly variable due to the heterogeneity of land and is influenced by surface cover type, such as vegetation cover and soil moisture. Quantification of emissivity is achieved by considering the ratio of energy emitted

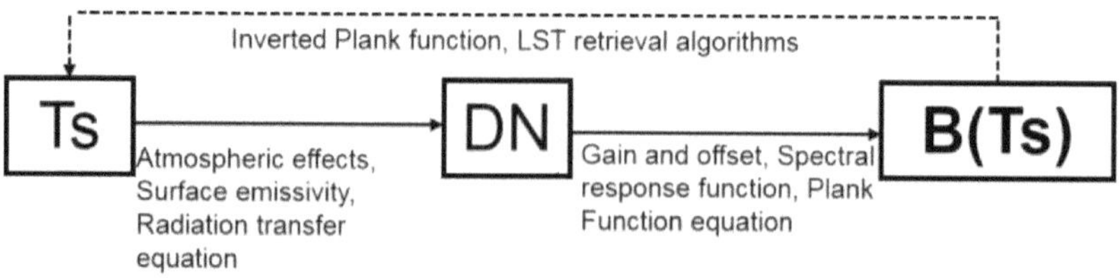

FIGURE 5.1 Transformation of surface temperature, scaled at-sensor radiance brightness temperature and estimated surface temperature.

by a surface regarding the energy emitted by a blackbody at the same temperature. However, calculations are complicated because natural surfaces do not behave like a blackbody and thus need correction using typical emissivity values. These corrections are done through complex algorithms, alongside extensive validation and verification, resulting in a final product that can be used by a researcher.

Over the past decades, the scientific community has successfully developed a variety of new theories and influential LST retrieval algorithms (e.g., Li et al., 2013) including the single-channel algorithm (SCA) (e.g., Hook et al., 1992), the day/night algorithms (e.g., Wan and Li., 1997), the mono-window algorithms (MWA) (e.g., Qin et al., 2001b), the generalized split-window (SWA) algorithm (e.g., Wan and Dozier,1996), machine learning (ML) empirical methods (e.g., Khoob, 2008), and physics-base methos (Lan et al., 2022). Many studies have been carried out, and different approaches have been proposed to derive LSTs from satellite TIR data, using the variety of methods to deal with the emissivity and atmospheric effects (Hook et al., 1992; Wan and Dozier, 1996; Wan and Li, 1997; Qin et al., 2001b). The advantage of the SCA algorithm is its simplicity because it only needs two parameters for LST retrieval: the ground emissivity (ε) and the water vapor content (w) (Yu et al., 2014). However, SCA and MWA algorithms show a significant overestimation of LSTs (Zhou et al., 2015). The third one can be carried out by utilizing two thermal bands, which is also known as the SWA algorithm (Wan and Dozier, 1996). Generally, the first two algorithms depend on a single TIR channel, while the other three algorithm require two adjacent TIR channels or more. In the following, the contents are cited in the paper by Pu (2023). In Section 5.3, we will introduce and discuss the five types of LST retrieval algorithms in detail.

5.2 LST RETRIEVAL BASIC THEORETICAL BACKGROUND

All objects with temperatures greater than absolute zero emit radiation; in accordance with blackbody theory, the thermal radiation amount emittance from an object of blackbody in thermal equilibrium at wavelength λ and temperature T can be expressed as the Planck's radiance function:

$$B_\lambda(T) = \frac{C_1}{\lambda^5 (e^{C_2/(\lambda T)} - 1)} \tag{5.1}$$

where $B_\lambda(T)$ is the spectral radiance of the blackbody in W/(m^2.sr.μm); λ the is wavelength in meters; C_1 (=1.19104356 X 10^{-16} Wm^{-2}) and C_2 (= 1.43876869 X 10^{-2} mK) are two the constants; and T is temperature in Kelvins.

In the absence of atmospheric effects, T (LST) of a ground object can be theoretically determined, if the emitted at sensor spectral radiance $B_\lambda(T)$ is measured (e.g., by thermal sensor) by inverting the Planck's radiance function as follows:

$$T = \frac{C_2}{\lambda \cdot \ln\left[\dfrac{C_1}{\lambda^5 \cdot B_\lambda(T)} + 1\right]} \tag{5.2}$$

Because most natural objects are gray bodies (non-blackbodies; e.g., vegetation, water bodies, and soil, etc.), the emissivity ε, which is defined as the ratio between the radiance of an object and that of a blackbody at the same temperature, must be taken into account (Li et al., 2013). The spectral radiance of a gray body is given by the spectral emissivity multiplied by Planck's Law as shown in Eq. (5.1). Obviously, if the atmosphere exerts no influence on the measured radiance, LST (i.e., T) can be retrieved by making temperature the subject of Eq. (5.1) once the emitted radiance and emissivity are known. The wavelength λ_{max} of the peak monochromatic radiance at a given temperature (T) is given by Wien's displacement law:

$$T_{\lambda max} = 2897.9 / T \tag{5.3}$$

where T is the temperature in K and λ is the wavelength nanometer.

According to this equation, the wavelength λ_{max} at which maximum emission occurs varies roughly from 11.6 μm to 8.8 μm if the LST ranges from 250 K to 330 K with the average temperature of the Earth being approximately 288 K. The wavelength region between 8 and 13 μm coincides within a clear window in the atmosphere, which is most transparent to TIR radiation (Li et al., 2013). In cases where the temperature of the surface exceeds 330 K, the wavelength peak moves to shorter and shorter wavelengths, for example for a wildfire (about 800 K), the maximum emission is around 3.6 μm in the mid-infrared (MIR) region (around AVHRR channel 3), which also coincides with a clear window in the atmosphere.

Eqs. (4.1)–(4. 3) are actually approximations to the theoretical radiation transfer equation in which monochromatic quantities are replaced with channel effective values, but these approximations or simplifications require several important preconditions: (1) the integral of a product is assumed to be equal to the product of the integrals. This assumption is true only if the variables are constant within the integration limits, which is rarely the case. (2) Either the surface is assumed to be Lambertian or the downward atmospheric and solar diffuse radiation are assumed to be isotropic in the calculation of the downward radiations reflected by the surface (e.g., $\left[(1-\varepsilon(\lambda))(Latm\downarrow(\lambda)+Lsun\downarrow(\lambda))\right]$ in Eq. (4.1)). In practice, these conditions are never fulfilled. However, because the surface-reflected downward atmospheric thermal radiation term is much smaller than the surface thermal emission, and the surface-reflected diffuse solar radiation term is much smaller than the surface-reflected direct solar term, this simplification of Eqs. (4.1, 4.2) is reasonable and does not create significant errors (Li et al., 2013).

The theoretical basis for the remote sensing of LST is that total radiative energy emitted by ground surface objects increases rapidly with increase in temperature. The spectral distribution of the energy emitted by a ground object also changes with temperature. According to Wien's Displacement Law for the Earth with an ambient temperature of 300 K, the peak of its spectral radiance occurs at about 9.6 μm. Therefore, theoretically, the thermal energy in relation to the physical temperature of the ground surface can be remotely observed by using airborne or spaceborne TIR sensors operating at a wavelength around 10 μm, which has been defined as thermal channels in remote sensing systems (e.g., Landsat TM). The obtained temperature of the ground surface objects at the satellite level is called the brightness temperature (Reutter et al.,

1994). On the other hand, the spectral characteristics of the atmosphere indicate that there is an atmospheric window in the spectral region 8–14 μm, where atmospheric absorption is minimum and through which the radiance energy of a ground surface can transmit without great losses. Thus, ground surface temperature can be remotely sensed using the channels within the atmospheric window at the thermal spectral wavelength.

5.3 METHODS AND TECHNIQUES FOR RETRIEVING LST

Generally, with the availability of thermal sensing data, such as Landsat Thematic Mapper (TM6), Advanced Spaceborne Thermal Emission and Reflection Radiometer (ASTER) thermal infrared (TIR) images, and the Advanced Very High Resolution Radiometer (AVHRR) thermal channels, the study of LST from thermal images has been a topic of great interest in the remote sensing community for the past four decades. Thermal data, covering TIR spectral range 8–12 μm, can be acquired in single broadband (e.g., Thermal Airborne Broadband Imager, TABI) or multi-band images (e.g., five ASTER TIR channel images). Technically, for retrieving LST from thermal images, five types of methods have been developed: the single infrared channel method (SCA), the mono-window algorithm (MWA), the split-window method (SWA), the machine learning (ML) empirical method, and the physics-based method (PBM), depending on the number of bands used. The single-channel method or mono-window algorithm usually requires a good radiative transfer model and atmospheric profiles that must be provided by either satellite soundings or conventional radiosonde (Qin et al., 2001b). The split-window method (SWA) corrects for atmospheric effects based on the differential absorption in adjacent bands (Qin et al., 2001a). The physics-based method (PBM) requires atmospheric transmittance, path radiance, and downward sky irradiance, estimated using a radiative transfer mode. Because of the difficulties in correcting for atmospheric absorption, atmospheric emission, and surface emissivity, the development of an accurate LST algorithm and correct application are not an easy job. In the following, the five types of LST retrieval of methos are introduced.

5.3.1 Mono-Window Algorithms (MWAs)

The procedure of LST retrieved from the TM(ETM+)6 thermal channel can be described in the flowchart (Figure 5.2). By the flowchart, input data are scaled thermal radiance (i.e., digital number). The MWA needs three parameters for retrieving LST: atmospheric transmittance (τ), land surface emissivity (ε), and effective mean atmospheric temperature (T_a in K). After determining the three parameters and with input of TM(ETM+)6 BT data, we can calculate the pixel-based LST with the MWA.

Based on Qin et al. (2001b) and also on Hang and Rahman (2018), Wang et al. (2019), and Sekertekin and Bonafoni (2020), a typical MWA used for retrieving LST from Landsat 5/7 TM(ETM+)6 thermal image can be expressed as:

$$T_s = \frac{1}{C_6}\left\{a_6 \cdot (1 - C_6 - D_6) + \left[b_6 \cdot (1 - C_6 - D_6) + C_6 + D_6\right] \cdot T_6 - D_6 \cdot T_a\right\} \quad (5.4)$$

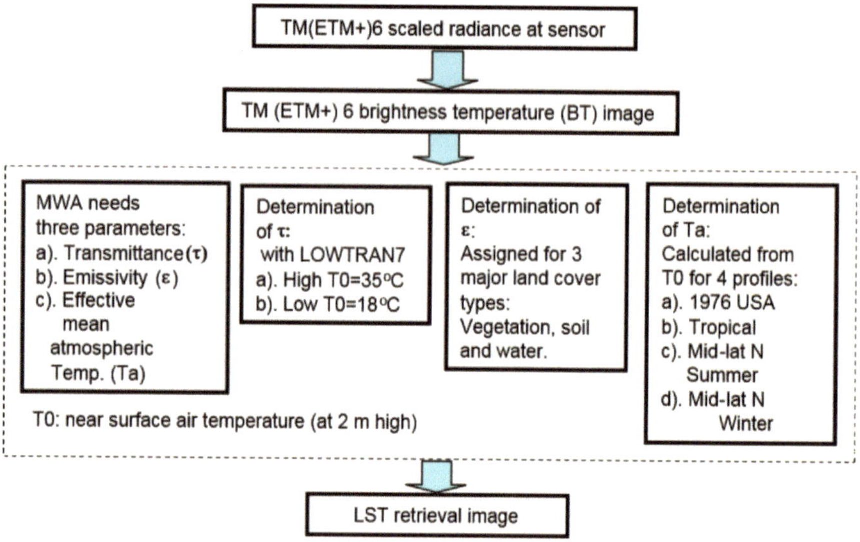

FIGURE 5.2 A flowchart of retrieving LST from the TM6 or ETM+6 thermal channel.

where $\begin{cases} C_6 = \varepsilon_6 \cdot \tau_6 \\ D_6 = (1-\tau_6)\cdot[1+(1-\varepsilon_6)\cdot\tau_6] \\ a_6 = -67.355353 \\ b_6 = 0.458606 \end{cases}$. Therefore, LST can be calculated from the TM(ETM+)6 thermal image with the MWA algorithm if $\varepsilon_6, \tau_6, and\, T_a$ are known.

5.3.1.1 Calculation of Brightness Temperature

Based on Schneider and Mauser (1996) and the Handbook of Landsat7 products, the TM6 or ETM+6 (L-1G For ETM+6) thermal data can be converted from scaled radiance (or digital number, *DN*) to thermal radiance (L_6) at sensor with the following formula:

$$\begin{aligned} L_6 &= 0.1238 + 0.005632156 \cdot DN_6, \quad mw/(cm^2 \cdot sr \cdot nm) \quad for\ TM6; \\ L_6 &= 0.3200 + 0.003705882 \cdot DN_6, \quad mw/(cm^2 \cdot sr \cdot nm) \quad for\ ETM{+}6 \end{aligned} \tag{5.5}$$

Once the spectral radiance (L_6) is computed, the brightness temperature at sensor can be directly calculated by using the following approximation of formula:

$$\begin{aligned} T_6 &= 1260.56 \big/ \ln(1 + 60.776 / L_6) \quad for\ TM6; \\ T_6 &= 1282.71 \big/ \ln(1 + 66.609 / L_6) \quad for\ ETM{+}6 \end{aligned} \tag{5.6}$$

5.3.1.2 Determination of τ

To determine the transmittance τ_6, Qin et al. (2001b) used LOWTRAN7 to simulate the linear relationship between atmospheric transmittance τ_6 and water vapor content w within a range of 0.4–3.0 g/cm^2 under the two profiles: low and high temperature profiles. The atmospheric temperature near the surface for the high temperature profile is defined as 35°C and for the low one 18°C. The estimation equations of transmittance with water vapor content are listed in Table 5.1.

5.3.1.3 Determination of ε

We can assign emissivity values for the seven land cover types for the TM(ETM+)6 band through referring to publications by Sabins (1986) and the spectral library at UC Santa Barbara (http://www.icess.ucsb.edu) (Table 5.2).

5.3.1.4 Determination of T_a

Qin et al. (2001b) derived the simple linear relations for approximation of the effective mean atmospheric temperature T_a from near surface air temperature T_0 (at 2 m high) for the four standard atmospheric profiles as follows:

- 1976 USA Ta = 25.9396 + 0.88045 • T0
- tropical Ta = 17.9769 + 0.91715 • T0

TABLE 5.1
Relationships Between Transmittance and Water Vapor Content in the Atmosphere

Profiles	w (g/cm^2)	Transmittance estimation equation
High air temperature (35°C)	0.4–1.6	$\tau_6 = 0.974290 - 0.08007w$
	1.6–3.0	$\tau_6 = 1.031412 - 0.11536w$
Low air temperature (18°C)	0.4–1.6	$\tau_6 = 0.982007 - 0.09611w$
	1.6–3.0	$\tau_6 = 1.053710 - 0.14142w$

TABLE 5.2
Emissivity Assigned for the Seven Land Cover Types

Land cover type	ε_6
dense vegetation covering 90–100%	0.985
75% vegetation, 25% bare soil	0.975
50% vegetation, 50% bare soil	0.965
bare soil (wet)	0.970
75% bare soil (dry)	0.960
bare soil (dry) or rock area	0.945
sea/lake	0.990

- mid-Lat-N-Summer Ta = 16.0110 + 0.92621 • T0
- mid-Lat-N-Winter Ta = 19.2704 + 0.91118 • T0

5.3.1.5 Application of the MWA to ETM+6

To test the MWA, we calculated LST from the one ETM+6 thermal image that was acquired on October 25, 2001 at 10:30 am local time from Lake Tahoe, California/Nevada, USA. Figure 5.3a is ETM+6 scaled radiance while Figure 5.3b presents LST results retrieved from the ETM+6 thermal image. From Figures 5.3b, the minimum/maximum/average temperatures in °C are 1.1/45.7/19.1.

The MWA algorithm with input of parameters' values, either simulated or assigned, can be employed in most of the low- to middle-latitude regions of the Earth. Currently, total atmospheric water vapor (g/cm^2), measured from ground, for the two profiles, summer and winter, is 0.4–3.0. Input of scaled thermal radiance of TM(ETM+)6 is necessary for using the algorithm. Use of a fixed value of emissivity over the entire scene of TM(ETM+)6 image may not be reasonable.

In addition, according to Qin et al. (2001a), Ottlé and Stoll (1993), and Pu et al. (2006), the observed radiance of thermal sensor can be expressed as

$$B(T) = \varepsilon \cdot \tau \cdot B(T_s) + (1-\tau)[1+(1-\varepsilon)\tau]B(T_a) \tag{5.7}$$

where $B(T)$ is the observed thermal radiance of brightness temperature T at sensor in w/(m^2.sr.μm); $B(Ts)$ is the land surface thermal radiance at surface temperature T_s in w/(m^2.sr.μm); B(T_a) is the thermal radiance of effective mean atmospheric temperature T_a in w/(m^2.sr.μm); ε is the ground surface emissivity; and τ is atmospheric transmittance.

Per Eq. (5.7), if we know $B(T_s)$ and $B(T_a)$ we have $ALST = (B - CT_a + D)$. Therefore, LST can be calculated from the MWA algorithm as:

$$LST = (B - CT_a + D)/A \tag{5.8}$$

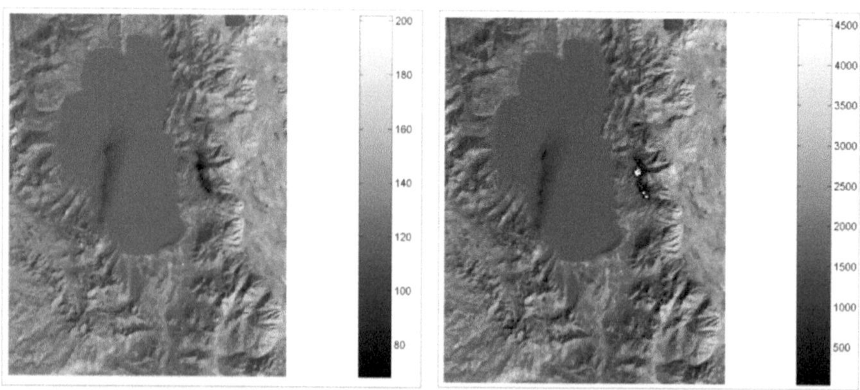

FIGURE 5.3 LST retrieved from the ETM+6 thermal channel, 10/25/2001, at 10:30 am, Lake Tahoe, California/Nevada, USA. (a) ETM+6 DN image; (b) retrieved LST image (°C × 100) with Min/Max/Average: 1.1/45.7/19.1°C.

$$\text{where} \begin{cases} A = a \cdot \varepsilon \cdot \tau \\ B = a \cdot T_s + b \cdot \varepsilon \cdot \tau - b \\ C = (1-\tau) \cdot \left[1 + (1-\varepsilon) \cdot \tau\right] \cdot a \\ D = (1-\tau) \cdot \left[1 + (1-\varepsilon) \cdot \tau\right] \cdot b \end{cases} \tag{5.9}$$

Using Eq, (5.7) with input of single thermal channel T_s BT in K, if ε, τ, and T_a are known, we can retrieve LST from the thermal image data.

5.3.2 Single-Channel Algorithms (SCAs)

According to Jiménez-Munoz and Sobrino (2003) who have developed a generalized single-channel method in order to retrieve LST from only one thermal channel, many researchers have used this SCA to retrieve LST from a single TIR channel, in which the LST is given by the following equation:

$$T_S = \gamma\left[\varepsilon^{-1}\left(\psi_1 L_{sensor} + \psi_2\right) + \psi_3\right] + \delta \tag{5.10}$$

with

$$\gamma = \left\{\frac{C_2 L_{sensor}}{T_{sensor}^2}\left[\frac{\lambda^4}{C_1} L_{sensor} + \lambda^{-1}\right]\right\}^{-1} \approx \frac{T_{sensor}^2}{b_r L_{sensor}}$$

$$\delta = -\gamma L_{sensor} + T_{sensor} \approx T_{sensor} - \frac{T_{sensor}^2}{b_r} \tag{5.11}$$

and where L_{sensor} is the at sensor radiance in W m^{-2} sr^{-1} μm^{-1}; T_{sensor} is the at-sensor brightness temperature in K; λ is the effective wavelength (11.457 μm for band TM6); C_1 = 1.19104 10^8 W μm^4 m^{-2} sr^{-1} and C_2 = 14387.7 K; and b_r is the 1324 K for TM6. The atmospheric functions ψ_1, ψ_2, and ψ_3 can be obtained as a function of the total atmospheric water vapor content (w) according to the following equations particularized for TM6 data:

$$\begin{bmatrix} \psi_1 \\ \psi_2 \\ \psi_3 \end{bmatrix} = \begin{bmatrix} 0.04918 & 0.02916 & 1.01523 \\ -0.38333 & -1.50294 & 0.20324 \\ 0.00918 & 1.36072 & -0.27514 \end{bmatrix} \begin{bmatrix} w^3 \\ w \\ 1 \end{bmatrix} \tag{5.12}$$

5.3.3 Split-Window Algorithms (SWAs)

5.3.3.1 General SWAs

Applying an LST retrieval algorithm from TIR remote sensing data, especially with SWA, is a well-established and validated method. However, no single algorithm performs best under all conditions, even if it generally achieves good accuracy. This suggests that a more stable and robust LST retrieval algorithm may be gotten by integrating various individual LST retrieval algorithms. Generally, the SWA algorithms can be divided into two sub-types: linear and non-linear SWA subtypes. A linear SWA algorithm can be developed utilizing the differential absorption in two adjacent TIR channels *i, j* (e.g., AVHRR channels 4 and 5) in the spectral wavelength 10–12.5 μm through linearizing the radiation transfer equation (RTE) with respect to the temperature and wavelength. This algorithm expresses the LST as a simple linear combination of the two brightness temperatures T_i and T_j measured in the two TIR channels (McMillin, 1975). A typical linear split-window algorithm (Li et al., 2013) can be expressed as

$$LST = a_0 + a_1 T_i + a_2 (T_i - T_j) \tag{5.13}$$

where, a_k (k = 0, 1, and 2) are coefficients that depend primarily on the spectral response function of the two channels $g_i(\lambda)$ and $g_j(\lambda)$; the two channel emissivities ε_i and ε_j, the WV, and the VZA. Thus:

$$a_k = f_k(g_i, g_j, \varepsilon_i, \varepsilon_j, WV, VZA) \tag{5.14}$$

It should be pointed out that the accuracy of this type LST retrieval method is dependent on the correct choice of the coefficients a_k. They are pre-determined either by regressing the simulated satellite data with a set of atmosphere parameters and surface parameters or empirically by comparing the satellite data against in situ LST measurements (Li et al., 2013).

The errors introduced by linearizing of the RTE in Eq. (5.13) by some approximations used in the development of SWA algorithms, e.g., approximating the transmittance as a linear function of the WV, result in large errors in LST retrieval under warm and cold atmospheric conditions. To improve the accuracy of LST retrieval, non-linear SWA algorithms have been developed as follows (Li et al., 2013):

$$LST = c_0 + c_1 T_i + c_2 \left(T_i - T_j\right) + c_3 (T_i - T_j)^2 \tag{5.15}$$

where c_k (k = 0–3) are coefficients pre-determined by regressing Eq. (5.15) on simulated satellite data with a set of atmosphere parameters and surface parameters, similar to the a_k in Eq. (5.13).

Many similar forms for non-linear SWA algorithms have been developed in the literature in recent decades. Similar to the linear SWA algorithms, some of these non-linear SWA algorithms incorporate the LSEs into the coefficients c_k, some use both the LSE and the WV, and some also incorporate the VZA. Table 5.3 shows a list of widely used candidate SWAs algorithms for readers' convenience. In the table, T_{11} and T_{12} represent the channels centered at approximately 11 and 12 mm, respectively. For example, data from NOAA-AVHRR channels 4 and 5 operating respectively at 10.5–11.3 mm and 11.5–12.5 mm. Landsat 8 channels 10 and 11 operating respectively at 10.60–11.19 mm and 11.50–12.51 mm can be used to estimate ground surface objects' temperature.

TABLE 5.3
A List of Spite-Window Algorithms (SWAs) (Zhou et al., 2019b; Ma et al., 2000)

Model	Formula	Reference
OV1992	$T_s = A_0 + A_1T_{11} + A_2(T_{11} - T_{12})$	Ottlé and Vidal-Madjar (1992)
FO1996a	$T_s = A_0 + A_1T_{11} + A_2(T_{11} - T_{12}) + A_3(T_{11} - T_{12})^2$	Francois and Ottlé (2002)
PR1984	$T_s = A_0 + A_1T_{11} + A_2(T_{11} - T_{12}) + A_3T_{11}\varepsilon_{12} + A_4(T_{11} - T_{12})(1-\varepsilon_{11}) + A_5T_{12}\Delta\varepsilon$	Price (1984)
UC1985	$T_s = A_0 + A_1T_{11} + A_2(T_{11} - T_{12}) + A_3(1-\varepsilon)$	Ulivieri and Cannizzaro (1985)
BL1990	$T_s = A_0 + (A_1 + A_2\frac{1-\varepsilon}{\varepsilon} + A_3\frac{\Delta\varepsilon}{g^2}(T_{11} - T_{12}) + (A_4 + A_5\frac{1-\varepsilon}{\varepsilon} + A_6\frac{1-\Delta\varepsilon}{\varepsilon^2})(T_{11} - T_{12})$	Becker and Li (1990)
PP1991	$T_S = A_0 + A_1\frac{T_{11}-T_0}{\varepsilon_{11}} + A_2\frac{T_{12}-T_0}{\varepsilon_{12}} + A_3\frac{1-\varepsilon_{11}}{\varepsilon_{11}} + T_0$	Platt and Wright (1991)
VI1991	$T_s = A_0 + A_1T_{11} + A_2(T_{11} - T_{12}) + A_3\left(\frac{1-\varepsilon}{\varepsilon}\right) + A_4\frac{\Delta\varepsilon}{\varepsilon}$	Vidal (1991)
UL1994	$T_s = A_0 + A_1T_{11} + A_2(T_{11} - T_{12}) + A_3(1-\varepsilon) + A_4\Delta\varepsilon$	Ulivieri et al. (1994)
WA2014	$T_s = A_0 + \left(A_1 + A_2\frac{1-\varepsilon}{\varepsilon} + \frac{\Delta\varepsilon}{\varepsilon^2}\right)(T_{11} - T_{12}) + \left(A_4 + A_5\frac{1-\varepsilon}{\varepsilon} + A_6\frac{\Delta\varepsilon}{\varepsilon^2}\right)(T_{11} - T_{12}) + A_7(T_{11} - T_{12})^2$	Wan (2014)
FO1996b	$T_s = A_0 + (A_1\omega + A_2\omega^2 + A_3)T_{11} + (A_4\omega + A_5\omega^2 + A_6)T_{12} + A_7\omega + A_8\omega^2$	Francois and Ottlé (2002)
SO1991	$T_s = A_0 + A_1T_{11} + [A_2\omega + A_3 + (A_4\omega + A_5)(1-\varepsilon_{11}) + (A_6\omega + A_7)\Delta\varepsilon](T_{11} + T_{12})$ $\frac{1-\varepsilon_{11}}{\varepsilon_{11}}T_{11}[A_8\omega + A_9(A_{10}\omega + A_{11})\Delta\varepsilon] - \frac{1-\varepsilon_{12}}{\varepsilon_{12}}T_{12}[A_{12}\omega + A_{13}(A_{14}\omega + A_{15})\Delta\varepsilon]$	Sobrino et al. (1991)
UL1985	$T_s = A_0 + A_1T_{11} + (A_2\omega + A_3)(T_{11} + T_{12}) + (A_4\omega + A_5)(1-\varepsilon) + (A_6\omega + A_7)\Delta\varepsilon$	Ulivieri and Cannizzaro (1985)

(Continued)

TABLE 5.3 (*Continued*)
A List of Spite-Window Algorithms (SWAs) (Zhou et al., 2019b; Ma et al., 2020)

Model	Formula	Reference
CC1997	$T_s = 0.39 \times T_{11}^2 + 2.3 \times T_{11} - 0.78 \times T_{11}T_{12} - 1.34 \times T_{12} + 0.39 \times T_{12}^2 + 0.56$	Coll and Caselles (1997)
KE1992	$T_s = T_{11}(0.5 \times P_v + 3.1) + T_{12}(0.5 \times P_v - 2) - 5.5 \times P_v + 3.1$	Kerr et al. (1992)
UL1994b	$T_s = A_0 + A_1T_{11} + (A_2w + A_3)(T_{11} + T_{12}) + (A_4w + A_5)(1-\varepsilon) - \Delta\varepsilon(A_6w + A_7)$	Ulivieri et al. (1994)
SO1996	$T_s = -A_0 + T_{11} + A_1(T_{11} - T_{12}) + A_2(1-\varepsilon) - A_3\Delta\varepsilon$	Sobrino et al. (1996)
CO1994	$T_s = A_0 + A_1T_{11} + A_2(T_{11} + T_{12}) + A_3(T_{11} + T_{12})^2 + [(A_4\omega + A_5)T_{11} + (A_6\omega + A_7)]$ $+(1-\varepsilon) - [(A_g\omega + A_g)T_{11} + [(A_{10}\omega + A_{11})]\Delta\varepsilon$	Coll et al. (1994)
SR2000	$T_s = A_0 + A_1T_{11} + A_2(T_{11} - T_{12}) + A_3(T_{11} - T_{12})^2$	Sobrino and Raissouni (2000)
MT2002	$T_s = A_0 + A_1T_{11} + A_2(T_{11} - T_{12}) + A_3(T_{11} - T_{12})^2 + (A_4\omega + A_5)(1-\varepsilon)$	Ma et al. (2002)
BL1995	$T_s = A_0 + A_1\omega + [A_2 + (A_3\cos\theta + A_4) + (1-\varepsilon_{11}) + (A_5\omega + A_6)\Delta\varepsilon](A_{11} + A_{12}) +$ $[A_7 + A_g\omega + (A_9 + A_{10}\omega)(1-\varepsilon_{11}) - (A_{11}\omega + A_{12})\Delta\varepsilon](A_{11} + A_{12})$	Becker and Li (1995)
GA2008	$T_s = A_0 + A_1T_{11} + A_2(T_{11} - T_{12}) + A_3(T_{11} - T_{12})^2 + (A_4 + A_5\omega + A_6\omega^2)(1-\varepsilon) + (A_7 + A_g\omega)\Delta\varepsilon$	Galve et al. (2008)
JM2014	$T_s = A_0 + T_{11} + A_1(T_{11} - T_{12}) + A_2(T_{11} - T_{12})^2 + (A_3\omega + A_4)(1-\varepsilon) - (A_5\omega + A_6)\Delta\varepsilon$	Jiménez-Muñoz et al. (2014)

Note: subscripts 11 and 12 denote the channels centered at approximately 11 and 12 μm, respectively, while T_{11} and T_{12} and ε_{11} and ε_{12} are the their associated BTs and LESs; $\varepsilon = (\varepsilon_{11} + \varepsilon_{12})/2$, $\Delta\varepsilon = (\varepsilon_{11} - \varepsilon_{12})$, and A_i are the coefficients; T_0 in PP1991 is the 273.15 K; ω is the column water vapor content; and θ is the VZA.

Based on the validation results of verifying most algorithms in Table 5.3, Zhou et al. (2019b) assessed the algorithms for the GLASS LST product with global simulation datasets. They found that the multialgorithm ensemble approach yields an accuracy better than 1.0 K when LSEs yield uncertainties of –0.02 to 0.02 and water vapor yields an uncertainty of –1.0 to 1.0 g·cm^{-2}. Thus, the algorithms can theoretically satisfy the target accuracy of the GLASS LST product, namely, 2.0 K for MODIS LST and 3.0 K for AVHRR LST. For the GLASS-AVHRR LST product, its accuracy is 2.89 K when validated against the measurements from the Baseline Surface Radiation Network (BSRN). Nevertheless, the GLASS-AVHRR LST still requires more in-depth comparison to better understand its quality and applicability due to the lack of historical, in situ measured LST data. For the GLASS-MODIS LST, a validation based on the measurements from six grassland/cropland surface radiation network (SURFRAD) sites indicated that the mean RMSE values are 1.82–2.15 K at nighttime for Terra/Aqua MODIS in 2003 and 2013; this accuracy is similar to the official MODIS LST product. Validation at an arid shrubland site (i.e., Desert Rock, NV) of SURFRAD showed that an accuracy yields a systematic underestimation of LST, as does the official MODIS product, due to large variations in surface spectral emissivity. Further comparison confirms the agreement between the GLASS-MODIS LST and official MODIS LST.

5.3.3.2 A Case of SWA Study

AVHRR thermal channels Ch4 (10.5–11.3 μm) and Ch5 (11.5–12.5 μm) can be used to retrieve LST. The viewing direction of the remote sensing determines the pathway of the radiance through the atmosphere to reach the sensor. Because the zenith angle of viewing (ZAV, θ) can be up to 75° for an AVHRR sensor, we should consider the effect of large θ on atmospheric transmittance. AVHRR data's nominal spatial resolution is 1.1 km at nadir. The procedure of the SWA LST retrieved from the two AVHRR thermal channels can be presented in the flowchart (Figure 5.4). From the flowchart, input data have been converted into BT at sensor.

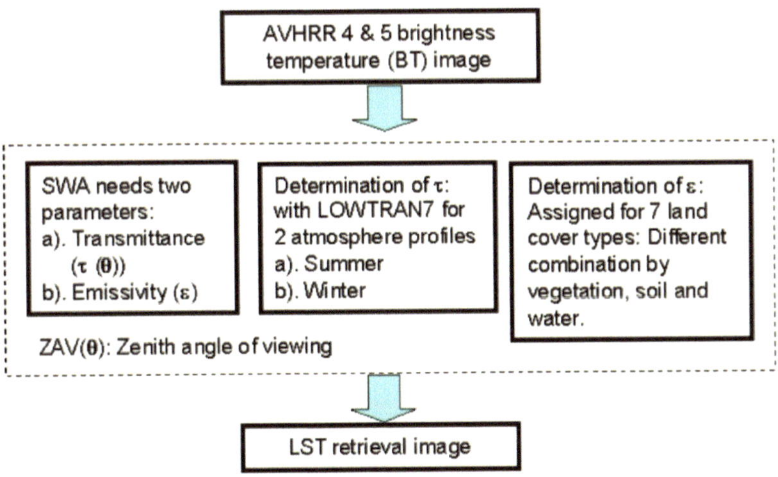

FIGURE 5.4 A flowchart of retrieving LST from the two AVHRR thermal channels 4 and 5.

The SWA needs two parameters for retrieving LST: atmospheric transmittance (**τ(θ)**) and land surface emissivity (**ε**). After determining the two parameters and with input of AVHRR Ch4 and Ch5 BT data, we can calculate the pixel-based LST with the SWA.

5.3.3.2.1 A Split-Window Algorithm (SWA)

Based on Qin et al. (2001a), an SWA used for retrieving LST from NOAA/AVHRR, two thermal channels 4 and 5 can be written as:

$$T_s = A_0 + A_1 T_4 - A_2 T_5 \tag{5.16}$$

where

$$\begin{cases} A_0 = E_1 a_4 - E_2 a_5 \\ A_1 = 1 + A + E_1 b_4 \\ A_2 = A + E_2 b_5 \end{cases} \begin{cases} A = D_4 / E_0 \\ E_0 = D_5 C_4 - D_4 C_5 \\ E_1 = D_5 (1 - C_4 - D_4) / E_0 \\ E_2 = D_4 (1 - C_5 - D_5) / E_0 \end{cases}$$

$$\begin{cases} C_4 = \varepsilon_4 \cdot \tau_4(\theta) \\ D_4 = (1 - \tau_4(\theta)) \cdot [1 + (1 - \varepsilon_4) \cdot \tau_4(\theta)] \\ C_5 = \varepsilon_5 \cdot \tau_5(\theta) \\ D_5 = (1 - \tau_5(\theta)) \cdot [1 + (1 - \varepsilon_5) \cdot \tau_5(\theta)] \end{cases} \text{and} \begin{cases} a_4 = -62.239281 \\ b_4 = 0.430589 \\ a_5 = -66.540666 \\ b_5 = 0.465845 \end{cases}.$$

Therefore, LST can be calculated from the two AVHRR thermal channels using the SWA algorithm if $\varepsilon_4, \tau_4(\theta), \varepsilon_5, and\, \tau_5(\theta)$ are known, and we can get LST.

5.3.3.2.2 Determination of τ

To determine the $\tau_4(\theta) and\, \tau_5(\theta)$, Qin et al. (2001a) first simulated transmittance at ZAV = 10° for the two thermal channels: $\tau_4(10) and\, \tau_5(10)$ with LOWTRAN7, then calculate transmittance differences with ZAV. Therefore, $\tau_4(\theta) and\, \tau_5(\theta)$ can be expressed as:

$$\begin{cases} \tau_4(\theta) = \tau_4(10) - \Delta\tau_4(\theta) \\ \tau_5(\theta) = \tau_5(10) - \Delta\tau_5(\theta) \end{cases} \tag{5.17}$$

where $\Delta\tau_4(\theta) and\, \Delta\tau_5(\theta)$ are the atmospheric transmittance differences of AVHRR channels 4 and 5 and θ is the ZAV in degrees. According to Qin et al. (2001a), the ZAV correction equations for AVHRR channels 4 and 5 are

$$\begin{cases} \Delta\tau_4(\theta) = -2.3994 \cdot 10^{-3} + 2.2976 \cdot 10^{-5} \cdot \theta^2 \\ \Delta\tau_5(\theta) = -3.2766 \cdot 10^{-3} + 3.1454 \cdot 10^{-5} \cdot \theta^2 \end{cases} \tag{5.18}$$

The $\tau_4(10) and \tau_5(10)$ can be calculated from the estimation equations in Table 5.4, separated for the two atmospheric profiles: summer and winter.

5.3.3.2.3. Determination of ε

We assigned emissivity values for the seven land cover types for the two AVHRR thermal channels through referring to publications by Sabins (1986) and the spectral library at UC Santa Barbara (http://www.icess.ucsb.edu) (Table 5.5).

5.3.3.2.4 Application of SWA

In this case, we calculated LST from the two AVHRR thermal channels 4 and 5. The AVHRR thermal data were acquired on September 2, 1999 at approximately 3:00 pm local time from San Francisco, California, USA. Figures 5.5a and 5.5b are AVHRR BT data of channels 4 and 5, respectively, while Figure 5.5c presents LST results retrieved from the two thermal channels.

From Figure 5.5c the minimum/maximum/average temperatures in °C are 7.6/51.6/25.4.

TABLE 5.4
Relationships Between Transmittance at ZAV = 10° and Water Vapor Content in the Atmosphere

w, g/cm2	Estimation equations summer profile	w, g/cm2	Estimation equations winter profile
0.4–1.6	$\tau_4(10) = 0.979160 - 0.062918w$ $\tau_5(10) = 0.968144 - 0.098942w$	0.4–1.6	$\tau_4(10) = 0.983311 - 0.072444w$ $\tau_5(10) = 0.981868 - 0.121979w$
1.6–3.0	$\tau_4(10) = 1.035378 - 0.097514w$ $\tau_5(10) = 1.026468 - 0.135133w$	1.6–3.0	$\tau_4(10) = 1.058059 - 0.121354$ $\tau_{5(}10) = 1.048364 - 0.163678w$
3.0–5.0	$\tau_4(10) = 1.098068 - 0.118847w$ $\tau_5(10) = 1.034865 - 0.139598w$	3.0–4.4	$\tau_4(10) = 1.111348 - 0.140080w$ $\tau_5(10) = 1.033785 - 0.160330w$
5.0–6.4	$\tau_4(10) = 1.060569 - 0.111773w$ $\tau_5(10) = 0.866450 - 0.105842w$	4.4–5.4	$\tau_4(10) = 1.071156 - 0.131166w$ $\tau_5(10) = 0.879292 - 0.125053w$
		5.4–6.4	$\tau_4(10) = 0.964106 - 0.111430w$ $\tau_5(10)0 = 0.681518 - 0.088431w$

TABLE 5.5
Emissivity Assigned for the Seven Land Cover Types

Land cover type	ε_4	ε_5
dense vegetation covering 90–100%	0.990	0.994
75% vegetation, 25% bare soil	0.980	0.984
50% vegetation, 50% bare soil	0.970	0.974
bare soil (wet)	0.970	0.974
75% bare soil (dry)	0.960	0.964
bare soil (dry) or rock area	0.950	0.954
sea/lake	0.995	0.999

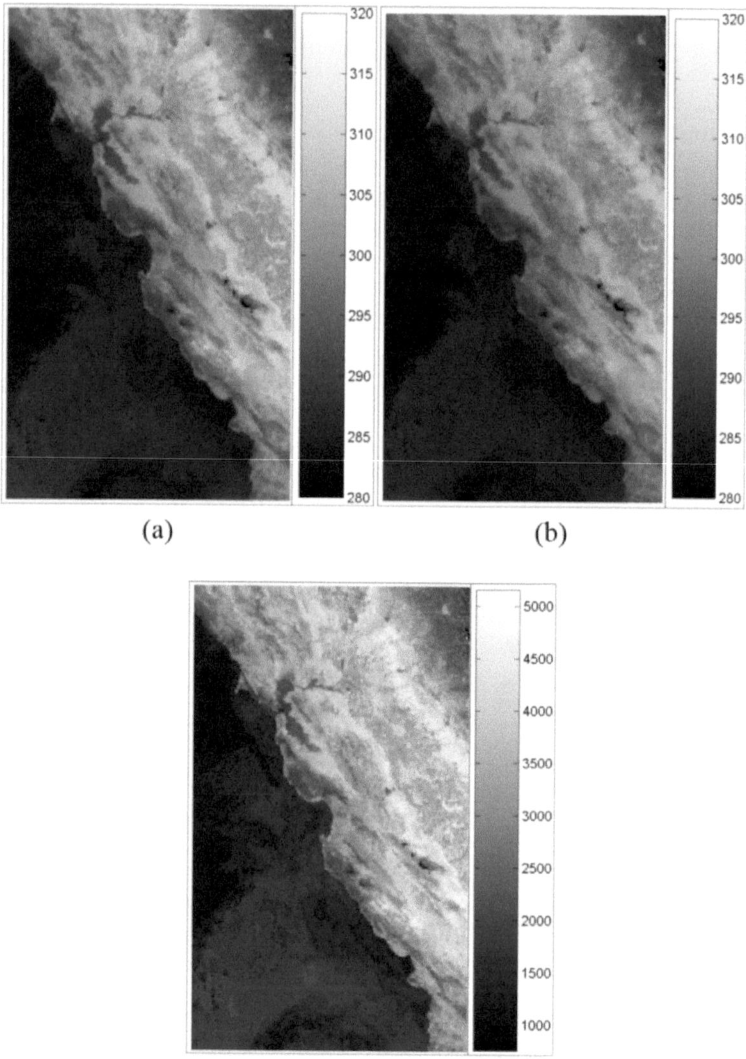

FIGURE 5.5 LST retrieved from the two AVHRR thermal channels, 09/2/1999, around 3:00 pm, San Francisco, California, USA. (a) AVHRR Ch4 BT image; (b) AVHRR Ch5 BT image; and (c) retrieved LST image (°C × 100) with Min/Max/Average: 7.6/51.6/25.4°C.

5.3.3.2.5 Requirements of the Current Version of the SWA

The SWA algorithm with input of parameters' values, either simulated or assigned, can be employed in the low- to middle-latitude regions of the Earth. Currently, total atmospheric water vapor (g/cm^2), measured from the ground, for the two profiles, summer and winter, is 0.2–6.4. Input of scaled brightness temperatures of both AVHRR channels (Ch 4 and Ch5) is necessary for using the algorithm. Use of a fixed value of emissivity over the entire scene of AVHRR image may not be reasonable.

5.3.4 Machine Learning (ML) Empirical Methods

In the past few decades, ML has been applied to the retrieval of geophysical parameters, which has considerably outperformed traditional models. The main advantages of artificial neural network (ANN) methods, some of the major ML methods, over conventional retrieval methods, are their ability to learn complex patterns, generalization to noisy environments, and incorporation of both knowledge and physical constraints. For example, Aires et al. (2002) developed a regularized neural network to retrieve atmospheric profiles and LSTs from the high-resolution infrared atmospheric sounding interferometer (IASI) spaceborne instrument. The surface temperature for tropical situations yields an RMS error of 0.4 K for instantaneous retrievals. Wang et al. (2010) attempted to establish ANNs to simultaneously retrieve the LST, LSE, and atmospheric profiles from hyperspectral TIR data. The preliminary results demonstrated that ANNs can be used to simultaneously retrieve the LST, LSEs, and atmospheric profiles from hyperspectral TIR data with an acceptable accuracy for some applications. Similarly, Mao et al. (2008) also used ML methods to achieve good results in atmospheric profiles/LST/LSE retrieval. Recently, some of the related studies with ML/DL methods focused on reconstructing the LST of satellite datasets, forecasting daily LST from time series data, and fusing multisource data to estimate subpixel LST.

However, it is difficult to determine the architectures and learning schemes for an ANN, which are directly associated with its ability to learn and generalize. Although one or two hidden layers are recognized to be enough for most application problems (Sontag, 1992; Aires et al., 2002), a number of experiments are still needed to determine what architecture-related parameters will improve the accuracy, such as the number of input and hidden nodes, the initial weight range, the activation functions, the learning rate and momentum coefficient, and the terminating criterion. Until now, no ANN architecture is universally accepted for a particular problem.

5.3.5 Physics-Based Method (PBM)

Many hyperspectral infrared sensors onboard satellites provide land surface data such as the infrared atmospheric sounding interferometer (Lan et al., 2022), and the accurate retrieval of the LST from hyperspectral infrared data is being widely discussed (Zhou et al., 2008). Based on the radiative transfer equation (RTE), the LST and land surface emissivity (LSE) are retrieved (Li et al., 2013). Many methods have been developed to acquire LST information from the at-sensor radiance after accurate atmospheric correction. Retrieving LST information without accurate atmospheric correction is a key issue (Li et al., 2013). Nowadays, physical retrieval methods (Masiello et al., 2018) provide solutions to obtain the LST without accurate atmospheric correction from hyperspectral TIR data. The physical retrieval method is mostly based on an RTE linearized using the tangents around the initial estimates of the LST, LSE, and atmospheric parameters. Principal component analysis (Masiello et al., 2018) in combination with the Tikhonov regularization method can be utilized to reduce the number of unknowns and stabilize the ill-posed problem. In this subsection, a new two-step physical method was introduced to retrieve the land surface temperature (LST), land surface emissivity (LSE), atmospheric equivalent temperature (T_a), and water vapor content (w) from infrared atmospheric sounding interferometer (IASI) observations without any prior land surface knowledge and atmospheric information. At the same time, it uses a simple way to relinearize the RTE to obtain an LST with a good accuracy from IASI data. The results indicate that the root-mean-square error (RMSE) of the simulated LST is ~1K based on an initial LST estimate with an RMSE of 2 K (1.9 K). The physical retrieval method is introduced later (Lan et al., 2022).

Following the basic concept of the two-step physical retrieval method (Ma et al., 2000), we only consider the inversion of the surface temperature but not the retrieval of the atmospheric profile. Therefore, we only pay attention to the effect of the atmosphere on the inversion of the surface temperature. The parameters for atmospheric equivalent temperature (T_a) and water vapor content (w) were utilized to simplify the retrieval method.

A linearized form of the RTE equation was first obtained. The at-sensor radiance L_λ at wavelength λ can be expressed as follows (Li et al., 2013):

$$l_\lambda = \varepsilon_\lambda B_\lambda^{T_s} \tau_\lambda + l_\lambda^{up} + \left(1-\varepsilon_\lambda\right) L_\lambda^{down} \tau_\lambda \tag{5.19}$$

where, ε_λ is the LSE; $B_\lambda^{T_s}$ is the Planck function B at surface temperature T_s at wavelength λ; τ_λ is the atmospheric transmittance; L_λ^{up} is the atmospheric upwelling radiance; and L_λ^{down} is the atmospheric downwelling radiance. Equation (5.19) can be further approximated as (Qin et al., 2001a)

$$B_\lambda^{T_{b\lambda}} = \varepsilon_\lambda B_\lambda^{T_s} \tau_\lambda + \left(1-\tau_\lambda\right) B_\lambda^{T_{a\lambda}} + \left(1-\varepsilon_\lambda\right) \tau_\lambda \left(1-\tau_\lambda^{50}\right) B_\lambda^{T_{a\lambda}} \tag{5.20}$$

where $B_\lambda^{T_{b\lambda}}$ is the Planck function B at the brightness temperature T_{b_λ}; $B_\lambda^{T_{a\lambda}}$ is the Planck function B at the atmospheric equivalent temperature T_{a_λ}; and τ_λ^{50} is the atmospheric transmittance at the viewing zenith angle of 50°. According to the first-order variation of (Anderson et al., 2008), the following equation is obtained:

$$\begin{aligned}\delta B_\lambda^{T_{b\lambda}} &= \delta B_\lambda^{T_{a\lambda}} \left(1-\tau_\lambda^{50}\tau_\lambda - \varepsilon_\lambda \tau_\lambda + \varepsilon_\lambda \tau_\lambda^{50} \tau_\lambda\right) \\ &+ \delta\varepsilon_\lambda \left(B_\lambda^{T_s} \tau_\lambda - \tau_\lambda \boldsymbol{B}_\lambda^{T_{a\lambda}} + B_\lambda^{T_{a\lambda}} \tau_\lambda \tau_\lambda^{50}\right) + \delta B_\lambda^{T_s} \tau_\lambda \varepsilon_\lambda \\ &+ \delta\tau_\lambda \left(\tau_\lambda B_\lambda^{T_s} - \tau_\lambda^{50} \boldsymbol{B}_\lambda^{T_{a\lambda}} - \varepsilon_\lambda \boldsymbol{B}_\lambda^{T_{a\lambda}} + \varepsilon_\lambda \tau_\lambda^{50} B_\lambda^{T_{a\lambda}}\right) \\ &+ \delta\tau_\lambda^{50} \left(\varepsilon_\lambda \boldsymbol{B}_\lambda^{T_{a\lambda}} \tau_\lambda - \tau_\lambda B_\lambda^{T_{a\lambda}}\right)\end{aligned} \tag{5.21}$$

where τ_λ and τ_λ^{50} are approximated as $a_\lambda \cdot e^{\left(-\frac{k_\lambda w}{\cos\theta}\right)}$ and $a_\lambda \cdot e^{\left(-\frac{k_\lambda w}{\cos 50^0}\right)}$, respectively; w is the water vapor content; θ is the viewing zenith angle; and a_λ and k_λ are the fitting and extinction coefficients, respectively, which are wavelength dependent. Based on $\alpha_\lambda = (\partial B_\lambda/\partial Ta_\lambda/\partial B_\lambda/\partial Tb_\lambda)$ and $\beta_\lambda = (\partial B_\lambda/\partial T_s/\partial B_\lambda/\partial Tb_\lambda)$, Eq. (5.21) can be written as

$$\delta Tb_\lambda = W_{T_{a\lambda}} \delta T_{a\lambda} + W_{\varepsilon\lambda} \delta\varepsilon_\lambda + W_{T_{s\lambda}} \delta T_S + W_{w\lambda} \delta w \tag{5.22}$$

where

$$W_{T_{a\lambda}} = a_\lambda\left(1-\tau_\lambda^{50}\tau_\lambda - \varepsilon_\lambda\tau_\lambda + \varepsilon_\lambda\tau_\lambda^{50}\tau_\lambda\right) \tag{5.23a}$$

$$W_{\varepsilon\lambda} = \frac{1}{\dfrac{\partial B_\lambda}{\partial T_{b\lambda}}}\left(\boldsymbol{B}_\lambda^{T_s}\tau_\lambda - \tau_\lambda B_\lambda^{T_{a\lambda}} + B_\lambda^{T_{a\lambda}}\tau_\lambda\tau_\lambda^{50}\right) \tag{5.23b}$$

$$W_{T_{s\lambda}} = \beta_\lambda\left(\varepsilon_\lambda\tau_\lambda\right) \tag{5.23c}$$

$$W_{w\lambda} = -\frac{K_\lambda / \cos\theta}{\partial B_\lambda / \partial T_{b\lambda}}\left(\varepsilon_\lambda B_\lambda^{T_s}\tau_\lambda - \tau_\lambda^{50} B_\lambda^{T_{a\lambda}}\tau_\lambda - \varepsilon_\lambda B_\lambda^{T_{a\lambda}}\tau_\lambda + \varepsilon_\lambda\tau_\lambda^{50} B_\lambda^{T_{a\lambda}}\tau_\lambda\right)$$
$$-\frac{K_\lambda / \cos 50^0}{\partial B_\lambda / \partial T_{b\lambda}}\left(\varepsilon_\lambda B_\lambda^{T_{a\lambda}}\tau_\lambda\tau_\lambda^{50} - \tau B_\lambda^{T_{a\lambda}}\tau_\lambda^{50}\right). \tag{5.23d}$$

Finally, for a given observation with N channels (wavelength λ_i, i = 1, . . . , n), the group of equations (n equations) can be presented in matrix form as:

$$\delta \boldsymbol{y} = \boldsymbol{W}\delta \boldsymbol{x} \tag{5.24}$$

where

$$\delta y = \begin{bmatrix} \delta Tb_{\lambda 1} \\ \vdots \\ \delta Tb_{\lambda N} \end{bmatrix} \tag{5.25a}$$

$$W = \begin{bmatrix} W_{T_{a\lambda 1}} & \cdots & 0 & W_{\varepsilon_{\lambda 1}} & \cdots & 0 & W_{T_{s\lambda 1}} & W_{W_{\lambda 1}} \\ \vdots & \ddots & \vdots & \vdots & \ddots & \vdots & \vdots & \vdots \\ 0 & \cdots & W_{T_{a\lambda N}} & 0 & \cdots & W_{\varepsilon_{\lambda N}} & W_{T_{s\lambda N}} & W_{W_{\lambda N}} \end{bmatrix} \tag{5.25b}$$

$$\boldsymbol{\delta x} = \begin{matrix} \delta T_{a_{\lambda 1}} & \cdots & 0 \\ \vdots & \ddots & \vdots \\ 0 & \cdots & WT_{a\lambda N} \\ \delta\varepsilon_{\lambda 1} & \cdots & 0 \\ \vdots & \ddots & \vdots \\ 0 & \cdots & \delta\varepsilon_{\lambda N} \\ \delta T_s & \cdots & \delta T_s \\ \delta w & & \delta w \end{matrix} \tag{5.25c}$$

The aforementioned linearized form of Eq. (5.24) is always underdetermined regardless of the number of channels (2n + 2 unknowns for n channels). To make the group of equations solvable, it is necessary to reduce the number of unknowns. Because there is a strong correlation between the channels for T_a and LSE, principal component analysis is adopted in this study to reduce the dimensions of T_a and LSE by using eigenvalues and eigenvectors (Singh, 1993). Accordingly, Eq. (5.24) can be expressed as follows:

$$\delta y = W \delta x = WVf = W^D f \tag{5.26}$$

where the terms V and f are the eigenvector matrix and coefficient vector of δx, respectively. The eigenvector of the LST and w is assumed to be a unit vector. Therefore, the corresponding coefficient vector f contains m coefficient vectors of Ta, t coefficient vectors of LSE, one for LST, and one for w. To stabilize this ill-posed problem, the Tikhonov regularization estimate (Tikhonov, 1963) can be used, which is the solution to the following minimization problem:

$$\text{argmin} \| \delta y - W^D f \|_2^2 + \gamma^2 \| Lf \|_2^2 \tag{5.27}$$

where, γ is the regularization parameter based on the L-curve method (Hansen and O'Leary, 1993) and L is a side constraint matrix. The second term in Eq. (5.27) is called 'regularizer' or 'side constraint' and includes prior knowledge about the expected behavior of f. When the initial estimates of the atmospheric and surface parameters have been updated using the Tikhonov regularization, the solution of the equation can be further optimized with iterative methods. The discrepancy principle (DP) iteration algorithm (Huang and Li, 2000) is utilized with the regularized solution as the initial estimate to obtain the final maximum likelihood solution of the LST. Compared with the two-step physical method (Masiello et al., 2018) that considers the stratified atmospheric profile, here the introduced physical method adopts T_a and w to retrieve the LST. The RTE was linearized using the tangents around the initial estimates of the LST, LSE, T_a, and w, which represents a simpler LST retrieval.

Among the five types of LST retrieval methods/algorithms, the advantage of the SCA algorithm is of simplicity because it only needs two parameters for LST retrieval: the ground emissivity (ε) and the water vapor content (w) (Yu et al., 2014). However, SCA and MWA algorithms show a significant overestimation of LSTs (Zhou et al., 2015). The third one, SWAs, can be carried out by utilizing two thermal bands (Wan and Dozier, 1996). They are mostly based on the absorption difference between two adjacent thermal bands within the range of 10–12.5 μm based on the linearization of the radiative transmittance equation in terms of temperature or wavelength, and they are relatively simple and robotic. The ML methods are able to learn complex patterns, generalization to noisy environments, and incorporation of both knowledge and physical constraints, but intensive computational resources and complex structure of algorithms are mostly required and they are generally a black box model. The PBM methods do not need a priori atmospheric corrections but are dependent on the initial guess. Generally, the first two algorithms depend on a single TIR channel, while the other algorithms require two adjacent TIR channels or over. Table 5.6 provides a summary of the five types of methods/algorithms from their characteristics and descriptions, advantages and limitations, major factors, and references.

5.4 EMISSIVITY RETRIEVAL METHODS AND TECHNIQUES

LSE is a key parameter in retrieving LST from TIR remote sensing data. Depending on the spectral channels of the sensor and the available temporal sampling, there

TABLE 5.6
A Summary of Methods/Algorithms Frequently Used in Retrieving LST

The type of DLST methods	Characteristic and description	Advantage and limitation	Major factor	Typical reference
Mono-window algorithm (MWA)	Use radiative transfer model and atmospheric profiles to retrieve LST	(1) Simplicity, (2) applicable to sensors with only one TIR channel, (3) require a priori knowledge of the pixel emissivity, (4) require accurate atmospheric profiles, (5) the uncertainty of atmospheric profiles may have strong effects on the accuracy of LST retrieval	Atmospheric transmittance (τ), LSE (ε) and effective mean atmospheric temperature (Ta in K)	(Qin et al., 2001b; Sobrino et al., 2004; Zhou et al., 2012)
Single infrared channel method (SCA)	Use radiance measurements in one IR-window channel and correct the atmospheric effects to determine LSTs	(1) Simplicity, (2) applicable to sensors with only one TIR channel, (3) require a priori knowledge of the pixel emissivity, (4) require accurate atmospheric profiles and a good RTM to estimate atmospheric quantities, (5) the uncertainty of atmospheric profiles may have strong effects on the accuracy of LST retrieval	LSE (ε) and atmospheric water vapor content (w)	(Jiménez-Muñoz and Sobrino, 2003; Li and Meng, 2018; Sekertekin and Bonafoni, 2020)
Split-window method (SWA)	Based on the absorption difference between two adjacent thermal bands within the range of 10–12.5 μm based on the linearization of the radiative transmittance equation in terms of temperature or wavelength	(1) Simplicity and robustness, (2) use adjacent channels with different properties to calculate atmospheric attenuation, (3) eliminates atmospheric effect based on differential absorption between the two bands for the LST estimation, (4) does not require high-precision atmospheric profiles	Atmospheric transmittance (τ) and LSE (ε)	(Francois and Ottle, 2002; Qin et al., 2001a; Wan et al., 2004)

(Continued)

TABLE 5.6 (*Continued*)
A Summary of Methods/Algorithms Frequently Used in Retrieving LST

The type of DLST methods	Characteristic and description	Advantage and limitation	Major factor	Typical reference
Machine learning (ML) empirical methods	Most ML methods are non-parameter methods. ANN's backpropagation algorithm is often used to train the multi-layer perceptron neural network model with input of various features and output of probability of LST	(1) Be able to learn complex patterns, generalization to noisy environments, and incorporation of both knowledge and physical constraints; (2) intensive computational resources, complex structure of algorithms, and a black box model; (3) the model structure is complex and corresponding tools/software for different CNN architectures are often not available and accessible	Test and find a set of ideal architecture parameters, e.g., the number of input and hidden nodes, the initial weight range, the activation functions, the learning rate and momentum, and the terminating criterion	(Del Frate and Solimini, 2004; Khoob, 2008; Shen et al., 2020)
Physics-based method (PBM)	Mostly based on an RTE linearized using the tangents around the initial estimates of the LST, LSE, and atmospheric parameters, then principal component analysis in combination with the Tikhonov regularization method can be utilized to reduce the number of unknowns without significant loss of accuracy	(1) Do not need a priori atmospheric corrections; (2) PCA and Tikhonov regularization can be used to make the solution more stable and accurate; (3) simultaneously retrieve the atmospheric profiles, the LST, and the LSEs; (4) complexity; (5) require adequate number of channels; (6) the solution is strongly dependent on the initial guess	How to make good initial guesses of the atmospheric profiles, the LST, and the LSEs is critical	(Ma et al., 2000; Masiello and Serio, 2013; Masiello et al., 2018)

are various LSE estimation methods, including the classification-based emissivity method (CBEM), NDVI-based emissivity method (NBEM), day/night temperature independent spectral-index (TISI), and temperature emissivity separation method (TES) (Gillespie et al., 1998; Li et al., 2013). The following sections will introduce the four types of methods.

5.4.1 Classification-Based Emissivity Method (CBEM)

This method assumes that similarly classified land covers exhibit very similar LSEs. In order to estimate LSEs from conventional static land cover classes and dynamic information, Snyder et al. (1998) proposed the 'classification-based emissivity' method, which is utilized to develop an emissivity knowledge base. The LSE can be assigned from look-up tables based on conventional land cover classification information.

Generally, the CBEM is the simplest method in terms of processing LSE, and it can provide accurate LSEs for LST retrieval as long as the land surfaces are accurately classified and each class has well-known LSEs (Gillespie et al., 1996). In Snyder et al. (1998), each pixel is classified into one of 14 emissivity classes; biospherically distinct cover types, which model estimates shown to have nearly the same emissivity, e.g., evergreen needle forest and green deciduous forest, are merged into one class. Furthermore, some conventional classes are separated, e.g., growing crops and idle crops, because they show different emissivities. For data at coarse resolutions, the CBEM can be applicable using a linear mixing model (Sun and Pinker, 2003). However, the LSE may be less accurate because it is difficult to estimate the weights of each component (endmember) within a coarse pixel. The LST retrieval accuracy is determined primarily by the accuracy of the LSEs (Li et al., 2013). As described by Snyder et al. (1998), the accuracy of CBEM can degrade due to uncertainties in the soil moisture, vegetation phenology change, and the appearance of snow and ice. In addition, classifications based on VNIR data are generally not well correlated to the LSE in the TIR channel (Li et al., 2013). Finally, the CBEM is most suitable for spectral regions with low spectral contrast LSEs, for example around 11 μm and 12 μm for most territorial surfaces. Otherwise, a large retrieval error may be created by variations in the emissivity variation. All of these uncertainties may prevent an accurate estimation of the LSE from CBEM, thus degrading the accuracy of LST retrieval (Li et al., 2013).

5.4.2 NDVI-Based Emissivity Method (NBEM)

The NBEM is based on a statistical relationship between the NDVI derived from the VNIR bands and the LSE in the TIR channels. Van de Griend and Owe (1993) first found a very high correlation between the LSE in the TIR channels covering 8–14 μm and the logarithmic NDVI. Valor and Caselles (1996) applied and validated this method for different atmospheric environments and different areas, e.g., mid latitudes and tropical latitudes with an error of 0.6% on the estimated emissivity. According to Van de Griend and Owe (1993), after logarithmic transformation, results established a high correlation ($R^2 = 0.94$) between measured ε an NDVI. The correlation can be expressed as:

$$\varepsilon = a + b \ln(NDVI) \tag{5.28}$$

where a = 1.0094 and b = 0.047 derived from regression analysis. However, the relationship is only valid for areas with characteristics similar to the Botswana region, but it can be adopted to other areas. Valor and Caselles (1996) proposed an operational model of this method that is suitable for other areas with several soil and vegetation types and changing vegetation cover. They defined the effective emissivity of a heterogeneous surface as the sum of the emissivities of its simple components as follows:

$$\varepsilon = \varepsilon_0 + d\varepsilon \tag{5.29}$$

where ε_0 is the fraction of emissivity corresponding to the direct radiation from the different components and $d\varepsilon$ corresponds to the fraction of emissivity caused by internal reflections between the wall and the ground (cavity effect).

$$\varepsilon_0 = \varepsilon_t P_t + \varepsilon_G P_G + \varepsilon_s P_s \tag{5.30}$$

Subscripts t, G, and s denote the top, ground, and side elements of the heterogeneous system and P_t, P_G, and P_s are the fractions viewed by the sensor. Shadows are neglected. Using a single emissivity ε_V and vegetation cover fraction $P_V = P_t + P_s$ for the top and the side of vegetation, ε_0 can be written as:

$$\varepsilon_0 = \varepsilon_V P_V + \varepsilon_G \left(1 - P_V\right) \tag{5.31}$$

where P_V is the vegetation proportion obtained according to Carlson and Ripley (1997):

$$P_V = \left[\frac{NDVI - NDVI_{min}}{NDVI_{max} - NDVI_{min}}\right]^2 \tag{5.32}$$

The term $d\varepsilon$ in Eq. (5.29) takes into account effect of the geometrical distribution of the natural surfaces and also the internal reflections. For plain surfaces, this term is negligible, but for heterogeneous and rough surfaces, such as forest, this term can reach a value of 2%. A good approximation for this term can be given:

$$d\varepsilon = \left(1 - \varepsilon_s\right)(1 - P_V)F\varepsilon_V \tag{5.33}$$

where F is a shape factor that assumes different geometrical distribution being 0.55. The shape factors depend on the geometrical structure. For vegetation modeled as infinite parallel rows, the shape factors can be calculated from the length, height, and separation of vegetation (Valor and Caselles, 1996). When NDVI values are greater than 0.2 but lower than 0.5, the pixels are assumed to be a mixture with bare soil and vegetation, and the emissivity is determined using the expression after taking Eqs. (5.29), (5.31), (5.33):

$$\varepsilon = mP_V + n \tag{5.34}$$

with

$$m = \varepsilon_V - \varepsilon_s - (1 - \varepsilon_V) F \varepsilon_V \tag{5.35a}$$

$$n = \varepsilon_s - (1 - \varepsilon_s) F \varepsilon_V \tag{5.35b}$$

When the NDVI obtained is higher than 0.5, the pixel is considered as fully vegetated, and a constant value for the emissivity is assumed, typically 0.99.

Because of NBEM's simplicity, this method has already been applied to various sensors with access to VNIR data. Similar to the CBEM, an accurate atmospheric correction is unnecessary. However, the NDVI thresholds that indicate bare soil and full vegetation cover, the vegetation fraction, any cavity effects, and the LSEs for bare soil and full vegetation must be known in advance. Since NDVI is used as a proxy for the fraction of vegetated surface within the pixel, it can be replaced by more accurate estimates of this variable. Nevertheless, one of the drawbacks of this method is the lack of continuity in the LSE values of regions transitioning from bare soil to full vegetation cover, because the LSEs in those regions are calculated using different formulae, and the determination of the soil emissivity may be the primary source of error in this method.

5.4.3 Day/Night Temperature Independent Spectral-Index (TISI)

Becker and Li (1995) and Li and Becker (1990) first developed a TISI-based method to perform spectral analysis in the TIR spectral region. They assumed that the $TISI_{ij}$ (i is the MIR channel and j is the TIR channel) in the daytime without the contribution of solar light is the same as the $TISI_{ij}$ in the nighttime, and Li and Becker (1993) further developed a day/night TISI-based method to first extract the bidirectional reflectivity in MIR channel i by eliminating the emitted radiance during the day in this channel by comparing the $TISI_{ij}$ in the daytime and the nighttime. The idea of the TISI method (Becker and Li 1995) is to construct indices that are (nearly) independent of LST and directly related to relative emissivity. The method uses the power-law approximation of Planck's function (Slater, 1980):

$$B_\lambda(T) \cong \alpha_\lambda(T_0) T^{n\lambda(T_0)} \tag{5.36}$$

where α_λ and $n\lambda$ are the wavelength dependent on stands for modest temperature variations around a reference temperature T_0; T_0 in Eq. (5.36) and its first derivative with respect to T is exact and can be derived for α_λ and $n\lambda$:

$$n_\lambda(T_0) = \frac{c_2}{\lambda T_0}\left[1 + \frac{1}{\exp(c_2 / \lambda T_0) - 1}\right] \tag{5.37}$$

$$\alpha_\lambda\left(T_0\right)=\frac{B_\lambda\left(T_0\right)}{T_0^{n\lambda(T_0)}} \tag{5.38}$$

Eq. (5.19) gives the at-sensor radiance L_λ in IR-channel i. Atmospherically corrected at-ground radiance R_i, e.g., for AVHRR channels 4 and 5, as well as for channel 3 at nighttime, is given by

$$R_i=\varepsilon_i B_i\left(T_s\right)+\left(1-\varepsilon_i\right)L_{i,atm}^{\downarrow} \tag{5.39}$$

where, $L_{i,atm}^{\downarrow}$ is channel i downwelling atmospheric irradiance divided by π. Channel-specific constants α_i and n_i can be determined using a linear regression to solve Eq. (5.36) at T_0 or from Eqs. (5.37), (5.38). Inserting Eq. (5.36) with channel-specific constants into Eq. (5.39) yields:

$$R_i=\alpha_i T_{i,G}^{n_i}=\varepsilon_i\alpha_i T_s^{n_i}C_i \tag{5.40}$$

where, $C_i=1+\frac{\left(1-\varepsilon_i\right)L_{i,atm}^{\downarrow}}{\varepsilon_i B_i\left(T_s\right)}$.

In order to obtain TISI, two channel radiances R_i and R_j given by Eq. (5.40) are ratioed with adequately chosen powers a_i and a_j:

$$\frac{R_i^{a_i}}{R_j^{a_j}}=\frac{\varepsilon_i^{a_i}\alpha_i^{a_i}T_s^{n_i a_i}C_i^{n_i}}{\varepsilon_j^{a_j}\alpha_j^{a_j}T_s^{n_j a_j}C_j^{a_j}}=\frac{\varepsilon_i\alpha_i C_i}{\varepsilon_j^{n_{ij}}\alpha_j^{n_{ij}}C_j^{n_{ij}}} \tag{5.41}$$

where $a_i = 1$ and $a_j = n_i/n_j = n_{ij}$ in order to verify $a_i n_i - a_j n_j = 0$, which makes the ratio nearly independent of temperature. The TISI are then expressed as:

$$TISI_{ij}=\frac{\alpha_j^{n_{ij}}R_i}{\alpha_i R_j^{n_{ij}}}=\frac{\varepsilon_i}{\varepsilon_j^{n_{ij}}}\cdot C_{ij} \tag{5.42}$$

where $C_{ij}=C_i\left(C_j^{n_{ij}}\right)^{-1}$, which depends on emissivity, atmosphere, and surface condition. Strictly, $C_{ij} = 1$ only for $T_s \to\infty$ or for $\varepsilon_i = \varepsilon_j = 1$, but for a large range of surface temperature and emissivity $C_{ij} = 1$ is a good approximation in spectral range 8–12 μm.

Without further assumptions TISI only facilitates the calculation of emissivity ratios. Dash et al. (2002) made an application of Eq. (5.41) derive angular emissivity via bidirectional reflectivity using day and night data. They used radiances at ground level for AVHRR channels 3, 4, and 5 that are given by Eq. (5.39), accounting for daytime for channel 3 solar-reflection; then ε_3 was obtained as a directional emissivity using Kirchoff's law. From ε_3 and brightness temperature data in channel 4 and 5, ε_4 and ε_5

can be readily obtained. The detailed information to obtain directional emissivity for individual channels can be seen in Dash et al. (2002). Once the bidirectional reflectivity in an MIR channel is retrieved, the directional emissivity in that MIR channel can be estimated. Finally, based on the concept of the TISI, the LSEs in the TIR channels can be obtained from the two channel TISI and the emissivity in the MIR channel.

5.4.4 Temperature Emissivity Separation Method (TES)

The TES was first developed by Gillespie et al. (1996) to separate the LST and the LSE to be estimated using ASTER TIR data. This method relies on an empirical relationship between the spectral contrast and the minimum emissivity to increase the number of equations so that the undetermined retrieval problem becomes deterministic (Li et al., 2013). The methodology is based on three modules, the normalized emissivity method (NEM) obtains preliminary values of LST and LSE, the ratio module estimates the emissivity normalized spectrum (β spectrum), and the maximum minimum difference (MMD) transforms the β spectrum into the absolute LSE and re-calculates the LST using an empirical relation between the minimum emissivity and the spectral contrast. See the following paragraphs for a brief explanation of each module.

NEM module: this module estimates normalized emissivities and provides an initial estimate of LST. The emissivity $\varepsilon(\lambda)$ is initially set to 0.97, in order to be within ±0.03 for typical land surfaces, e.g., vegetation, snow, soil, rocks, etc. (Schmugge et al., 1998). The reflected sky irradiance is removed iteratively. The temperature passed to the next ratio module should be within ±3K at 340K and within ±2K at 273 K (Dash et al., 2002).

Ratio module: the temperature from the NEM module is used to calculate emissivity ratios β (Eq. (5.43)); these are an unbiased measure of emissivity spectral shape. The emissivity ratios β_i are calculated by dividing the atmospherically corrected radiances by the average of the channels:

$$\beta_i = \frac{R_i \overline{B(T)}}{\bar{R} B_i(T)} \tag{5.43}$$

where $\bar{R} = 1/N\sum_1^N R_i$ and $\overline{B(T)} = 1/N\sum_1^N B_i(T)$.

For emissivities from 0.7–1.0, the ratios β_i are generally within 0.7–1.4.

Maximum-minimum difference (MMD) module: the maximum–minimum difference between the emissivity ratios β_i is given by MMD = max(β_i) – min(β_i). An empirical relation between minimum emissivity and MMD is:

$$\varepsilon_{min} = 0.994 - 0.687(MMD)^{0.737} \tag{5.44}$$

The relation was determined by regression analysis using emissivities measured in the laboratory and in a field experiment (Salisbury and D'Aria, 1992). It was observed that MMD ranges from 0.0 to 0.4 in most cases. From ε_{min} given by Eq. (5.44), the emissivites in all channels can be calculated by pixel-by-pixel basis:

$$\varepsilon_i = \beta_i \left(\varepsilon_{min} / \min(\beta_i) \right) \tag{5.45}$$

Numerical modelling showed that TES can recover LST to within ~1.5 K and emissivity to within 0.015 (Gillespie et al., 1998). The main advantage of the TES is that it combines attractive features of three precursors and uses an empirical relationship between the range of emissivities and the minimum emissivity in the N channels to retrieve the LST and LSEs. However, some reports have indicated that the TES method exhibited significant errors in the LST and LSEs of surfaces with low spectral contrast emissivity (e.g., water, snow, vegetation) and under hot and wet atmospheric conditions (e.g., Gillespie et al., 1996; Sawabe et al., 2003). The uncertainties in the atmospheric corrections will result in a large apparent emissivity contrast. This effect is more serious over graybody surfaces (Hulley and Hook, 2011). To minimize atmospheric correction errors, Hulley (2013) improved the TES method by using a water vapor scaling (WVS) approach proposed by Tonooka (2001).

5.5 LIMITATIONS AND FUTURE DIRECTIONS

Currently, existing satellite TIR remote sensors are feeling the drawback of TIR sensors' configuration. For accurate LST retrieval to address local climate applications over urban environments, a multispectral TIR remote sensing configuration may be required. A minimum configuration with only three TIR bands would be necessary for LST/LSE retrieval. This minimum configuration should include one TIR band in the 8–9 μm spectral region and two TIR bands in the 10–12 μm spectral window, enabling use of both TES and SWA algorithms for LST retrieval. Accurately using the TES algorithm, atmospheric correction is critical for successful LST retrieval. It is therefore important to complement the TIR sensor with appropriate bands for characterizing atmosphere (e.g., water vapor retrieval).

For a successful LST retrieval, atmospheric corrections are critical and difficult to implement. The presence of the atmosphere between the surface and the sensors significantly affects the radiances measured by a sensor at the TOA. These affected radiances result primarily from emission/reflection at the ground surface modified by the effects of the attenuation and emission of the atmosphere. Therefore, the atmospheric corrections consist of correcting the radiance measured by the sensors for the effects of the attenuation, emission, and emission-reflection of atmosphere. Accurately correcting for the atmospheric effects requires accurate knowledge of the atmospheric vertical profiles of water vapor and temperature, both of which are highly variable in space and time.

Validation of LST retrieved from satellite measurements at the scale of imaging pixels is challenging. The major problems of validating LST retrieved from satellite sensors are due to the difficulty of obtaining in situ LST measurements and of collecting representative in situ LST data at the scale of a single pixel. In general, temperatures over the land surfaces vary strongly in space and time (Prata et al., 1995), and it is not unnormal for the LST to vary by more than 10 K over just a few centimeters of distance or by more than 1 K in less than a minute over certain cover types (Li et al., 2013). Collecting and appropriately scaling the satellite derived LSTs to those measured at ground level in a heterogenous landscape, especially at large scale, is difficult and is almost impossible. The strong spatial heterogeneity and temporal variation of the LST makes it difficult to collect ground-based measurements and validation while several relatively homogeneous land surface types, such as lakes, deserts, and densely vegetated land can use direct comparisons of in situ LST measurements with LSTs retrieved from the satellite data (Hook et al., 2007). A complementary approach is to use sites that are spatially homogenous in terms of emissivity using the radiance-based validation approach (Hulley et al., 2009). For example, sand dunes can be these type of sites, which are referred to as pseudo-invariant sites (Hulley et al., 2009). Furthermore, how to conduct a representative measurement of the LST in a complex heterogeneous surface is also a critical question and challenge. Spatial scaling methods should be paid attention to when developing to relate the measurements at different scales and helping validate the retrieved LST.

By reviewing literature in the past several decades, the most important problem in LST validation might be the accuracy and representativeness of the ground-truth LST at the satellite pixel scale, especially in a heterogenous landscape. Although ground-based validation is considered to be the most direct and reliable validation technique, measurements of the in situ LST are limited by the difficulty of how to find a homogeneous region as large as the satellite pixel size and by the difficulty and associated costs of sampling measurements over heterogeneous landscapes. The first difficulty might be overcome by improving the spatial resolution of TIR data, while the second difficulty will require the development of a few new sampling techniques for ground-level LST measurements, such as a wireless-net observation system or multiple-scale observation methods with corresponding new instruments. Attempts to use the LSTs retrieved from satellite sensos to indirectly validate the LSTs retrieved from space from other satellite sensors must devote more attention to improving the accuracy of the retrieval of LST dealing with the scale mismatch issue in both space and time. For cross-validation techniques, an appropriate scaling method to remove the mismatch effects of spatial, temporal, and angular influences on the LST must be developed. Therefore, LST validation is still an ongoing research direction.

More specific future research directions of developing novel methods and techniques may focus on developing or refining existing advanced models and methods to improve the accuracy of LST retrieval. As reviewed in Section 4.4, while most researchers developed and used various MWA, SCA, SWA, and PBM methods to retrieve LSTs, a few of them have utilized ML methods. DL methods are very useful in reconstructing LST products when data missing occurs. For example, Wu et al. (2019a) proposed a fully convolutional network with a multiscale feature connected architecture for reconstructing geostationary satellite LSTs with large missing regions. Compared with traditional reconstruction methods, the one proposed shows a better performance with varied land-cover types from different satellites and different seasons. Therefore, it provides a new framework and advances capabilities for reconstructing other remotely sensed data products with spatiotemporal dynamics applications. Such a capability will be beneficial for monitoring land-surface and ecological dynamics. Jia et al. (2021) also developed a time series prediction framework for urban LST under data missing, such as that caused by cloud, snow, and other interference factors. It is helpful to solve the problem of the impact of data loss on surface temperature prediction. In this study, the geographically weighed regression (GWR) method is used to realize the simulation of LST based on the current date. The DL prediction network was constructed to predict the spatial distribution of LST on the next observation date. For LST prediction using DL methods, the result of the prediction on LST space distribution was relatively good. The purpose of this study is to make up for the missing data and to be applied to the prediction of the spatial and temporal distributions of LST. Those advanced and novel methods and techniques have demonstrated their excellent performance and robustness. However, more effort is still needed in he future research.

5.6 SUMMARY

In the beginning, the importance and necessity of the LST retrieved from TIR sensor data is introduced. This is because LST is one of the most important parameters in the physical processes of surface energy and water balance at local and global scales. To better understand and accurately evaluated various LST retrieval methods, some basic theoretical backgrounds are discussed. In Section 5.3, five types of LST retrieval algorithm/methods are introduced and discussed, including (1) the single-channel method (or mono-window algorithm, MWA), which usually requires a good radiative transfer model and atmospheric profiles that must be provided by either satellite soundings or conventional radiosonde; (2) the split-window method (SWA), which corrects for atmospheric effects based on the differential absorption in adjacent bands; (3) the ML methods, which are able to learn complex patterns, generalization to noisy environments, and incorporation of both knowledge and physical constraints; and (4) the physics-based method (PBM), which requires atmospheric transmittance, path radiance, and downward sky irradiance, estimated using a radiative transfer mode. In Section 5.4, four frequently used LSE methods are introduced. They include, depending on the spectral channels of the sensor and the available temporal sampling, (1) the classification-based emissivity method (CBEM); (2) the NDVI-based emissivity method (NBEM); (3) the day/night temperature independent spectral-index (TISI); and (4) the temperature emissivity separation method (TES). In the last section, some limitations and future directions are identified and discussed, including issues configurating TIR remote sensing sensors, correcting atmospheric effects, validating retrieved LST products, and developing or refining existing advanced models and methods to improve the accuracy of LST retrieval.

6 Thermal Infrared Remote Sensing Data Downscaling

Land surface temperature (LST), retrieved from moderate resolution or downscaled from coarse thermal infrared (TIR) data, represents a crucial parameter in Earth's surface processes. Addressing diverse research and application needs requires accurate and reliable methods for downscaling Land surface temperature (DLST) from moderate or coarse-resolution TIR sensors/systems. This chapter reviews and discusses various DLST methods, starting with the rationale for performing DLST processes from moderate or coarse-resolution TIR sensor data (Section 6.1). It also explores the current utilization of satellite TIR remote sensing sensors/systems for DLST processes (Section 6.2), methods and techniques employed in DLST (Section 6.3), the spatial scaling effects within DLST processes (Section 6.4), and outlines existing limitations and future research directions (Section 6.5). Throughout this chapter, the primary content draws heavily on insights from Pu and Bonafoni (2023).

6.1 INTRODUCTION

Thermal infrared (TIR) remote sensing, covering TIR range 3–14 μm (typical 8–14 μm), can be acquired by electro-optical sensors in either a single broadband or multispectral or even hyperspectral images. Land surface temperature (LST) retrieved from modern remote sensing TIR data or downscaled from coarse resolution thermal data is one of the key Earth surface processes parameters from local to regional to global scales (Zhan et al., 2013; Mao et al., 2021). Usually, coarse remote sensing sensors/systems can acquire TIR data at a low spatial resolution but high temporal resolution compared to optical (visible-near infrared (VNIR) and short-wave infrared (SWIR)) data. Due to the issue of the trade-off between spatial and temporal resolutions in modern thermal remote sensing technologies, it is extremely difficult to acquire both high temporal and spatial resolution TIR imagery (Zhan et al., 2013; Xia et al., 2019; Mao et al., 2021). Given that LST plays a critical role in controlling physical and biological processes on Earth's surface, it is desirable to have LST data at both high spatial (i.e., tens of meters or even a few meters) and high temporal resolutions for many application purposes, such as mapping surface evapotranspiration in agricultural and arid regions (Anderson et al., 2012; Olivera-Guerra et al., 2017; Liu et al., 2018b), studying urban thermal environments (Stathopoulou and Cartalis, 2009; Sobrino et al., 2012; Zakšek and Oštir, 2012; Wang et al., 2020b), and characterizing urban heterogenous landscapes (Yang et al., 2019; Zawadzka et al., 2020).

DOI: 10.1201/9781003276173-6

To overcome the dilemma (i.e., low spatial resolution but high temporal resolution) of most available satellite TIR remote sensing sensors, during the last four decades, a lot of studies have been conducted to investigate downscaling LST (DLST) processes (from a relatively lower/coarser spatial resolution to a relatively finer resolution) to generate high spatial LST data to meet many application purposes. Such investigations of DLST techniques/methods of coarse TIR data have attracted a lot of interest from researchers. Sequentially, there are more and more research projects/studies that have been done and their analysis results and novel findings were reported in the current literature. This review focused on DLST methods and techniques by (1) examining high temporal resolution but low spatial resolution TIR sensors and systems frequently used for coarse resolution TIR data spatial sharpening and spectral unmixing studies; (2) introducing, discussing, and characterizing different techniques/methods for downscaling LST from coarse/native resolutions to finer target resolutions; (3) assessing spatial scaling effect on DLST results and review approaches to measure and reduce the scaling effects on DLSTs; and (4) identifying limitations and constraints existing in current DLST processes and pointing to future research directions for improving DLST results and promoting applications of TIR remote sensing at finer resolutions in different areas. In general, DLST processes can provide a better resolution dataset of LST at both finer spatial and temporal resolutions and thus it is widely thought of as an effective tool to utilize subpixel thermal energy for many applications. In current literature, the relevant studies on DLST techniques/methods have used different terms from DLST but have the same or similar meaning, which include the frequently used disaggregation of LST, spatial resolution improvement of LST, spatial enhancement of LST, spatial (thermal) sharpening, subpixel temperature estimation, thermal spectral unmixing, and component temperature retrieval. In this chapter, DLST will be a representative term for all of them even though the various terms may be directly cited in specific studies reviewed.

6.2 SATELLITE TIR REMOTE SENSING SENSORS/SYSTEMS

Satellite TIR remote sensing sensors' data that are currently available for downscaling processes include a moderate to coarse spatial resolution and 35 days to 15 min temporal resolution (Table 6.1). The platforms of modern TIR remote sensing sensors/systems consist of different polar orbits and geostationary obits. Table 6.1 summarizes TIR remote sensing sensors/systems currently used for DLST processes from sensors' names and characteristics including band setting and information sources, etc.

6.2.1 Polar Orbits (I): Moderate Resolution TIR Sensors

Moderate resolution satellite TIR sensors currently used for DLST processes include those onboard Landsat series and Terra with spatial resolution better a few 100 m. The Thematic Mapper (TM) onboard Landsat 4/5, the Enhanced Thematic Mapper Plus (ETM+) onboard Landsat 7, and the Thermal Infrared Sensor (TIRS) onboard Landsat 8 measure land surface thermal energy in one or two thermal bands, while

TABLE 6.1
A List of Moderate to Coarse Spatial but High Temporal Resolution Thermal Remote Sensing Sensors/Systems Frequently Used for Downscaling LST Processes (Pu and Bonafoni, 2023)

Satellite	Sensor/ system	Revisit period*	No. of TIR bands	TIR spectral range, wavelength (μm)	Spatial resolution (m)	Operation	Information source**
Polar orbits (I): moderate resolution (pixel size < 1000 m)							
Landsat-4,-5	TM	16 D	1	10.4–12.5	120	1982–2011	https://earthexplorer.usgs.gov/
Landsat-7	ETM+	16 D	1	10.4–12.5	60	Since 1999	https://earthexplorer.usgs.gov/
Landsat-8	TIRS	16 D	2	10.60–11.19, 11.50–12.51	100	Since 2013	https://earthexplorer.usgs.gov/
Terra	ASTER	On-demand	5	8.125–8.475, 8.475–8.825, 8.925–9.275,10.25–10.95, 10.95–11.65	90	Since 1999	https://earthexplorer.usgs.gov/
Polar orbits (II): coarse resolution (pixel size ≥ 1000 m)							
Terra/Aqua	MODIS	1–2 D	2	10.78–11.28, 11.77–12.27	1,000	Since 1999	https://modis.gsfc.nasa.gov/
NOAA	AVHRR	0.5 D	2	10.5–11.3, 11.5–12.5	1,100	Since 1979	https://www.avl.class.noaa.gov/release/data_available/avhrr/index.htm
Envisat	AATSR	35 D	2	11.0, 12.0	1,000	Since 2002	http://www.atsrsensors.org/aatsr.htm
ERS	ATSR	35 D	2	11.0, 12.0	1,000	Since 1991	http://www.atsrsensors.org/aboutATSR.htm
Sentinel-3(A/B)	SLSTR	1 D	2	10.85, 12.02	1,000	Since 2016	https://sentinel.esa.int/web/sentinel/home
Fengyun-3(A/B/C)	VIRR	5.5 D	2	10.3–11.3, 11.5–12.5	1,100	Since 2008	https://fy4.nsmc.org.cn/data/en/instrument/VIRR.html
Geostationary orbits							
GOES	Imager	3 H	2	10.2–11.2. 11.5–12.5	4,000	Since 1975	https://www.goes.noaa.gov/
Meteosat-8, MSG	SEVIRI	15 M	2	9.8–11.8, 11.0–13.0	3,000	Since 2005	https://www.eumetsat.int/seviri
INSAT	VHRR	< 1 D	1	10.5–12.5	8,000	Since 2003	https://www.mosdac.gov.in/insat-3a
Fengyun-4A	AGRI	15 M	4	8.0–9.0, 10.3–11.3, 11.5–12.5, 13.2–13.8	4,000	Since 2016	https://fy4.nsmc.org.cn/nsmc/en/theme/FY4A_instrument.html

Note: *: D-day, H-hour, M-minute (limited in the table); ** Accessed on November 14, 2021.

the Advanced Spaceborne Thermal Emission and Reflection Radiometer (ASTER) sensor carried by Terra satellite senses land surface TIR energy in five multispectral thermal bands (Table 6.1). In recent years, many DLST processing studies have indicated that the Landsat series sensors and ASTER TIR data have been popularly used for DLST processes to create high (e.g., 30 m resolution) and very high (better than 30 m resolution) VHR (e.g., Jiang and Weng, 2013; Pu, 2021), and, coupling some coarse spatial resolution but high temporal TIR sensors (e.g., Moderate Resolution Imaging Spectroradiometer (MODIS)), high spatiotemporal Landsat-like time series TIR data (e.g., Herrero-Huerta et al., 2019; Yao et al., 2020). For example, Jiang and Weng (2013) downscaled Landsat TM 120 m LST to 30 m fine resolution by using a vegetation cover-based thermal sharpening method. Bonafoni (2016) conducted downscaling Landsat LST over the urban area of Florence, Italy from a native 120 resolution to 30 fine resolution with TM optical data derived scaling factors. Essa et al. (2012) downscaled ETM+ 60 LST to 30 m resolution with 15 spectral indies as scaling factors. Zawadzka et al. (2020) proposed a DLST method for downscaling Landsat 8 TIRS derived 100 m resolution LST to 2–4 m by using scaling factors extracted from very high resolution hyperspectral aerial imagery and large-scale topographic maps. In assessing scaling effect in DLST processes at different at high and very high spatial resolutions, Pu (2021) conducted downscaling ASTER 90 m resolution LST to 4–30 m finer spatial resolutions. In addition, Herrero-Huerta et al. (2019) and Yao et al. (2020) downscaled TM/TIRS 120 m and 100 m LSTs coupled with MODIS 1,000 m LST to create high spatiotemporal Landsat-like 30 m resolution time series LSTs. There are many other similar studies on downscaling the moderate resolution TIR sensors' data to high and very high resolutions including—but not limited to—DLSTs from TM 120 m to finer resolutions by Guo et al. (2014); from ETM+ 60 m to finer resolutions by Deng and Wu (2013a); from TIRS 100 m to finer resolutions by Wang et al. (2017a); from ASTER 90 m LST to finer resolutions by Gustafson et al. (2003) and Xu et al. (2020).

Given the fact that the moderate resolution TIR sensors' data can be downscaled to high and VHR resolutions, most of the DLSTs downscaled from the TIR sensors' data are used for mapping and characterizing thermal environments with heterogenous landscapes, such as heterogenous urban landscapes. For example, Huang et al. (2013) and Yao et al. (2020) utilized DLSTs at high resolutions, downscaled from the Landsat TIR sensors' LST to characterize and monitor urban heat islands (UHIs). Zawadzka et al. (2020) generated the VHR DLSTs for characterizing heterogenous urban landscapes.

6.2.2 Polar Orbits (II): Coarse Resolution TIR Sensors

Typical high temporal resolution (0.5–2 days) but low spatial resolution (~1000 m) sensors/systems, popularly used for DLST processes, specify the NASA-MODIS sensor onboard the Terra/Aqua satellites and the NOAA-Advanced Very High-Resolution Radiometer (AVHRR). Other polar orbits coarse resolution TIR sensors with less investigation for DLST processes also include the European Space Agency's Along-Track Scanning Radiometer (ATSR), Advanced ATSR (AATSR), and the Sea and Land Surface Temperature Radiometer (SLSTR) onboard Sentinel-3(A/B) and the

Visible and Infrared Radiometer (VIRR) onboard Fengyun-3 (B/C) (Table 6.1). The coarse resolution TIR sensors' data can be used to downscale to different moderate resolutions (mostly a few 100 m). In the current literature, it is easily found that there are a lot of studies on DLST processes with the coarse resolution TIR sensors' data, especially the MODIS coarse resolution LST data, for which researchers either directly used real MODIS LST data to downscale to moderate resolutions or used MODIS-like LST data simulated by aggregating moderate resolution sensors' data to downscale. For example, Mukherjee et al. (2014) used the original 1000 m MODIS LST data to downscale to 250 m LST and the Landsat TM upscaled 960 m LST data to downscale to 480 m and 240 m LSTs over heterogeneous landscapes. Sharma et al. (2020) also utilized both real MODIS 1000 m LST data and aggregated Landsat 8 LST data at 1,000 m resolution to downscale to 400, 300, 200, and 100 m resolutions. In addition, to create high spatiotemporal time series TIR data, the coarse resolution MODIS data are coupled with moderate resolution sensors (e.g., TM, TIRS, and ASTER) by using spatiotemporal-based models (discussing them in Section 6.3.3). For instance, Moosavi et al. (2015) fused multitemporal MODIS and Landsat TIRS TIR data to predict daily LST at 100 m resolution. To monitor an environmental process, Wu et al. (2015) generated daily high spatial resolution LSTs by fusing ASTER and MODIS LST products. There were some studies on downscaling other coarse resolution TIR sensors' data to moderate resolutions, including AVHRR (e.g., Matson and Dozier, 1981; Stathopoulou and Cartalis, 2009; Chybicki and Łubniewski, 2017).

The moderate resolution LSTs downscaled from the coarse resolution TIR sensors are mostly used for mapping, monitoring, and characterizing large landscapes and some cases at a regional scale. For example, such DLSTs from the coarse TIR sensors' data were used to map and monitor soil moisture (e.g., Amazirh et al., 2019; Bai et al., 2019), agriculture landscapes (e.g., Jeganathan et al., 2011; Eswar et al., 2016), and evapotranspiration (e.g., Olivera-Guerra et al., 2017; Liu et al., 2018b).

6.2.3 Geostationary Orbits TIR Sensors

Geostationary orbits TIR sensors can provide TIR image data within the same region (dish area) of the Earth at high temporal resolution (e.g., 15 min) (Mao et al., 2021). They include Imager onboard Geostationary Operational Environmental Satellite (GOES), SEVIRI onboard Meteosat Second Generation (MSG), AGRI onboard Fengyun-4A, etc. They are currently used for LST downscaling from very coarse native spatial resolution (a few kilometers but very high temporal resolution) to general coarse and moderate resolution LSTs (most in time series). Detailed information on commonly used geostationary orbits TIR sensors in LST downscaling processes is summarized in Table 6.1. There were many studies on DLST processes to generate MODIS-like spatial resolution LSTs for monitoring the Earth surface thermal budget including urban thermal environments. The DLSTs disaggregated from geostationary orbits= TIR sensors are mostly at high temporal resolutions up to 15 min and cover regional or continental areas (e.g., SEVIRI covering most European and GOES-East most South and North Americans' areas). For example, Weng and Fu (2014) downscaled GOES Image 4 km LST to 1 km resolution to create time series LST products for monitoring of UHI in the Los Angeles region by extracting and using key diurnal temperature cycle parameters. Jiang et al. (2015) used the disaggregated GOES Image LST data to MODIS-like 1 km resolution LSTS for assessing heat wave health risks at a regional scale. To create time series LST data for performing an urban heat island diurnal cycle analysis, Zakšek and Oštir (2012) downscaled SEVIRI onboard MSG LST data with 3 km × 5 km to 1 km spatial resolution and 15 min temporal resolution LSTs. Sismanidis et al. (2017) also disaggregated ~4 km SEVIRI LST DLST to 1 km MODIS-like LST data to create a diurnal LST product. There are more studies on disaggregating geostationary orbits TIR sensors' data to regular coarse (~ 1 km) and moderate resolution LST data (a few of 100 m) during the last two decades, including GOES TIR data (e.g., Kustas et al., 2003; Agam et al., 2007; Inamdar and French, 2009); MSG- SEVIRI TIR data (e.g., Bechtel et al., 2012), and Fenyen-4A/SGRI TIR data (e.g., Zhu et al., 2018b).

6.3 TECHNIQUES AND METHODS OF DOWNSCALING LST

6.3.1 Reference Data

To assess the performance of a new method/technique to downscale coarse resolution TIR sensors' data to finer resolutions, reference data are necessary for evaluating quality and accuracy of DLST results. According to different reference data sources, there are two approaches that can be adopted to collect the reference data, the field-based (in-situ, UAV-based LST, and TIR radiance measurements) approach and TIR/LST image-based approach. In addition, simulated data can be also used as the reference data that allow us to evaluate the strengths and weaknesses of a DLST method in some aspects. However, since fewer studies with the simulated data as reference data are reported in the relevant literature, they were not reviewed in this chapter.

6.3.1.1 Field (In-Situ)-Based Approach

Based on requirements for assessing the performance of a new downscaling method and validating its DLST results, sampling methods (a random or systematic or typical sampling method) and samples (usually a set of plots distributed in areas of different land use/cover types) are first selected and deployed in a project area, then a set of in-situ samples of LST/TIR radiance are measured by adopting one field LST/TIR radiance measurement approach and required instrument. In practice, the field collected reference data are usually used for verifying DLST results at a high or very high resolution (e.g., better than 15 m) and a local scale (i.e., covering a relatively small study area), and thus the field-based approach is not suitable for validating DLST results covering a large or a regional area. For example, Zhou et al. (2016c) downscaled ASTER 90 m LST data to 15 m high resolution and Landsat TIRS 100 m LST data to 30 m resolution. The DLST results were verified with field-measured LST data. To validate the DLST results at very high resolution (2 m) from upscaled moderate resolutions (40 m, 30 m, and 20 m), Bonafoni and Tosi (2017) first verified the retrieved LST from 2 m resolution airborne TIR data with the ground-measured LST data, then used the verified LST image to validate the DLST results. In recent years, a flexible UAV-carried sensor (LST thermometer or TIR radiometer) may be used for collecting the reference (imaging) data of LST/TIR radiance for DLST validation purposes (e.g., Wang et al., 2020).

Field LST measurements do not frequently necessarily represent LST values over a pixel area due to large spatial variations of LST at a pixel scale, depending on the pixel size (spatial resolution) and thermal spatial variation within the pixel area and on spatial heterogeneity of landscapes within the pixel area. This is because natural land cover materials and types and their corresponding LST and LSE (land surface emissivity) values are quite variable at scales of a few of 100 m^2 or of km^2 (Li et al., 2013). Homogeneous and flat Earth surfaces, such as inland water, sand, snow, and ice, which can be relatively easily measured and characterized, can serve as validation sites/areas. Thus, the field-based approach might not be suitable for validating DLST results covering heterogenous surface and variable terrain areas, especially with a large pixel size.

6.3.1.2 Image-Based Approach

In LST/TIR data downscaling processes, the use of the image-based approach for assessing the DLST methods and validating DLST results perhaps is one of the most popular and effective approaches. This is because (i) the TIR/LST images, including airborne or satellite sensors' images, frequently have the same spatial or temporal or both spatiotemporal characteristics as those DLST results, and (ii) the most of moderate to coarse resolution TIR/LST images are easy to access and available for most users. Therefore, not like the field-based approach, verifying DLST results covering a heterogenous and variable terrain surface and at moderate to coarse resolutions is reliable and accurate with the image-based approach via comparing the DLST results with appropriate scale/resolution (spatial and temporal) images. To fit the spatial/temporal resolutions of DLST results, if the spatial/temporal resolutions of the images used for validating the DLSTs do not match those of DLST results, the images for validation purposes can be upscaled to the required resolutions for validating the DLST results. For example, Zhang et al. (2019) and Li et al. (2018b) downscaled 1,000 m or 990 m resolution MODIS LST to 90 m resolution LST, then validated the DLSTs at 90 m resolution with the ASTER 90 m resolution LST product. Wang et al. (2020) and Luo et al. (2021) also disaggregated 1,000 m resolution MODIS LST to 100 m resolution LST then validated the DLST at 100 m resolution with Landsat 8 TIRS 100 m LST data. Stathopoulou and Cartalis (2009) downscaled 1,000 m resolution AVHRR data to 120 m DLST and validated with Landsat 120 m resolution TM LST. Jeganathan et al. (2011) first downscaled MODIS 1 km LST product to 810 m, . . . , and 180 m resolutions, then upscaled ASTER 90 m LST products to the upscaled corresponding resolutions LSTs to verify the DLSTs from MODIS. When the acquisition time for the images for validation purposes is different from that for the coarse resolution LST image, it is necessary for the former images to normalize the possible LST difference caused by the time difference of acquiring images by adopting some normalization approaches (e.g., one introduced in Pu, 2021).

To test a newly developed DLST method/technique, many studies use upscaled (aggregated) TIR or LST data at lower resolutions to downscale to different finer resolution DLSTs then validate them with corresponding upscaled TIR or LST data.

For example, Chen et al. (2014a) first upscaled ASTER 90 m LST to 720 m resolution then downscaled to 90 m resolution LST and validated it with ASTER 90 m LST. Liu et al. (2018a) first aggregated ETM+ 60 m LST to 960 m, 240 m, and 120 m for site 1 (an agriculture area) and TM 120 m LST to 960 m for site 2 (an urban area) then disaggregated to 240 m, 120 m, and 60 m for site 1 and to 480 m, 240 m, and 120 m for site 2. They validated the DLSTs at different resolutions with corresponding resolution upscaled ETM+ LSTs from 60 m and TM LSTs from 120 m, respectively. There are a lot of other studies using upscaled lower resolution LSTs to downscale to different finer resolution LSTs with DLST methods and then validate them with corresponding upscaled LSTs, such as using Landsat 8 TIRS LST by Xia et al. (2018) and ASTER LST by Sattari et al. (2018). With the same sensor's data for testing and validating downscaling processes, one unique advantage is that DLST data and the images for a validation purpose have an identical image acquisition time.

6.3.2 Types and Selection of Scaling Factors

Scaling factors (SFs), used as independent variables for kernel-based DLST processes, are extracted from relatively high resolution multi-/hyper-spectral optical sensors' data, mostly covering visible, NIR, and MIR spectral ranges and also from no-remote sensing ancillary data (e.g., DEM data). In the relevant DLST processes literature, there are many different terms but with the same meaning and/or function as SFs, frequently including kernels, predictors, independent variables, explanatory variables, etc. Although different terms have been used in different studies, in this chapter, the SF is adopted.

6.3.2.1 Types and Characteristics of SFs

In the kernel-based DLST processes, there are five types of SFs frequently seen in the literature, including (1) original spectral bands, (2) spectral (vegetation) indices, (3) land use land cover (LULC) types or their fraction images, (4) surface thermal related indices, and (5) terrain surface indices. Table 6.2 summarizes characteristics and references of the five types of SFs. Selected spectral band images in digital number or reflectance may directly be used as SFs for downscaling LST processes. There are some studies on DLST processes that directly used original spectral bands selected from optical bands as SFs, such as selecting original spectral bands from multispectral imagery by Tom et al. (1985), Li et al. (2018), and Hutengs and Vohland (2016) and from hyperspectral imagery by Ghosh and Joshi (2014).

In the kernel-based DLST processes, various spectral indices or vegetation indices (SIs or VIs) are commonly used as SFs. The SIs (VIs) are constructed with two to four typical blue, green, red, NIR, SWIR 1 (~1.6 μm), and SWIR 2 (~2.2 μm) bands selected from multi-/hyper-spectral imagery (Table 6.3). Table 6.3 presents a summary of 15 SIs frequently used as SFs in the most DLST processes, including their definitions (formulas), functions/characteristics, and references. The physical characteristics of most SIs are related to surface features, such as biomass, vegetation cover, soil moisture, and impervious areas, which directly or indirectly denote surface thermal characteristics. Compared with directly using original spectral original bands, benefits of using these SIs may include being easy to use and more enabling to denote surface thermal characteristics and thus to reflect spatiotemporal variation of LST. Therefore, SIs (e.g., NDVI) were used in some early classic DLST methods (e.g., DisTrad and TsHARP) and more recent studies have used different SIs (Table 6.3) as SFs in the kernel-based DLST techniques and methods. For example, Guo et al. (2014) and Lillo-Saavedra et al. (2018) used Landsat TM/TIRS LST data directly to downscale to finer resolution LST with their optical multispectral data derived SIs and other SFs. Eswar et al. (2016) and Wu et al. (2019b) used Landsat ETM+ and TIRS LST data first to upscale to lower resolution LSTs and then to downscale to finer resolution LSTs with SIs and other SFs extracted from Landsat optical multispectral data. By using ASTER optical data SIs and other SFs, Jeganathan et al. (2011), Chen et al. (2014b), and Pereira et al. (2018) also used ASTER LST data first to upscale to lower resolution LSTs and then to downscale to finer resolution LSTs.

Another commonly used type of SFs for the kernel-based DLST processes includes fraction of vegetation cover (F_{vc}), fraction or abundance images of endmembers, and LULC map, which are all extracted from high resolution optical sensors' data. A vegetation canopy cover or a fraction of vegetation cover (F_{vc}) within a pixel area in percentage (%) or [0, 1] is commonly used as an important SF for some classic and modern DLST processes. F_{vc} may be calculated via NDVI with different definitions/formulas, such as $(NDVI-NDVI_{min})/(NDVI_{max}-NDVI_{min})$ (Amazirh et al., 2019) and $1-((NDVI_{max}-NDVI)/NDVI_{max}-NDVI_{min}))^{0.625}$ (Agam et al., 2007). Fraction images of endmembers or LULC types can be calculated through a spectral unmixing analysis (either linear spectral mixture or multiple endmember linear spectral mixture or extracted from rule images of LULC classification results created with supervised classifiers (e.g., maximum likelihood classifier and support vector machine, SVM). Per LULC map, it is mostly created from high resolution optical multispectral imagers (satellite or airborne sensor's data). Since all the three subtypes of fraction and LULC maps as SFs have close relationships with LST, there are a lot of studies including them as important SFs for the kernel-based DLST processes. For example, Liu et al. (2018b) and Amazirh et al. (2019) downscaled 1,000 m LST data to 60 m and 100 m LSTs by using F_{vc} extracted from Landsat optical 30 m NDVI data to 30 m. To test the performance of a new DLST method, Liu and Su (2016) first used Landsat TM/ETM+ LST to upscale to 960 m and then downscale to 480 m, 240 m, 120 m, and 60 m LSTs with F_{vc} and other SFs extracted from 30 m resolution Landsat optical data. Gustafson et al. (2003) and Pu (2021) downscaled 90 m ASTER LST product to finer resolution LSTs with abundance images of LULC types by spectral unmixing analysis and rule images created with supervised SVM classifiers. Deng and Wu (2013b) used a thermal spectral unmixing model to downscale TM 120 m LST to 30 m LST with fraction images of LULC types extracted from the IKONOS image. Yang et al. (2011) and Sismanidis et al. (2016) used LULC map (created from optical high resolution satellite sensors' data) and other SFs to disaggregate coarse resolution LST data to finer resolution LSTs.

TABLE 6.2
Summary of Types and Characteristics of Different Scaling Factors (Kernels) (Pu and Bonafoni, 2023)

Type of scaling factors	Scaling factor (SF)	Characteristic and description	Reference
Spectral band	Spectral band	Spectral bands in digital number or reflectance are directly selected from multi-/hyper-spectral optical sensors	(Tom et al., 1985; Ghosh and Joshi, 2014; Li et al., 2018)
Spectral index (SI)	Spectral index	Using two or more multi-/hyper-spectra bands to construct a ratio or normalized difference ratio or other forms of SIs by an arithmetic operation. See Table 4 for list of SIs frequently used for scaling factors	(Pu, 2021; Mao et al., 2021)
LULC type or its fraction	Fraction of vegetation cover (Fvc)	A vegetation canopy cover within a pixel area in percentage (%) or [0, 1], frequently created with NDVI (e.g., $Fvc = (NDVI–NDVI_{min})/(NDVI_{max}–NDVI_{min})$ or $Fvc=1–((NDVI_{max}-NDVI)/(NDVI_{max}-NDVI_{min}))^{0.625}$)	(Agam et al., 2007; Amazirh et al., 2019)
	Fraction (abundance) of endmembers (or LULC types)	Using spectral unmixing methods to create fraction/abundances of endmembers or LULC types. Using some supervised classifiers to created rule images (similar to fractions) of endmembers or LULC types	(Gustafson et al., 2003; Deng and Wu, 2013b; Bonafoni and Tosi, 2017; Pu, 2021)
	LULC type	Using supervised classifiers to create LULC types from high resolution optical multi-/hyper-sensors' data	(Yang et al., 2011; Sismanidis et al., 2016; Li et al., 2019; Wu and Li., 2019)
Surface thermal related index	Albedo	Extracted from finer resolution optical visible and NIR band images	(Dominguez et al., 2011; Jiang et al., 2015; Yang et al., 2019)
	emissivity	Converted and mapped from LULC type map that is created with supervised classifiers and high resolution optical imagery	(Inamdar and French, 2009; Jiang et al., 2015; Mitraka et al., 2015; Agathangelidis and Cartalis, 2019)
Terrain surface index	Digital elevation model	Elevation values extracted from DEM model at a better than 30 m resolution	(Sismanidis et al., 2015; Jiang et al., 2015; Hutengs and Vohland, 2016; Wu and Li., 2019)
	Slope	Slope values extracted from DEM model at a better than 30 m resolution	(Guo et al., 2014; Li et al., 2019; Wu and Li., 2019)
	Aspect	Aspect values extracted from DEM model at a better than 30 m resolution	(Sismanidis et al., 2015; Li et al., 2019; Wu and Li., 2019)
	Solar incident angle	Extracted from DEM data at a better than 30 m resolution and with the same coarse TIR imaging acquisition time in a year	(Li et al., 2018; Hutengs and Vohland, 2016)

TABLE 6.3
A Summary of Spectral Indices (SIs) Frequently Used as Scaling Factors in DLST Processes (Pu and Bonafoni, 2023)

Index	Functions and characteristics	Formula	References
AVI	Anomaly Vegetation Index, denoting variation of NDVI in an area	$NDVI - \overline{NDVI}$	Wu et al. (2019b)
BSI	Bare Soil Index, reflecting variation of soil moisture	$(B_3 + B_6 - B_4 - B_1) / (B_3 + B_6 + B_4 + B_1)$	Wu et al. (2019b)
DVI	Difference Vegetation Index, reflecting differences of biomass and vegetation cover	$B_4 - B_3$	Wu et al. (2019b)
EVI	Enhanced Vegetation Index, compared to NDVI, EVI improving sensitivity to high biomass regions and vegetation monitoring	$2.5(B_4 - B_3) / (B_4 + 6B_3 - 7.5B_1 + 1)$	Bala et al. (2020)
IBI	Improved New Building Index, see NDBI	$\left(NDBI - \frac{SAVI + MNDWI}{2}\right) / \left(NDBI + \frac{SAVI + MNDWI}{2}\right)$	Xu (2008)
IVI	Improved New Vegetation Index, characterizing soil, vegetation, and impervious surface	$\left(SAVI - \frac{NDBI + MNDWI}{2}\right) / \left(SAVI + \frac{NDBI + MNDWI}{2}\right)$	(Liu et al., 2012; Wu et al., 2019b)
MNDWI	Modified NDWI, high linearly related to LST	$(B_2 - B_5) / (B_2 + B_5)$ SAVI-(NDBI+MNDWI)/2	Xu (2008)
MSAVI	Modified Soil-Adjusted Vegetation Index, see SAVI	$\left[2B_4 + 1 - \sqrt{(2B_4 + 1)^2 - 8(B_4 - B_3)}\right] / 2$	Wu et al. (2019b)
NDBI	Normalized Difference Built-up Index, high correlation with impervious surface area and less sensitive to seasonal change	$(B_5 - B_4) / (B_5 + B_4)$	Zha et al. (2003)
NDMI	Normalized Difference Moisture Index, stronger linearly related to LST, sensitive to vegetation/soil moisture	$(B_4 - B_5) / (B_4 + B_5)$	Lillo-Saavedra et al. (2018)

(Continued)

TABLE 6.3 *(Continued)*
A Summary of Spectral Indices (SIs) Frequently Used as Scaling Factors in DLST Processes (Pu and Bonafoni, 2023)

Index	Functions and characteristics	Formula	References
NDVI	Normalized Difference Vegetation index, well-documented inverse relationship between LST and NDVI and positive relationship between NDVI and soil moisture	$(B_4 - B_3)/(B_4 + B_3)$	Fassnacht et al. (1997)
NDWI	Normalized Difference Water Index, high linear correlation with LST	$(B_2 - B_4)/(B_2 + B_4)$	(Guo et al., 2017; Bala et al., 2020)
NMDI	Normalized Multi-band Drought Index, monitor the water content of soil and vegetation by using absorption properties in NIR band and absorption differences between NIR and SWIR bands	$\left[B_4 - (B_5 - B_6)\right]/\left[B_4 + (B_5 - B_6)\right]$	(Wang and Qu, 2007; Yang et al., 2011)
RVI	Ratio Vegetation Index, considering the inverse relationship between chlorophyll absorption of red band and high reflectance of NIR band for healthy plant canopies	B_3 / B_4	Wu et al. (2019b)
SAVI	Soil-Adjusted Vegetation Index, characterizing interaction of soil properties and vegetation systems	$\frac{(B_4 - B_3)(1+L)}{B_4 + B_3 + L} where, L = 0.5$	van Leeuwen and Huete (1996)

Note: 1–6 in B1–6 represent blue, green, red, NIR, SWIR1 (1.6 μm), and SWIR2 (2.2 μm) bands in order

The surface thermal related indices may include surface albedo and emissivity, which are often used as SFs. The albedo is often extracted from visible, NIR, and SWIR bands of high-resolution multispectral sensor data, such as from Landsat 8 OLI bands 2–7 (in a linear transformation) (Yang et al., 2019). High resolution spectral emissivity can be obtained via assigning its value to each surface cover type by referring to typical emissivities of various common materials over the thermal spectral range 8–14 μm recommended by Lillesand et al. (2008) or other references. The surface thermal related indices have been applied as SFs in the DLST processes. For instance, Dominguez et al. (2011), Jiang et al. (2015), and Yang et al. (2019) used the albedo as one of SFs, extracted from different sources (NASA-ATLAS, MODIS, and Landsat 8), for the DLST process. Similarly, Inamdar and French (2009), Mitraka et al. (2015), Chybicki and Łubniewski (2017), and Agathangelidis and Cartalis (2019) utilized the emissivity as one of the SFs in their DLST processes.

In the kernel-based DLST processes, researchers also often use DEM data derived ancillary SFs including elevation, slope, aspect, and solar incident angle for their DLST processing studies. This is because the DEM derived SFs directly reflect the surface thermal energy budget and thus directly or indirectly influence spatial variation of LST in a study area. For example, Keramitsoglou et al. (2013) and Sismanidis et al. (2016) downscaled MSG-SEVIRI LST data (4 km × 5 km) to MODIS-like 1 km resolution LST using SFs composed by two groups of SFs: the static SFs including land cover, slope, aspect, and altitude maps and the dynamic SFs corresponding to emissivity and SIs maps. Guo et al. (2014), Hutengs and Vohland (2016), and Wu and Li (2019) also used terrain surface indices, land cover data, and other SFs to downscale lower resolution LSTs (MODIS and Landsat TM and ASTER) to finer resolution LSTs.

6.3.2.2 Scaling Factor Selection

The issue of the SF selection seems not to be very crucial compared to the feature/variable selection for application purposes of remote sensing. This may be that researchers who conduct their DLST processing studies may have elaborately selected what SFs are the most suitable for their DLST process projects based on their available datasets and research objectives. However, per this literature review, there still are many studies on DLST processes that consider the issue of SF selection to determine a final subset selected from all candidate SFs for the DLST processes. The key reasons for the SF selection are generally lowering SF dimensionality, reducing SF data redundancy and collinearity, and improving efficiency of DLST processing. Currently, there may be three main types of SF selection methods commonly adopted by researchers: (1) Random Forest (RF) algorithm, (2) stepwise regression procedure, and (3) Support Vector Machine-Recursive Feature Elimination (SVM-RFE). With the ranked important scores of each candidate SF, calculated with the RF algorithm, an optimal subset of SFs may be selected from all input SFs with a threshold of importance scores. For example, Li et al. (2019) and Yang et al. (2017) used the RF algorithm to calculate importance scores of input SFs (SIs, spectral bands, terrain surface variables, etc.) to select an important subset of SFs to efficiently downscale 1,000 m MODIS LST data to relatively higher resolution LSTs. Forward stepwise regression procedure may be used to build robust DLST models. In this procedure, SFs are added in a selected subset in an order of one by one when an SF is added, then the subset increases the most significantly explained model variance (R^2). For instance, Bechtel et al. (2012) used the procedure to find an optimal SF subset for their downscaling LST scheme. The SVM-RFE SF selection method is using RFE that is an embedded ranking-based feature selector (Tuia et al., 2009; Zhang et al., 2017), which chooses a fixed number of high-ranking SFs for subsequent analyses. Therefore, the SVM-RFE SF algorithm, like the RF algorithm, may be used to select an optimal subset of SFs based on their high-ranking values for DLST processes. For instance, Ebrahimy and Azadbakht (2019) applied this SF selection method to choose the three similar high-ranked SFs, namely DEM, NDVI, and a red band, for their downscaling MODIS 960 m LST data to 240 m resolution LST.

6.3.3 DLST Methods and Algorithms

To obtain finer spatiotemporal resolution LST data through downscaling coarse spatial resolution but high temporal resolution LST data, many DLST investigators have developed and improved different advanced DLST methods and techniques. Generally, there are two categories of DLST methods and techniques: kernel-driven process methods and fusion-based process methods. Some researchers also consider the combined (hybrid) categories with the two basic categories. However, by understanding and utilizing the two basic categories, the hybrid category of DLST methods should be easy to understand and master. The kernel-driven processes are the most frequently used thermal downscaling methods, such as classic-based, regression-based, machine-learning-based, TIR spectral unmixing-based, etc. (see their detailed introductions later). The category of methods relies on relationships between the kernels (SFs) and LSTs to obtain high-resolution LSTs by performing a spatial sharpening process or a TIR spectral unmixing process. The kernels or SFs are extracted from high resolution sensors' data and no-remote sensing ancillary data (see Section 6.3.2). The fusion-based process methods are based on fusion algorithms rather than on kernels to obtain spatial details from known fine spatial but low temporal resolution LSTs (e.g., Landsat series) and low spatial but high temporal resolution LSTs (e.g., MODIS), which mostly create high resolution spatiotemporal time series LST products (e.g., Landsat-like LST product). Figure 6.1 illustrates the flow diagram of the two basic categories of DLST process methods. Table 6.4 also summarizes the characteristics, advantages/limitations, and references of the five general types of DLST process methods popularly used during the last four decades. In the following subsections, the five types of DLST methods are introduced and discussed.

6.3.3.1 Classic-Based Methods

The earliest studies on the thermal spatial sharpening process were done by such as Tom et al. (1985) and Nishii et al. (1996), who used SFs extracted from relative high resolution optical images to enhance or improve relatively low resolution TIR band images. However, the typical and popularly used classic-based downscaling LST (DLST) processing methods are Disaggregation of Radiometric Temperature (DisTrad) (Kustas et al., 2003) and Temperature Sharpening (TsHARP) (Agam et al.,

2007), and the classic-based methods also include their variants (e.g., TsHARP with local variant (Jeganathan et al., 2011)).

With DisTrad in Kustas et al. (2003), if T_{coarse} and $NDVI_{coarse}$ *and* T_{fine} and $NDVI_{fine}$ represent an image in coarse and fine resolutions, respectively, the first step is to select a subset of pixels from the scene where NDVI is most uniform within the coarse aggregated pixels. Next, a least-squares fit performed relating T_{coarse} and $NDVI_{coarse}$ values associated with this subset of 'uniform' pixels using the following expression.

$$\hat{T}_{coarse} = a + b\,NDVI_{coarse} + cNDVI^2_{coarse} \tag{6.1}$$

A linear equation (c = 0) between T_{coarse} and $NDVI_{coarse}$ may be more suitable in some cases, particularly when the range between T_{coarse} and $NDVI_{coarse}$ values is limited. Spatial variability due to soil moisture effects can be incorporated into the sharpened thermal image by adjusting each fine resolution pixel by the deviation from the regressed temperature observed at coarse resolution:

$$\Delta\hat{T}_{coarse} = T_{coarse} - \hat{T}_{coarse} \tag{6.2}$$

Then, for each *ith* fine resolution pixel within the coarse resolution pixel (where i = 1, 2, 3 . . .), $\hat{T}_{fine}$ is computed via

$$\hat{T}_{fine}(i) = \hat{T}_{coarse}\left(NDVI_{fine}(i)\right) + \Delta\hat{T}_{cooarse} \tag{6.3}$$

where the first term on the right-hand side is evaluated using Eq. (6.1) with the $NDVI_{fine}$ value at that pixel, and the second term accounting for deviations from the mean is evaluated from Eq. (6.2).

Note that in the DisTrad when Eq. (6.1) changes its form into $\hat{T}_{coarse} = f\left(NDVI_{coarse}\right)$, the DisTrad becomes TsHARP technique (Agam et al., 2007) via:

$$f(NDVI) = \begin{cases} (a)\,a_0 + a_1\,NDVI & \textit{linear fit} \\ (b)\,a_0\,a_1\,NDVI + a_2\,NDVI^2 & \textit{polynomia fit} \\ (c)\,a_0 + a_1\left(1 - \left(\dfrac{NDVI_{max} - NDVI}{NDVI_{max} - NDVI_{Min}}\right)^{0.625}\right) & f_c\ \textit{transformation} \\ (d)\,a_0 + a_1(1 - NDVI)^{0.625} & \textit{sjmlified } f_c\ \textit{trans,} \\ (e) & 0 \end{cases} \tag{6.4}$$

Note that the fifth form (e) is included as a baseline case, representing no sharpening, i.e., no incorporation of high-resolution NDVI information.

The classic-based methods use a simple regression relationship between LST and an NDVI (DisTrad) or fraction of vegetation cover (F_{vc}, TsHARP) to spatially sharpen coarse resolution LST to a finer resolution. The mechanism underlying the

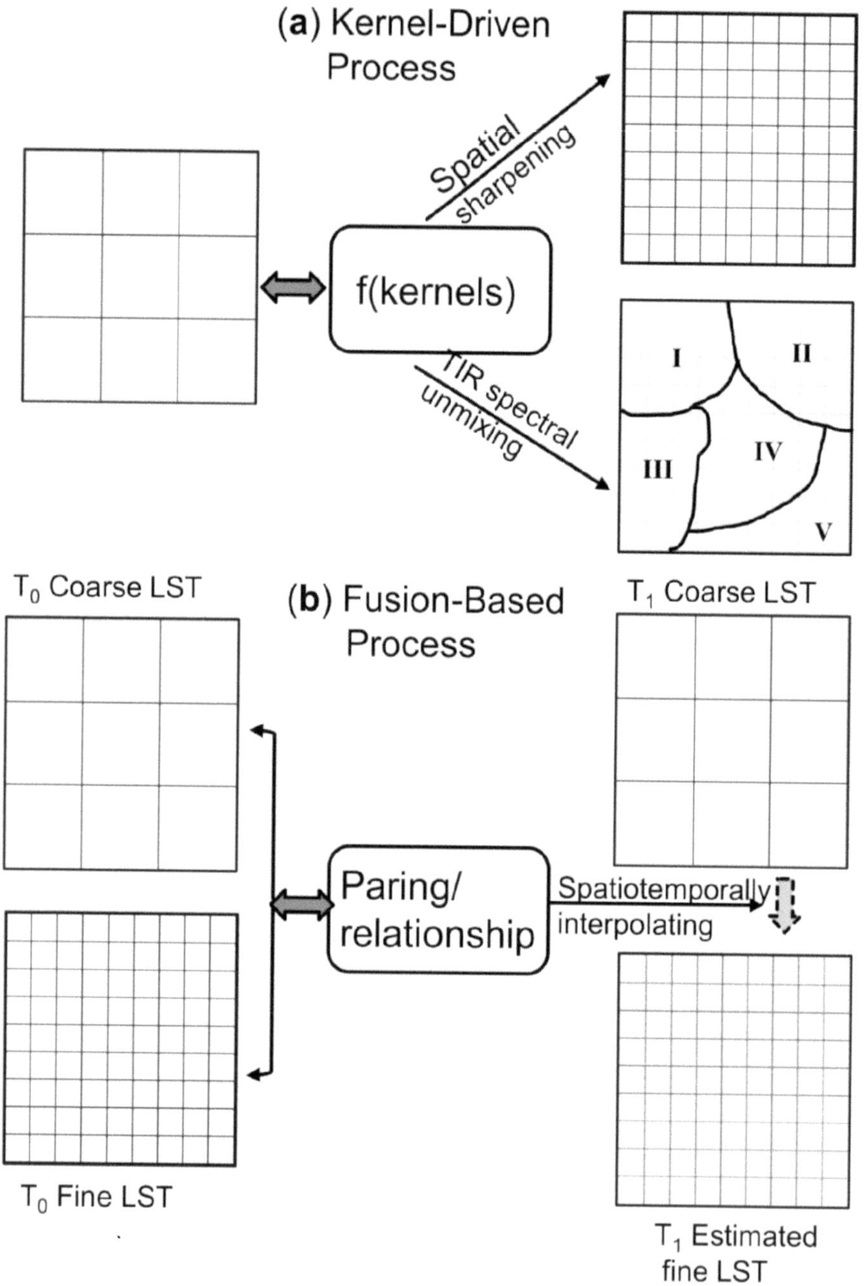

FIGURE 6.1 The flow diagram to illustrate the two basic categories of downscaling LST methods: (a) the kernel-driven process methods and (b) the fusion-based process methods (Pu and Bonafoni, 2023).

TABLE 6.4
A Summary of Methods/Algorithms Frequently Used in Downscaling LST (Pu and Bonafoni, 2023).

The type of DLST methods	Characteristic and description	Advantage and limitation	Major factor	Typical reference
Classic-based	Use a simple regression relationship between LST and an NDVI (DisTrad) or fraction of vegetation cover (Fvc, TsHARP) to spatially sharpen coarse resolution LST to finer resolution	DLST process is simple and running fast but not suitable for areas with sparse vegetation and no-vegetated areas; the relationships are site specific	Determine close relationships between LST and NDVI or Fvc.	(Kustas et al., 2003; Agam et al., 2007; Mukherjee et al., 2014)
Regression-based	Regression relationships between LST and SFs (kernels) within a study area are determined empirically at a coarse (thermal band) resolution and then applied to the finer resolutions to produce sharpened LSTs	This type of method is relatively easy to perform, and DLST results are satisfactorily accurate. But regression correlations between LST and SFs might be insufficient for some regions and significantly site specific	Extracting, selecting a number of effective SFs, and choosing an appropriate regression model.	(Maeda, 2014; Mukherjee et al., 2014; Luo et al., 2021)
Machine learning-based	By fitting nonlinear relationships between LST and SFs and reducing uncertainties by approximating relationships more accurately, they usually perform better compared to the regression-based methods and result in higher accuracies of DLST results	Nondeterministic reasoning, self-learning, self-organization, error tolerance, and excellent nonlinear approximation ability of ANN; rapidly processing high-dimensional data with SVM; and insensitivity to multicollinearity of independent variables of RF. But intensive computational resources, complex structure of algorithms, a black box model, etc.	Determination of representative training samples and structure parameters and selection of key SFs	(Yang et al., 2011; Ghosh and Joshi, 2014; Ebrahimy and Azadbakht, 2019)

(Continued)

TABLE 6.4 *(Continued)*
A Summary of Methods/Algorithms Frequently Used in Downscaling LST (Pu and Bonafoni, 2023).

The type of DLST methods	Characteristic and description	Advantage and limitation	Major factor	Typical reference
TIR spectral unmixing-based	The thermal component (TC) temperatures or radiances are obtained by solving the thermal spectral unmixing (TSU) model from coarse resolution thermal (or LST) data, and TC temperature can then be aggregated into different high resolution DLST maps	TC temperature or radiance instead of component fractions and emissivities can be directly obtained from coarse TIR data, less time consuming even with a large downscaling factor of 30, and higher downscaling accuracy	Define and determine typical thermal components from high resolution optical sensors' data	(Mitraka et al., 2015; Wang et al., 2020; Pu and Bonafoni, 2021)
Spatiotemporal fusion-based	First construct relationships at an initial time between fine resolution image LST (e.g., Landsat LST) and coarse resolution image LST (e.g., MODIS LST), then downscale a target time coarse resolution image LST to fine resolution LST when the target time fine resolution LST is not available	Can produce high spatiotemporal fine resolution (e.g., Landsat-like) LSTs. The effect of mixed pixels and uncertainty for the type of fusion-based methods need to be assessed	How to determine ideal pair of coarse-/fine- resolution sensors' data at appropriate time points and model spatial and temporal variations of LSTs between two time points.	(Gao et al., 2006b; Zhu et al., 2010; Xiao et al., 2019)

DLST processes relies on correlations between LST and vegetated SFs (NDVI and F_{vc}) that are directly related to the general cooling effect by vegetated surfaces due to shading, biomass, higher relative humidity, and evapotranspiration (Mao et al., 2021). Consequently, such classic-based methods for DLST process methods are limited to be applied in vegetated areas, agriculture areas in growing seasons, and thus, theoretically, they are not suitable for dessert or semi-arid dessert areas covered with sparse vegetation and non-vegetation. Since the classic methods are based on the simple (regression) relationships between LST and NDVI or F_{vc}, the relationships are site specific. Per an application perspective, currently, the classic-based methods are rarely solely used for downscaling coarse resolution LST to create finer resolution LST maps, instead of used as a standard or criterion for testing and evaluating the accuracy and performance of newly developed DLST process algorithms and methods. For example, with SFs of terrain features, land cover data, and surface reflectance, Hutengs and Vohland (2016) utilized a random forest (RF) regression approach in DLST processing. Compared to the DLST accuracy created with the classic TsHARP sharpening method, the RF model improved the DLST accuracy up to 19%. Ghosh and Joshi (2014) compared DLST results created using machine learning algorithms, boosting machine and SVM with those created by the DisTrad sharpening model and indicated that the machine learning algorithms have produced much better DLST results.

6.3.3.2 Regression-Based Methods

In downscaling LST processes, the regression-based methods are the most popularly used methods with different regression forms and a diversity set of SFs. This type of method includes linear and non-linear, single SF or multiple SFs regression or polynomial models (e.g., Chen et al., 2012; Gao et al., 2012; Ghosh and Joshi, 2014; Mukherjee et al., 2014; Bonafoni and Tosi, 2017; Gao et al., 2017; Wu and Li, 2019; Wang et al., 2020c; Zawadzka et al., 2020; Pu, 2021). Usually, a regression relationship in different forms of LST with SFs within a study area is empirically determined at a coarse (thermal band) resolution and then applied to a fine (optical band) resolution to generate a spatial sharpened thermal image. For example, Zawadzka et al. (2020) developed a method that used the Multivariate Adaptive Regression Splines (MARS) models associated with SFs extracted from very high resolution (VHR) hyperspectral aerial imagery and topographic maps to downscale the 100 m Landsat 8 TIRS data to 2–4 m resolutions. Their analysis results indicate that the MARS models could be used to create high resolution LST maps for studying urban thermal environments. Wu and Li (2019) established a nonlinear relationship of LST with input SFs by using a random forest regression model with multitype predictor variables (MTVRF) to downscale LST under various surface characteristics. The MTVRF model produced satisfactory results. Maeda (2014) applied simple and multiple regressions to quantify relationships between LST and SFs (e.g., altitude, land covers, NDVI, etc.) in order to downscale 1 km resolution LST to 250 m resolution. By considering geospatial and temporal effects on DLST results, Luo et al. (2021) utilized a geographically and temporally weighted autoregressive (GTWAR) model to disaggregate MODIS 1 km LST to 100 m resolution LST. Compared to the TsHARP, the geographically weighted regression (GWR), the geographically weighted autoregressive (GWAR), and the geographically and temporally weighted regression (GTWR) downscaling LST methods, the GTWAR was superior. Pereira et al. (2018) also used GWR but coupled with area-to-point kriging of regressed residuals to improve the DLST accuracy.

This type of method is relatively easy to perform, and usually their DLST results are satisfactorily accurate. However, the main limitations of this type are that regression correlations between LST and SFs might be insufficient for some regions (Wu and Li, 2019) and significantly site specific (Mukherjee et al., 2014).

6.3.3.3 Machine Learning (ML)-Based Methods

During the last decade, machine learning (ML)-based methods have been used to establish no-linear relationships between LST and SFs to downscale coarse resolution LSTs to finer resolutions. The ML-based methods may consist of various artificial neural networks (ANN), support vector machines (SVM), random forest (RF) models, the partial least square model, etc. (e.g., Yang et al., 2011; Ghosh and Joshi, 2014; Hutengs and Vohland, 2016; Bartkowiak et al., 2019; Ebrahimy and Azadbakht, 2019; Xia et al., 2019; Pu, 2021; Xu et al., 2021). The ML-based methods may also include some novel deep learning methods, such as convolutional neural networks, CNN) and recurrent neural networks (e.g., Yin et al., 2021). They usually perform better compared to the regression-based methods and result in a higher accuracy by fitting nonlinear relationships between LST and SFs (Li et al., 2019) and reducing uncertainties by approximating the relationships more accurately (Mao et al., 2021). For example, Yang et al. (2011) used ANN models trained with the 990 m resolution MODIS LST product and the corresponding input SFs (i.e., area ratios and endmember indices) to estimate the 90 m resolution DLST. The DLST was verified with ASTER LST product and was basically consistent with the latter. To address the issue of nonlinearity and spatial nonstationarity in DLST processes, Xu et al. (2021) proposed a multi-factor geographically weighted machine learning (MFGWML) method to downscale the 30 m LST data retrieved from Landsat 8 images to 10 m LST data mainly based on Sentinel-2A image derived SFs. MFGWML used three excellent machine learning algorithms (namely extreme gradient boosting, multivariate adaptive regression splines, and Bayesian ridge regression) and a geographically weighted regression algorithm to fit nonlinear relationships and allow for spatial nonstationarity. Their results indicate that MFGWML could characterize the local heterogeneity and obtain reliable and accurate DLST results. Ghosh and Joshi (2014) conducted a set of experiments to downscale LST retrieved from Landsat ETM+ thermal images with SFs extracted from EO1-Hyperion hyperspectral data over three different geographic locations with different landscape characteristics. They selected a partial least square regression and two machine learning algorithms, namely boosting machine (GBM) and SVM, for processing DLST. Compared to the DLSTs produced by the DisTrad sharpening model and the regression model, GBM and SVM have created much better DLST results. To disaggregate MODIS ~1 km LST product to ~250 m resolution, Hutengs and Vohland (2016) used a random forest (RF) regression approach with SFs of terrain features, land cover data, and surface reflectances. Compared with the DLST accuracy created with the commonly used

TsHARP sharpening method, the RF model improved the DLST accuracy up to 19%. With MODIS 1,000 m LST and Landsat LST datasets, Yin et al. (2021) developed a deep learning-based spatiotemporal temperature fusion network (STTFN) based on CNNs handling the non-linearity of LST temporal changes for creating finer spatio-temporal resolution LST products. Their experimental results demonstrate the better performance of the STTFN algorithm than other spatiotemporal-based methods (see detailed review in Section 6.3.3.5).

There are many advantages with different ML methods, including nondeterministic reasoning with a complex causality due to its self-learning/-organization, error tolerance, and excellent nonlinear approximation ability of ANN; rapid processing ability of high-dimensional data with SVM; and insensitivity to multicollinearity of independent variables to RF (Li et al., 2019). However, the major limitations for the ML technique-based methods may include intensive computational resources, complex structure of algorithms, a black box model, etc.

6.3.3.4 TIR Spectral Unmixing (TSU)-Based Methods

Compared to the spatial sharpening kernel-driven methods mentioned earlier (i.e., classic-, regression-, and ML-based DLST methods), there are fewer studies on DSLT processes based on thermal spectral unmixing (TSU). The earlier TSU-based DLST methods could be traced back to the early 1980s when Dozier (1981) and Matson and Dozier (1981) published their initial explored works based on the thermal spectral unmixing principle with moderate optical multispectral imagery. More recently, the TSU-based DLST methods can be described as follows: thermal component (TC) temperatures within a pixel area are decomposed based on multi-temporal, -spatial, -spectral, and/or -angular observations (Zhan et al., 2013). The TC temperatures or radiances at an initial high resolution, created by solving the TSU model from coarse resolution thermal (or LST) data, can then be used to produce different high resolution DLST maps.

The input is assumed as a radiance matrix **Rad** with n rows (samples) and m columns (band images), which is decomposed through the non-negative matrix factorization (NMF) as:

$$Rad_{n \times m} = F_{n \times r} \cdot P_{\mathrm{r \times m}} \tag{6.5}$$

where r is the number of the temperature components (TCs) that can be determined by the NMF itself by examining the explained variance and residuals. In this sense, the NMF is capable of determining the non-negative TCs and their number at the same time. Although, mathematically, the NMF is simple as a product of two matrices, the interpretation can be flexible in applications.

In spectral analysis, **F** can be the fraction (TC) map of r primary components, and **P** encodes the response of the components at each of the m input band images. Given any r, a positive solution of both **F** and **P** can be obtained by minimizing the unexplained or residual variance expressed as:

$$\min \, \| \boldsymbol{Rad\text{-}F \cdot p} \| \text{ with } \mathbf{F} > 0, \mathbf{P} > 0. \tag{6.6}$$

The solution can be derived through several iterative algorithms, among which the multiplicative update rule is the most popular approach.

The algorithm can be modified as a modified TSU model described as follows. Instead of using thermal radiance in the TSU model as in (Wang et al., 2020), we can directly use LST at a low resolution expressed as the thermal component temperature TC) change weighted by their fractions of spectral clusters F at a low resolution:

$$LST_{L(1\times n)} = TCV(1\times m) \bullet F_L(m \times n) + \varepsilon \tag{6.7}$$

with $\mathbf{min}\ \|\mathbf{LST} - \mathbf{TCV} \bullet \mathbf{F}\|$ subject to $F \geq 0$.

Where $\mathbf{LST}_{L(1\times n)}$ represents the LSTs of n low-resolution pixels; $\mathbf{TCV}_{(1\times m)}$ is the temperatures of m thermal components represented by m spectral clusters at a high resolution; $\mathbf{F}_{L(m\times n)}$ represents the fractions of spectral clusters (thermal components) at a low resolution; and residual ε represents the variance of LST unexplained by the spectral clusters. Per Eq. (6.7), if we know $\mathbf{LST}_L$ and $\mathbf{F}_L$, it is easy to solve the thermal component vector via a least-square estimation (LSE) of the model (Pu and Bonafoni, 2021). Consequently, $\mathbf{TCV}_{(1\times m)}$ can be obtained via the following LSE:

$$\mathbf{TCV}_{(1\times m)} = \mathbf{LST}_{L(1\times n)} \bullet \mathbf{F}^{T}_{L(n\times m)} \bullet (\mathbf{F}_{L(m\times n)} \bullet \mathbf{F}^{T}_{L(n\times m)})^{-1} \tag{6.8}$$

After obtaining $\mathbf{TCV}_{(1\times m)}$, we can map very-high-resolution LSTs based on their spectral cluster maps at high resolution (here, TCs and their temperatures can be assigned to the corresponding spectral clusters).

For example, Deng and Wu (2013b) proposed a VHR spectral unmixing and thermal mixing (VHR-SUTM) approach to disaggregate LST to a high resolution. The VHR-SUTM method may include: (1) spectral unmixing with IKONOS data to obtain fractions of land cover types that respond to unique thermal characteristics, and (2) thermal unmixing with lower resolution Landsat TM LST and corresponding land cover fractions to estimate high resolution LST pixel values. Their thermal spectral unmixed results indicate that the estimated LST map was high consistent with the corresponding resolution of retrieved LSTs. Li et al. (2021a) also developed a TSU-based method, named temperature downscale with data Fusion and spectral unmixing, to unmix Landsat 8 100 m LST to 16 m high resolution LST with surface feature abundances extracted from Gaofen-1 data. The results indicate that the TSU model can effectively estimate high resolution LST and reflect the spatial variation of the LST. Mitraka et al. (2015) developed a TSU-based method for downscaling low-resolution MODIS TIR data using SFs extracted from optical Landsat sensors to obtain time-series high spatial resolution LSTs. The results indicate that the proposed approach shows a high potential of being operationally used in the near future to produce high spatiotemporal resolution LSTs in cities. With VHR airborne LST and hyperspectral optical data (1.25 m resolution), Liu et al. (2016a) first upscaled both LST and optical data to 4 m, 8 m, 16 m, and 32 m resolution, then used a TSU model to downscale to 16 m, 8 m, and 4 m resolution LSTs. The analysis results show a good agreement between the DLST and the reference LST by using both visual interpretation and quantitative accuracy metrics.

With the TSU-based methods, one advantage is that high resolution thermal component (TC) temperature or radiance can be directly created from coarse TIR data (LST or radiance). For instance, Wang et al. (2020) proposed the TC-based spectral unmixing technique to create TC radiance. The TSU technique may be used directly to solve a system of linear spectral mixture models encoding the TC-LST (or thermal radiance) relationship at a coarse resolution (e.g., MODIS 1000 m LST), and then the solution is applied to a fine resolution to obtain fine resolution LST results. Their results indicate that downscaled LST could differentiate temperatures over major land types and capture both seasonal and diurnal LST dynamics. Pu and Bonafoni (2021) modified the TSU model developed by Wang et al. (2020) for directly downscaling the ASTER LST data to high and very high resolution LSTs. Their analysis results indicate that the modified TSU model for unmixing TC temperatures at an initial high resolution was effective and advanced in DLST processing compared with their previous works and those in current literature. In general, this type of DLST method is less time consuming even with a large downscaling factor of 30 (Wang et al., 2020) and outperformed classic downscaling techniques significantly.

6.3.3.5 Spatiotemporal Fusion-Based Methods

As conceptually illustrated by Figure 6.1b spatiotemporal fusion-based methods first construct relationships at an initial time between fine resolution (usually low temporal resolution) image LST (e.g., Landsat LST) and coarse resolution (usually high temporal resolution) image LST (e.g., MODIS LST), then downscale a target time coarse resolution image LST to fine resolution LST when the target time fine resolution LST is not available (Yoo et al., 2020). During the last two decades, the spatiotemporal fusion-based methods can be classified into two general types: (1) the methods based on the spatial and temporal adaptive reflectance fusion models (STARFM) and (2) the methods based on unmixing theory (Yang et al., 2020).

The STARFM algorithm (Figure 6.1b) can be summarized as the following Eq. (6.9):

$$L_{(x_{w/2},y_{w/2},t_0)} = \sum_{1}^{w}\sum_{1}^{w}\sum_{1}^{n} W_{ijk}\left(M_{(x_i,y_j,t_0)} + L_{(x_i,y_j,t_k)} - M_{(x_i,y_j,t_k)}\right) \tag{6.9}$$

where L and M are the Landsat and MODIS surface reflectance, respectively; w is the window size searching similar pixels; (x, y) is the location of the center focus pixel; (x_i, y_j) is the pixel location of Landsat and MODIS image pair; t_0 is the date for downscaling; t_k is the prediction date of MODIS and Landsat pairs; n is the number of input Landsat and MODIS data points; and W_{ijk} is the weight that determines the influence of neighboring pixels on the simulated reflectance of the central focus pixel.

The first type of method assumes that the ratio of coarse pixel value to neighboring similar pixels does not change over time, while the second type assumes that the value of each coarse resolution pixel is a linear combination of the responses of all endmembers within the coarse pixel. The two types of methods all use spatial relationships between coarse/native- and fine/target-resolution images acquired from different sensors and the temporal variation between the initial and target times to interpolate/predict fine-resolution image LST at the target time. The spatial resolution of the interpolated LST image corresponds to that of the fine-resolution image (e.g., 100 m LST Landsat TIRS data) while the temporal resolution of the interpolated LST image corresponds to that of the target time coarse resolution (e.g., daily MODIS LST data) (Xia et al., 2019). In short, the resultant fused LST data with the type of methods possess high spatiotemporal resolution characteristics of LSTs originally coming from fine spatial resolution sensors (e.g., Landsat TIRS) and high frequency sensors (e.g., MODIS). Per the first type of the methods, the most popularly used method for DLST processing is the STARFM, which originally fuses the daily MODIS and the 16-day Landsat surface reflectance data to produce a synthetic daily surface reflectance Landsat-like data (Gao et al., 2006). Later, the method was directly used to produced high temporal resolution and fine spatial resolution LST products. For example, Liu et al. (2018b) used the STARFM method to downscale MODIS 1,000 m LST to produce LST data at a fine resolution of 60 m and a fine temporal resolution of one day with Landsat TM data. More similar application works with the STARFM may also include those by Wu et al. (2015) and Liu and Weng (2018) etc. Since the STARFM might not be sensitive to heterogeneous landscapes, many researchers have improved the method, such as the enhanced STARFM (ESTARFM) and flexible spatiotemporal data fusion (FSDAF). The ESTARFM can increase the data fusion accuracy by considering heterogeneous landscapes, while FSDAF can not only consider heterogeneous landscapes but also maintain spatial details within a study area (Zhai et al., 2020). The unmixing-based spatiotemporal models, estimating high-resolution LST data based on a linear spectral mixing model, may include improved spatiotemporal Bayesian data fusion (ISTBDF) (Xue et al., 2019), improved spatial and temporal reflectance unmixing model (ISTRUM), and spatial temporal data fusion approach (STDAF) (Wu et al., 2012). For instance, Ma et al. (2018) downscaled coarse resolution MODIS LST data to create a synthesized time series of Landsat-like products by using the ISTRUM. This type of methods can predict an accurate estimation of LST by grouping similar pixels and calculating their average LST.

Different from kernel-driven methods, the spatial detailed information with this type of fusion-based methods is obtained from known fine resolution LST data rather than kernels. To further improve the prediction accuracy created with the spatiotemporal fusion-based methods, many researchers also made an effort to combine advantages of both kernel-driven and fusion-based process methods (i.e., detailed spatial information with a kernel-based method and high frequency time series LST products with a fusion-based method (Yoo et al., 2020)). One type of spatiotemporal fusion-based method considers the synergy of the kernel-driven and fusion-based approaches to produce time series high spatiotemporal Landsat-like LST products. For example, Xia et al. (2018) proposed two strategies for combining the regression-based method and fusion-based methods to create a high spatiotemporal resolution LST product, namely, the 'regression-then-fusion' (R-F) and 'fusion-then-regression' (F-R) methods. After testing Landsat 8 and ASTER data (first upscaling to coarse resolution then using R-F and F-R strategies to downscale to Landsat-like and ASTER-like LST products), their results clearly indicate that the R-F strategy performed better than the F-R strategy when the regression error at the start time was smaller than that at the target time, and otherwise, the F-R strategy was better than the R-F one. Xia et al. (2019) proposed a weighted combination of kernel-driven and fusion-based methods (CKFM) to further improve the spatiotemporal resolutions of time series LSTs. The CKFM is more accurate and robust compared to the kernel-driven method with an improvement of 0.1–0.6 K, and it may reconstruct more spatial details than those by a fusion-based method when it was used to downscale daily MODIS 1 km LSTs to 30 m resolution LSTs with Landsat 8 data.

By this literature review on applications of this type of methods, most investigations were found on producing high spatiotemporal resolution synthetic LST products (e.g., Landsat-like LST data series) by mostly fusing MODIS–Landsat data. More advanced spatiotemporal fusion-based methods include synergic methods by combining the kernel-driven process with the fusion-based process methods, which more emphasize on how to efficiently deal with spatial heterogeneity of landscapes (spatial variation of LST) and time change (temporal variation of LST) between any two time points. This is a key point for a research direction. In addition, it is necessary that the effect of mixed pixels with different LULC cover types on the high spatiotemporal LSTs and uncertainty for the type of fusion-based methods be assessed by considering the optical/thermal sensors' noises, LST retrieval errors, and spatial/temporal variations of LST induced by heterogenous landscapes, mixed pixels, and seasonal change (Mao et al., 2021).

6.4 SPATIAL SCALING EFFECT IN DOWNSCALING LST PROCESSES

Many DLST investigations have proved that a spatial scaling effect does exist in DLST processing, especially DLSTs at VHR resolutions. The spatial scaling effect in DLST processes in most studies may be considered as effects of spatial variations of spatial resolution and spatial extent in an image scene on DLST results at finer resolutions (Chen et al., 2012; Zhou et al., 2016). To more easily, specifically, and substantially understand DLST results, we can define the spatial scaling effect as an error in the downscaled LST process at a target (high) resolution, which may be understood as a difference between DLST and true/reference LST at high resolutions (Zhou et al., 2016; Pu and Bonafoni, 2021). In this section, we review and discuss the causes of the spatial scaling effect in DLST processes, the methods used to measure or assess it, and the techniques developed to reduce it.

6.4.1 Spatial Scaling Effect

Based on the literature review, the spatial scaling effect in DLST processes are mainly caused by a basic assumption in DLST processes, which assumes that the LST–descriptors relationship (linear or non-linear) at a native (coarse) resolution is scale-invariant at different DLSTs at finer resolutions (Kustas et al., 2003; Agam et al., 2007). The assumption is mostly problematic in real DLST processing cases due to spatial variation changes of the spatial resolution and spatial extent in spatial heterogenous environments (Chen et al., 2012; Ghosh and Joshi, 2014; Zhou et al., 2016). The variation of spatial resolution can be caused by the change of the pixel size that is associated with the change of spatial variation within a specific pixel area, while the variation of spatial extent can be related to different heterogeneous landscapes at a large scale within a scene of imagery. The former (spatial variation) may be directly or indirectly related to the spatial variation of bio- and physical surface materials (e.g., greenness patch size and bare soil patch size variations) within the different pixel size areas while the latter is due to existing different landscapes (e.g., with different LULC types: agriculture, grassland, urban land use) in a study area. Both variations in a scene of imagery directly result in the spatial scaling effects in DLST processes.

There are many studies that have demonstrated the spatial scaling effects in DLST processes caused by either or both the spatial variations discussed earlier. For example, Chen et al. (2012) studied the scaling effects of spatial variations of spatial resolution and spatial extent on DLST results by using relationships between vegetation index and LST at/within different resolutions and extents based on two scenes of ASTER data separately covering grassland and cropland. They found that the relationships between vegetation index and LST at different spatial resolutions were very different within different spatial extents. Zhou et al. (2016) directly proved that the scale effect was closely associated with the varying spatial distribution of the LST and corresponding SFs at the coarse and target resolutions. They utilized different linear regression models that were developed for different target resolutions to evaluate the spatial scale effect in DLST processes from moderate resolution sensors (e.g., ASTER TIR data and Landsat 8 TIRS data) to higher resolutions. Their results indicate that the scale effect depended on the variations of SFs, the phenology, and both native and target resolutions. There are many other investigations that proved the scaling effect caused by the variations of both spatial resolution and spatial extent, including Jeganathan et al. (2011), who analyzed the scaling effect through DLSTs from 1 km MODIS LST product downscaled to 810 m, 630 m, . . . , and 180 m resolution LSTs and indicated the scaling effect caused by different spatial resolutions and Ghosh and Joshi (2014) who also confirmed that the scale effect is induced by a spatial variation of the surface characteristics.

6.4.2 Assessing and Reducing Scaling Effect

As discussed earlier, although advanced statistical models and machine learning methods have improved DLST results at finer resolutions via reducing error (spatial effect) between downscaled LSTs and their reference LSTs, the spatial scaling effect still exists on DLST results at finer resolutions. Many researchers have made efforts to quantify the scaling effect in DLST processes by using different measurements and approaches. For example, Chen et al. (2012) used the slope of relationships of VI–LST at different spatial resolutions and within different spatial extents to quantify the scaling effect, while Zhou et al. (2016) also utilized the slope change regression models at different target resolutions and a native resolution. Jeganathan et al. (2011) confirmed the both slope and intercept in linear regression models depending on native and target resolutions. There were more approaches used to measure the scaling effect, including an error estimation method by Chen et al. (2014c), who used an equivalent random sample size and a semivariogram approach by Pu (2021) that when lag distances of spatial resolutions or spatial extents are longer than the range in a semivariogram, the scaling effect may be ignored, but, when the lag distance is shorter than the range of the semivariogram, the effect is statistically significant.

Given the fact that the spatial scaling effect always exists on DLST results in spite of how advanced the DLST methods are, how to develop some methods and techniques that can be used to directly reduce the effect on DLST maps is definitely significant in DLST processing in improving DLST results. However, only one initial effort in the literature was found to try to directly reduce the scaling effect, in which Pu and Bonafoni (2021) developed new approaches with a correction term expression that was used to correct the effect on the basis that a spatial scaling effect on DLST maps is associated with the degree of heterogeneity of surface features/materials within a pixel area at a specific resolution. Their experimental results indicate that the new approaches were effective in improving DLST maps by reducing the scaling effect at VHR resolutions.

6.5 LIMITATIONS AND FUTURE RESEARCH DIRECTIONS

6.5.1 Existing Limitations and Constraints

Many factors determine and influence the success and accuracy of DLST processes. The factors may be considered: reliability and accuracy of various scaling factors (SFs) determined and extracted from high resolution remote sensing and non-remote sensing data, variations of spatial resolution and spatial extent, generalization of DLST models/algorithms, and spatial scaling effect issue. The literature review indicates that existing methods/techniques used for accurate DLST processes under certain environmental settings and backgrounds frequently suffer from some limitations and/or constraints as one or more of the factors cannot meet its or their ideal requirement(s).

There always exists an uncertainty of DLSTs despite how advanced a kernel-driven or fusion-based process DLST model and algorithm applied. Such an uncertainty of DLST may be associated with (1) potential errors in determining, extracting, and selecting SFs (Yang et al., 2011, 2011 (2) errors caused by using a general global scaled model of DLST due to spatial variation changes of spatial resolution and spatial extent (Chen et al., 2012; Gao et al., 2017), and (3) errors caused by nonlinearity of temporal change of LST with SFs and LSTs in DLST processes. To improve efficiency of DLST processing, as discussed in Section 6.3.2, there are many types and characteristics of SFs that can be adopted in DLST processes, in which the potential errors in processing SFs may directly influence the accuracy (uncertainty) of a relationship of LST with SFs at a native (coarse) resolution. For example, Bechtel et al. (2012) used the stepwise procedure to find an optimal SF subset for their downscaling LST scheme, while Tuia et al. (2009) and Zhang et al. (2017) utilized an SVM-recursive feature elimination approach to choose an ideal subset of SFs based on their high-ranking values for their DLST processes. Based on this literature review, most existing kernel-driven DLST models are at a general global scale, which may be suffered from variations of spatial resolution and spatial extent and thus lead to a higher uncertainty of DLSTs (Chen et al., 2012). To optimize DLST accuracy and computational complexity, Gao et al. (2017) adopted an optimal moving-window size as a local window strategy for DLST processing and demonstrated the better performance of the local window strategy over a global window strategy. Currently, most fusion-based models (e.g., STARFM and ESTARFM) assume that the output high spatiotemporal resolution DLST (e.g., Landsat-like LST product) is linearly related to their input LSTs. Although surface cover types and landscape components do not change obviously within a certain time period, reflectance may change slowly over time (e.g., biophysical parameter change), which may result in corresponding LST quick temporal change in a nonlinearity (Mohamadi et al., 2019). To solve the problem, the fusion-based models need to consider the LST temporal change in a nonlinearity to improve the quality of output DLSTs (e.g., the SADFAT may consider using the MODIS annual cycle parameters, discussed by Weng et al. (2014)).

Existing literature review indicates that most DLST models/techniques lack generalization and most DLST studies are study site specific and thus difficult or impossible to extend outside of a study area. Currently, exploring and assessing new methods/techniques for downscaling DLST is mostly conducted in a relatively small test site (area). It is worth noting that only a few of efforts among those studies reviewed in this chapter have used more than one study area for DLST processing (e.g., Liu et al., 2019; Peng et al., 2019; Wu et al., 2019b; Wang et al., 2020; Pu and Bonafoni, 2021; Yin et al., 2021). The performance of the methods/techniques often varied by the characteristics of bio-physical conditions and environmental settings. For example, the methods using VIs, NDVI, or EVI as SFs (for DisTrad and TsHARP) might not work properly for areas with less or no vegetation cover (Mukherjee et al., 2014). Any developed DLST models/methods and findings derived from such a single and small test site may be unreliable and already biased. For this case, there may be two reasons to explain it: (1) a single or small test site cannot fully reflect and account for the spatial variations of bio-physical conditions and environmental settings that a newly developed method/algorithm needs to account for; (2) the collected thermal and optical data from such a single small test site might have been biased (due to a relatively narrow range of thermal component temperature and reflectance variation of surface cover types/materials) toward some favorable conditions. To ensure the generalized capability of DLST methods, one possible solution is to assess them over multiple areas/sites with different conditions and datasets and optimize the related parameters (Yoo et al., 2020).

After a relatively extensive literature review and compared with some traditional DLST methods, many advanced methods and techniques (e.g., machine learning methods) do improve LST downscaled results by reducing the effect on DLST maps to a certain extent (Chen et al., 2012; Pu, 2021). However, as discussed in Section 6.4, a certain spatial scaling effect still exists in DLST processes, especially at high resolutions. Therefore, it is interesting to us that developing a new approach to directly correct or compensate for the spatial effect on DLST maps is a research direction to improve DLSTs at finer resolutions.

6.5.2 Future Research Directions

According to this comprehensive review of the TIR remote sensing data downscaling investigations published during the last four decades, the future research directions of thermal downscaling may be drawn as follows: further reducing the uncertainty of DLST results, developing novel DLST models and algorithms, and directly reducing the spatial scaling effect in DLST processes.

It should be one general direction that continually developing some methods and techniques reduces the uncertainty of DLST results by considering selecting an optimal subset of SFs, establishing a DLST model with a localizing moving window strategy, and adopting a nonlinearity of temporal change when developing a fusion-based model. Currently, a few of studies have tested some novel methods and techniques and demonstrated that using a comprehensive and optimal subset of SFs (e.g., Tuia et al., 2009; Zhang et al., 2017), a localizing moving window strategy (Gao et al., 2017; Peng et al., 2019), and developing a fusion-based model with nonlinearity of temporal change over time (e.g., Weng et al., 2014 for SADFAT model) can result in significant improvement of DLST results (i.e., reducing uncertainty of DLSTs significantly). However, more efforts still need to be made in the future in developing such novel methods and techniques in downscaling coarse spatial LST to finer resolution LSTs.

More specific future research directions of developing novel methods and techniques may focus on developing or refining existing advanced models and methods, such as deep learning, local and temporal weighted autoregressive regression (GTWAR), and a weighted combination of kernel-driven and fusion-based methods (termed CKFM) to enhance the spatiotemporal resolutions of time series LSTs. As discussed in Section 6.3.3, while most researchers developed and used various statistical models for coarse resolution LST downscaling, a few of them have utilized machine learning methods. For example, a CNN model has been used in LST downscaling in an image fusion-based method by Yin et al. (2021), and a more advanced deep-learning model (i.e., generative adversarial network (GAN)) could improve the image pan-sharpening performance more than the CNN (Liu et al., 2018c; Zhang et al., 2021a). Luo et al. (2021) proposed a GTWAR approach to downscale the 1 km MODIS LST to 100 m. In this model, they considered the temporality, spatial heterogeneity, and spatial autocorrelation of the LST and SF data simultaneously. The GTWAR downscaling model could facilitate improvements in downscaling LST in time series. To further improve high spatiotemporal Landsat (or other fine spatial resolution)-like LST time series products by using the spatiotemporal fusion-based method, Xiao et al. (2019) proposed the CKFM model to enhance the spatiotemporal resolutions of time series LSTs by compensating the limitations kernel-driven process and fusion-based process methods separately in DLST processes. The kernel-driven process can offer spatial details from high resolution optical images, while the fusion-based process can create spatiotemporal information. Their testing results indicate that the CKFM is more accurate and robust than either the kernel-driven method or fusion-based method and it can reconstruct more spatial details than the fusion-based process method to generate a daily high spatial resolution LST series for various environmental studies.

Those advanced and novel methods and techniques have demonstrated their excellent performance and robustness. However, more efforts are still needed. For example, for the machine learning methods, their performances could be further enhanced if additional input data such as in-situ LST or other numerical model data are used together for a CNN and some up-to-date deep learning approaches to generate a more reliable DLST product in the future. Per the GTWAR model, the SFs are not comprehensive and optimal and the temporal change is too narrow. For the CKFM model, it may be modified with a temperature cycle model (e.g., diurnal/annual temperature cycle models) to further improve the high spatiotemporal resolution time series LST product. In developing a novel spatiotemporal fusion-based method, how to deal with spatial heterogeneity of landscapes (leading to spatial variation of LST) and temporal change (inducing temporal variation of LST) between any two time points should be a key point for a research direction.

Given that the scaling effect always exists in DLST processes, especially at very high resolutions, how to directly reduce the effect by developing some novel methods should be another direction and beneficial to many applications that need high resolution DLST data. The initial effort was made by Pu and Bonafoni (2021, 2022) who justified that the spatial scaling effect on DLST maps is related to the degree of heterogeneity of landscapes associated with spatial changes of surface features/materials within pixels at specific resolutions. This issue was proved early by Garrigues et al. (2006a, 2006b) by quantifying the heterogeneity within a pixel area for mapping NDVI and LAI with moderate resolution remote sensing data and non-linear estimation processes. Garrigues et al. (2008) also employed a geostatistical linear model to assess the landscape spatial heterogeneity with high resolution red and NIR band images. Based on the theory, Pu and Bonafoni (2021) developed a novel method with a correction term expression to directly correct the effect on DLST maps. The analysis results indicate that the new method could significantly reduce the root mean square error and the mean absolute error >30% and >33%, respectively and thus demonstrate that the method was effective in correcting the scaling effect on DLST maps at VHR resolutions.

6.6 SUMMARY

Land surface temperature (LST) retrieved from moderate resolution or downscaled from coarse thermal infrared (TIR) data is one of the key Earth surface processes parameters. Over the last four decades, most advanced remote sensing sensors/systems can acquire TIR data at a low spatial resolution but high temporal resolution. However, per different application purposes, both high spatial and temporal resolution TIR data are needed. In this chapter, an introduction to and discussion on TIR sensors/systems data and existing DLST methods and techniques was conducted, especially for the investigations on applications and research of DLST processes with low spatial resolution but high temporal resolution polar and geostationary orbit TIR sensors' data. Therefore, in the Introduction section, we indicate the significance of high resolution thermal remote sensing data in many applications and DLST. Then we introduce the process sensors/systems of moderate and coarse resolution polar orbit sensors currently used for DLST as well as geostationary orbit sensors. In the techniques and methods of DLST section, we first introduce methods of collecting reference data: the field-based (in-situ, UAV-based LST, and TIR radiance measurements) approach and TIR/LST image-based approach, then introduce approaches of types and selection of scaling factors, and finally introduce and discuss five types of DLST algorithms and methods: (1) classic-based methods, (2) regression-based methods, (3) machine learning (ML)-based methods, (4) thermal spectral unmixing (TSU)-based methods, and (5) spatiotemporal fusion-based methods. In the next section, the spatial scaling effect in downscaling LST processes was reviewed and assessing and reducing spatial scaling was introduced and discussed. And the last section reviewed existing limitations and constraints and pointed out future research directions.

APPENDIX

TABLE 6.5
Abbreviations and Acronyms Used in This Chapter.

AATSR	Advanced Along-Track Scanning Radiometer
AGRI	Advanced Geosynchronous Radiation Imager
ATSR	Along-Track Scanning Radiometer
ANN	artificial neural network (model)
ASTER	Advanced Spaceborne Thermal Emission and Reflection Radiometer, sensor
AVHRR	Advanced Very High Resolution Radiometer
CNN	convolutional neural networks
DisTrad	Disaggregation of Radiometric Temperature
Envisat	Environmental Satellite
ERS	European Remote Sensing Satellite
ESTARFM	enhanced spatial and temporal adaptive reflectance fusion model
ETM+	Enhanced Thematic Mapper Plus
FSDAF	flexible spatial and temporal adaptive reflectance fusion model
GAN	generative adversarial network
GOES	Geostationary Operational Environmental Satellite
INSAT	Indian National SATellite
LSE	land surface emissivity
LST	land surface temperature
LULC	land use/land cover
ML	machine learning
MODIS	MODerate-resolution Imaging Spectroradiometer
MSG	Meteosat Second Generation
NDVI	normalized difference vegetation index
NIR	near-infrared
RF	random forest
SEVIRI	Spinning Enhanced Visible and Infrared Imager
SF	scaling factor (kernel)
SLSTR	Sea and Land Surface Temperature Radiometer
STARFM	spatial and temporal adaptive reflectance fusion model
SWIR	shortwave infrared
SVM	support vector machine
TIR	thermal infrared (3–14 mm)

(Continued)

TABLE 6.5 *(Continued)*
Abbreviations and Acronyms Used in This Chapter.

AATSR	Advanced Along-Track Scanning Radiometer
TM	thematic mapper
TsHARP	Temperature Sharpening
UAV	unmanned aerial vehicle
UHI	urban heat island
VHRR	Very High Resolution Radiometer
VIRR	Visible and Infrared Radiometer
VNIR	visible-near infrared (0.4–1.0 mm)

7 Thermal Infrared Remote Sensing Applied in Hydrology

Remote sensing offers a powerful method for observing hydrological state variables over large areas. To address diverse research and application requirements, comprehensive reviews of thermal infrared (TIR) remote sensing studies in hydrology are essential. These studies encompass crucial tasks such as estimating surface soil moisture, evapotranspiration, and mapping ice/snow areas. This chapter provides a concise overview of the motivations behind conducting TIR remote sensing studies in hydrology (Section 7.1), the remote sensing of hydrometeorological state variables (Section 7.2), methods and techniques for soil moisture estimation from TIR data (Section 7.3), methods and techniques for evapotranspiration estimation from TIR data (Section 7.4) and the mapping of ice/snow areas using TIR images (Section 7.5). Each section introduces, reviews, and discusses these topics in detail.

7.1 INTRODUCTION

Remote sensing is the process of inferring surface parameters from airborne or spaceborne observations of the upwelling electromagnetic radiation from the land surface. This radiation includes both reflected and emitted energy by the land. The former specifies usually the reflected solar radiation while the latter is emitted radiation energy in both thermal infrared (TIR) and microwave portions of the spectra. Usually, the reflected solar radiation is used in hydrology for snow mapping, vegetation/land cover, and water quality studies. The emitted thermal radiation in the TIR spectral range is used for land surface temperature (LST) and in the microwave for soil moisture and snow studies. We will concentrate on the TIR radiation for surface temperature and energy balance in the hydrological studies. Remotely sensed observations can contribute to our knowledge of these quantities in relatively high spatial and temporal resolutions. Therefore, with remote sensing we can only observe the surface but can obtain the spatial variability and, if observations are made repeatedly, the temporal variability. In this chapter we will concentrate on those applications of TIR remote sensing that we believe are the most important in hydrology, including estimating soil moisture, evapotranspiration, mapping ice/snow, etc. A major focus of remote sensing research in hydrometeorology has been to develop approaches for estimating hydrometeorological states and fluxes. The primary set of state variables include land surface temperature, near surface soil moisture, and ice/snow cover. The hydrometeorological fluxes are primarily soil evaporation and plant transpiration or evapotranspiration and snowmelt runoff.

DOI: 10.1201/9781003276173-7

The methods have been used to quantify the components of the water and energy balance equation using TIR remote sensing. The water balance is commonly expressed as follows:

$$\frac{dS}{dt} = P - ET - Q \tag{7.1}$$

where dS/dt is the change in water storage in the soil and/or snow layer; P is the precipitation; ET is the evapotranspiration; and Q is the runoff. Because the energy and water balance at the land surface are closely linked, we can also consider the energy balance equation for most landscapes, except for tall forests, which is typically expressed as:

$$R_N - G = H + LE \tag{7.2}$$

where R_N is the net radiation; G is the soil heat flux; H is the sensible heat flux; and LE is the latent heat flux all in Wm^{-2}. The quantity R_N–G is commonly referred to as the available energy, and ET and LE represent the same water vapor exchange rate across the surface–atmosphere interface, except ET is usually expressed in terms of depth of water over daily and longer time scales, namely mm/day.

7.2 REMOTE SENSING OF HYDROMETEOROLOGICAL STATES VARIABLES

7.2.1 Land Surface Temperature (LST)

LST is the result of the equilibrium thermodynamic state dictated by the energy balance between the atmosphere, surface, and subsurface soil and the efficiency by which the surface transmits radiant energy into the atmosphere (surface emissivity). The surface emissivity depends on the composition, surface roughness, and physical parameters of the surface, e.g., soil moisture content. In addition, the emissivity generally will vary with wavelength for natural landscape surfaces. Thus, to make a quantitative estimate of the LST, we need to separate and know the effects of temperature and emissivity in the observed radiance. Airborne and spaceborne-based radiometers measure what is commonly called a 'brightness temperature' derived from the radiance reaching the sensors from ground surfaces. In general, the brightness temperature is affected by atmosphere and needs to be corrected for atmospheric attenuation (scattering and absorption), and the surface radiance considers the impact of surface emissivity before it can be regarded as an estimate of the LST.

The relationship between land surface and brightness temperature from an airborne or satellite-based sensor is usually expressed in terms of the radiation balance (Price, 1983):

$$L_{i,sensor} = L_{i,surface}\tau_i + L^{\uparrow}_{i,atm} \tag{7.3}$$

where L is the radiance from the ith waveband of the radiometer; $L_{i,sensor}$ is the at-sensor radiance; $L_{i,surface}$ is the surface radiance; $L^{\uparrow}_{i,atm}$ is the upwelling atmospheric radiance; and τ_i is the atmospheric transmission. Values of $L^{\uparrow}_{i,atm}$and τ_i can be calculated using atmospheric radiative transfer codes, such as MORTRANM 5 (Berk et al., 2005). This permits the upwelling radiance at the surface, which yields the LST, to be computed from the following expression:

$$L_{i,surface} = \varepsilon_i L_{BB}\left(\lambda_i, T_s\right) + \left(1 - \varepsilon_i\right) L^{\uparrow}_{atm} \quad (7.4)$$

where ε_i is the surface emissivity for the ith waveband; $L_{BB}\left(\lambda_i, T_s\right)$ is the Planck equation for the radiation from a blackbody at surface temperature T_s and λ_i is the central wavelength for the ith waveband of the radiometer. The value of $L^{\uparrow}_{atm}$ can also be determined from atmospheric radiative transfer codes. The remaining problem is to relate these radiances to the surface emissivity without direct knowledge of the land surface temperature (LST).

To retrieve LST from at sensor brightness temperature, due to the lack of adequate atmospheric profile observations, the development of alternative approaches such as so-called split window methods would be more operationally applicable (e.g., Price, 1984). These split window methods employ two channels at slightly different wavelengths,λ_1 and λ_2 in Eqs. (7.3) and (7.4) to essentially eliminate (using a few approximations) the need for estimating the atmospheric transmission and radiances. This approach has been used quite successfully over oceans where the spectral variation of the emissivity is small. Over land this is not the case and the split window methods are sensitive to spectral variation in the emissivities in the two wavebands; for example, at a brightness temperature 300 K, a difference $\varepsilon_1 - \varepsilon_2 \sim 0.1$ can yield an error in LST of ~2 K (Price, 1984). There has been much recent work on the use of these split window methods (see Chapter 5 for more split window methods) using the two thermal channels of the advanced very high-resolution radiometer (AVHRR). This approach is being used with the multi-spectral TIR data from the moderate resolution imaging spectroradiometer (MODIS) (e.g., Pan et al., 2020). They use a database of land surface emissivities based on land cover to correct for emissivity effects.

A recent example using MODIS LST products mosaicked in Northeast China in 2015 is displayed in Figure 7.1. The spatial distribution of land surface temperature (LST) reflects some significant temperature differences between farmland types and among different months in the 2015 grow season with the paddy field in blue with relatively low temperature and the rain-fed farmland in yellow in relatively high temperature. This LST type of spatially distributed information is very useful for evaluating spatial patterns of ET over large areas.

7.2.2 Surface Soil Moisture (SM)

Soil moisture (SM) is a critical factor in various hydro-meteorological studies and is used as a key variable in several applications such as drought severity and duration, irrigation scheduling, soil erosion, evapotranspiration, and agriculture management.

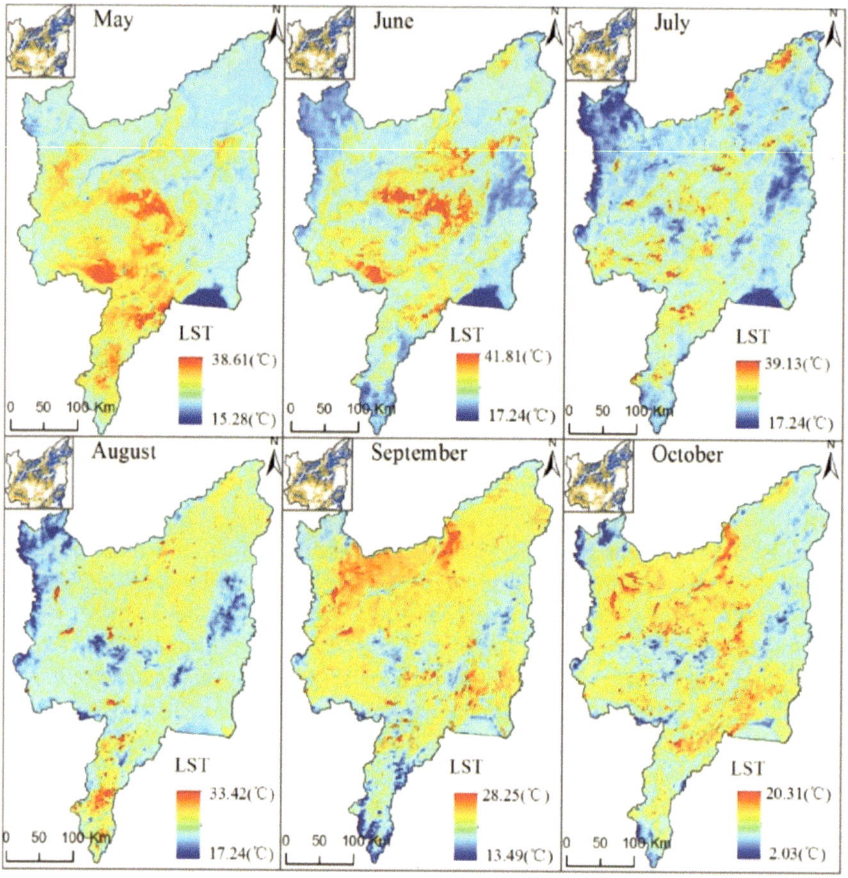

FIGURE 7.1 Maps of monthly LST throughout crop growth season, an example year in 2015. The small location map at the upper left corner of each figure represents the spatial distribution of paddy field and rain-fed farmland, in which the blue indicates paddy field and the yellow indicates rain-fed farmland.

Source: Pan et al., 2020.

Although direct measurement is the most accurate method for estimating soil moisture, this technique is expensive, time consuming, and only provides point measurements. Therefore, it may not represent the spatial distribution of SM and is not available for continuous spatial and temporal coverage at regional and global scales. Technological advances in airborne and satellite remote sensing have offered an alternative to field measurements of SM and enabled us to monitor it at higher temporal and spatial resolutions at lower cost and in a short time at regional and global scales. The development of airborne and satellite technology provides a powerful method for monitoring the spatial-temporal variations of regional SM and provides quantitative estimations.

Many experts have established different methods for retrieving SM based on the relationships between the SM and airborne and satellite-derived land surface parameters. Using the visible infrared band, the SM can be derived from the relationships between soil reflectance and different SMs or vegetation traits that occur under water stress. These methods can be divided into two categories, (1) the single spectral analysis method and (2) the vegetation index method. An optical remote sensing method known as surface temperature/vegetation index method is a promising approach to estimate SM as LST and vegetation have been found to have a complicated dependence on SM. The triangle method is proposed to extract the characteristics of land surface parameters based on integration between land surface temperature (LST) and multispectral indices. By referring to the relationship between LST and the normalized difference vegetation index (NDVI), temperature vegetation dryness index (TVDI) was proposed to examine soil moisture and monitor drought conditions (Le and Liou, 2021).

For thermal infrared data, the parameters related to the soil thermal properties are mainly used to derive the SM. This method includes two categories, (1) the thermal inertia method and (2) the temperature index method. Thermal inertia is the intrinsic property of materials and describes their impedance to variations of temperature. When soil moisture increases, the thermal inertia of soil increases, and there is a lower diurnal temperature fluctuation. Based on this principle, a few models for determination of thermal inertia have been developed (e.g., Xue and Cracknell, 1995; Tramutoli et al., 2000), and the thermal inertia method is routinely used to estimate soil moisture in bare soil and sparsely vegetated regions (e.g., Verhoef, 2004; Tramutoli et al., 2000). Combining visible and thermal infrared remotely sensed data can provide more information for estimating soil moisture than the single one. SM content controls soil density, heat capacity, thermal conductivity, and other characteristics, and soil thermal inertia is closely related to soil moisture (Cheruy et al., 2017). The apparent thermal inertia (ATI), which is simplified on the basis of thermal inertia, is simple to calculate and has been widely used to characterize SM (Lu et al., 2009). The calculation formula of the ATI model only involves surface albedo and a night–day surface temperature pair parameter, which can be directly derived from remote sensing data. Therefore, the ATI model has become one of the most commonly used methods for retrieving SM in bare soil and low vegetation areas by TIR technology.

Beyond the more traditional modelling approaches (e.g., linear and non-linear regression methods), recently developed machine learning (ML) methods offer some alternative approaches. The popularity of ML for the remote-sensing based estimation of biophysical parameters, such as estimate SM, has grown significantly over the last decade (Ali et al., 2015). ML refers to automatic or semi-automatic exploration and analysis of large datasets, in order to discover meaningful correlations, patterns, and rules among data. ML approaches are now commonplace and have been successfully applied for predicting SM using remotely sensed data in the semi-arid region. The flexibility of ML approaches is further highlighted by their potential to be used in various hydro-meteorological applications, from the prediction of SM to precipitation, temperature, or wind (Ahn and Palmer, 2016; Tyralis et al., 2019). Compared to more traditional approaches, these methods have two significant advantages: (1) they enable the construction of more objective, purely data-driven retrieval models, independent of necessary assumptions and (2) they allow the combination of data from different sources (like the combination of optical and TIR remote sensing), exploiting their relationship with a target variable. There are various ways to incorporate ML in the estimation approach. Often, it is applied for model inversion or downscaling purposes. For example, Moosavi et al. (2016) presented an example of how these concepts can be applied to MODIS and Landsat imagery for the downscaling of LST measurements and the estimation of high resolution (100 m) SM content by applying a support-vector-regression (SVR) and an adaptive neuro-fuzzy inference system for model inversion.

7.2.3 Ice/Snow Areas

The occurrence of precipitation in the form of snow typically causes a change in how a drainage basin responds to the input of water. There is such a vast difference in the physical properties of snow and other natural surfaces that the occurrence of snow on a drainage basin can cause significant changes in the energy and water budget. As an example, the relatively high albedo of snow reflects a much higher percentage of incoming solar shortwave radiation than snow-free surfaces (80% for relatively new snow as opposed to roughly 15% for snow-free vegetation). Snow cover can be detected and monitored with a variety of remote sensing devices. The greatest number of applications have been found in the VNIR region of the electromagnetic spectrum. Many users have used Landsat, SPOT, AVHRR, ASTER, and MODIS sensors' observations to map ice/snow areas.

Data from thermal satellite sensors, such as MODIS, have previously been used to map polynyas and leads. Some work has been done using thermal optical satellite data to estimate ice thickness in the Antarctic. For example, Ciappa and Budillon (2012) and Paul et al. (2015a, 2015b) used thermal MODIS data to estimate ice thickness of thin ice in polynyas in Terra Nova Bay and the Southern Weddell Sea. They used a surface energy balance model to estimate the ice thickness. In Rudjord et al. (2022), they presented an automatic method for retrieving the thickness of thin sea ice. The method relies on an approach described elsewhere where the ice thickness is estimated by considering the heat balance on the ice surface, but they here further developed and adapted for the Svalbard region. The region around Svalbard and the Barents Sea is a challenging region for mapping ice thickness. Due to ice exchange with the Kara Sea and the Arctic Ocean, the composition of the ice may at times be very complex, consisting of both local and imported ice of various origin and age.

Impassable and inaccessible of glaciers is one of the most important reasons for using satellite imagery instead of field operations. Moreover, many glaciers and ice sheets, are very large and this makes it difficult, if not impossible, to obtain data at a reasonable spatial and temporal resolution. In addition to the problems enumerated to the access of glacier areas, some properties of remote sensing techniques have led to the increasing application of them in glaciology. Some TIR remote sensing characteristics are as follows: wide coverage of satellite images, obtained repetitive satellite images with arbitrary return periods, and cost effectiveness of satellite images compared with field operations. These factors increased in glaciers naturally led to the development of suitable observations techniques, which even at the early stages already involved the acquisition of remotely sensed data. Glacier area is a fundamental requirement in a glacial inventory, and it is frequently used to determine total glacial budget in the world. In several climate models that calculate the energy fluxes according to the surface conditions, the glacier area and ice caps are of the most important factors. Since the glacier surface has a higher reflectance compared to the other objects around it, extracting the glacier area in aerial photo and remote sensing optical images is simpler than its other parameters. So, several methods have been invented to extract the glacier limited area by means of optical remote sensing imagery such as manual digitization of glacier perimeter, band rationing methods, image classification techniques (including supervised and unsupervised classification algorithms), and using vegetation indices (e.g., NDSI (Normalized Difference snow Index)). However, because the supraglacial and periglacial covers have the same reflectance characteristic (due to the fact that they originate from the same source) optical imagery can't separate them from each other. Hence, several methods have been developed to overcome this problem, which can be classified into three general categories: the geomorphometric-based methods, radar imagery-based methods, and thermal-based methods. Thermal techniques are based on the sufficiently lower temperature of the debris-covered glacier from adjacent periglacial debris outside the glacial margins. Some studies have shown that thermal-based methods are effective when thickness of debris cover did not exceed 40–50 cm. Several researchers have combined morphometric and temperature information to overcome this problem and their results were better than the other methods mentioned.

Rock glaciers and debris-covered glaciers are distinct types of ice-debris landforms and elements of the mountain cryosphere in which the high ice contents are not visible at the surface, creating challenges for remotely-sensed mapping of these features. Thermal remote sensing provides a set of variables that are useful for characterizing thermal and thermophysical surface characteristics of the natural and built environment. Brenning et al. (2012) explored the utility of remotely-sensed land surface temperature (LST) and derived apparent thermal inertia (ATI) for thermally characterizing and discriminating periglacial mountain environments, particularly rock glaciers and debris-covered glaciers. They demonstrated their approach utilizing ASTER daytime and nighttime data in a study area in the Andes of Central Chile as successful, where in-situ near-surface ground temperature (NSGT) measurements are available for comparison with remotely-sensed LST.

7.3 ESTIMATING SOIL MOISTURE FROM TIR REMOTE SENSING DATA

Based on the optical and thermal infrared remote sensing, many approaches have been developed by establishing relationships between SM and soil reflectivity or surface temperature/vegetation coverage and soil thermal properties. Although the signal penetration capabilities of optical methods are worse than those of thermal infrared and microwave methods, the reflectance domain of the optical methods is the most operational because the images are easily available over a broad range of ground resolutions. Combining visible and thermal infrared remotely sensed data can provide more information for estimating soil moisture than the single one. In this section, SM estimation methods relying on optical and thermal remote sensing are collated and summarized. Optical remote sensing spectral/vegetation index-based methods are first introduced. Next, the physical properties and advantages of triangle-trapezoid and thermal inertia methods in estimating SM are presented in detail. Finally, we introduce machine learning (ML)-based methods in SM estimation.

7.3.1 Spectral/Vegetation-Based Methods

Because some vegetation indices (VIs) are extremely sensitive to water stress, estimates by remote sensing have been widely used to detect drought conditions, such as the normalized difference vegetation index (NDVI) and enhanced vegetation index (EVI). Drought conditions can be reflected by vegetation growth, which can be directly represented by the VIs. By comparing the vegetation index at different times, the impacts of drought can easily be determined. However, the accuracy of using NDVI-based drought indices is often subject to changes in vegetation types and ecosystem differences. Many classical drought indices based on vegetation indicators were proposed and have been widely used in drought monitoring. To remove the influences of weather and geographic location differences on NDVI, Kogan (1990, 1995) developed the following Vegetation Condition Index (VCI, Table 7.1) based on statistical NDVI time-series data, where VCI is a modified NDVI vegetation condition index; NDVI is a weekly smoothed composite NDVI; $NDVI_{min}$ and $NDVI_{max}$ are the maximum and minimum values of NDVI, respectively, which can be obtained from historical data; *i* and *j* denote weeks and locations. VCI not only describes the surface vegetation cover and spatial-temporal variations but also quantifies the impacts of weather on vegetation type. Furthermore, VCI can be used to evaluate the effects of weather on different ecosystems.

Because NDVI's sensitivity to water differs, the relationships among different water absorption bands were applied to derive the SM. Gao (1996) presented a normalized difference water index (NDWI, Table 7.1) using the reflectance at 1.24 μm rather than 0.66 μm because the former is more sensitive to SM. More importantly, the NDWI is insensitive to the atmospheric conditions.

For a specific soil type, the soil line can be regarded as a line that characterizes the spectral behavior of non-vegetated pixels when the soil moisture content varies noticeably (Ghulam et al., 2007). The PDI (Table 7.1) describes the SM distribution in the NIR-red space. The points farther away from the coordinate origin indicate low soil moisture areas. In contrast, the closer points represent wet conditions. The PDI is suitable for bare soil but has many limitations in vegetated areas, especially for non-flat regions with different soil types. Ghulam et al. (2007) introduced fractional vegetation coverage (FVC) for the removal of vegetation information from mixed pixels and proposed the Modified Perpendicular Drought Index—MPDI (Table 7.1). f_v is the vegetation cover. By considering the SM status and vegetation growth, the MPDI can be applied to vegetated regions to obtain credible results. Because the vegetation index cannot immediately reveal information when the vegetation is restricted by water stress, a long period is required to show the water-limited features after some accumulation; thus, this method requires a certain time lag to show real-time SM conditions and thus VCI is necessary. Therefore, good quality satellite data acquired

TABLE 7.1
A Summary of Soil Moisture and Drought Indices Frequently Used in Soil Moisture Estimation (Edited from Sadeghi et al., 2017)

Index	Full name	Formula	References
VCI	Vegetation Condition Index	$VCI_{ij} = 100(NDVI_{ij} - NDVI_{min,j}) / (NDVI_{max,j} - NDVI_{min,j})$	Jiao et al. (2016)
SIWSI	Shortwave Infrared Water Stress Index	$SIWSI = (R_{SWIR} - R_{NIR}) / (R_{SWIR} + R_{NIR})$	Fensholt and Sandholt (2003)
ESMI	Ecological Safety Monitoring Index	$ESMI = NDVI/\alpha$	Ghulam et al. (2004)
PDI	Perpendicular Drought Index,	$PDI = (R_{red} + MR_{NIR})\sqrt{1+M^2}$	Ghulam et al. (2007)
MDPI	Modified Perpendicular Drought Index	$MDPI = (PDI - f_v PDI_V) / 1 - f_V)$	Ghulam et al. (2007)
SPDI	Shortwave-infrared Perpendicular Drought Index	$SDPI = (R_{SWIR} + MR_{NIR}) / \sqrt{1+M^2}$	Ghulam et al. (2007)
VCADI	Vegetation Condition Albedo Drought Index	$VCADI = (\alpha - i_w + s_w NDVI) / (i_d - i_w + (s_d - s_w) NDVI)$	Ghulam et al. (2007)
DDI	Distance Drought Index	$DDI = \sqrt{R_{red}^2 + R_{NIR}^2} / (1 + NDVI)$	Qin et al. (2010)
MSPDI	Modified Shortwave-Infrared Perpendicular Drought Index	$MSPDI = \dfrac{R_{\Sigma} + MR_{\Delta}}{1 - M^2}$	Feng et al. (2013)
VSDI	Visible and Shortwave-Infrared Drought Index	$VSDI = 1 - (R_{SWIR} + R_{red} - 2R_{blue})$	Zhang et al. (2013)
TSMI	Triangle Soil Moisture Index	$TSMI = c_0 + \Sigma_{i=1}^{10} c_i p_i$	Amani et al. (2016)
TVDI	Temperature/Vegetation Drought Index	$TVDI = (T_s - T_{s,min} / a + bNDVI - T_{s,max})$	Sandholt et al. (2002)
TMDI	Temperature-Soil Moisture Dryness Index	$TMDI = (T_s - T_{s,min}) / T_{s,max} - T_{s,min})$	Le and Liou (2022)
NDWI	Normalized Difference Water Index	$NDWI = (R_{NIR} - R_{SWIR}) / (R_{NIR} + R_{SWIR})$	Gao (1996)

(Continued)

TABLE 7.1 (*Continued*)

A Summary of Soil Moisture and Drought Indices Frequently Used in Soil Moisture Estimation (Edited from Sadeghi et al., 2017)

Index	Full name	Formula	References
NDLI	Normalized Difference Latent Heat Index	$NDLI = (R_{green} - R_{red}) / (R_{green} - R_{red} + R_{SWIR1})$	Le and Liou (2022)

Note: R_{red}, R_{blue}, R_{grren}, R_{NIR} and R_{SWIR}: reflectance for red, blue, green, NIR, and SWIR bands, respectively; α: broadband albedo; NDVI: normalized difference vegetation index, $R_{\Sigma} = R_{SWIR} + R_{red}$, $R_{\Delta} = R_{SWIR} - R_{red}$; M: slope of the soil line in the $R_{red} - R_{NIR}$ or $R_{SWI}R - R_{NIR}$ or $R_{\Sigma} - R_{\Delta}$ space; f_v: fractional vegetation cover; PDI_v: PDI of sole vegetation; i_d and s_d: intercept and slope of the dry edge in the α–NDVI space, respectively; i_w and s_w: intercept and slope of the wet edge in the α – NDVI space, respectively; p_i (I = 1 to 10): either a distance or an angle associated with the location of a random pixel in the $R_{red} - R_{NIR}$ space; c_i: regression coefficient of p_i; T_s: land surface temperature.

over a long period of many years must be obtained, and the obtained results only reflect the SM deficit situation and not the true soil moisture content.

Many research groups have successfully used the temperature vegetation dryness index (TVDI, Table 7.1) for SM status assessment with different satellite-based multispectral imageries. However, the TVDI has certain limitations. The normalized difference latent heat index (NDLI, Table 7.1) has already been proven as a potential index for representing the surface water availability by optimizing the spectral signatures of water at three bands, including green, red, and SWIR in the incorporation (Liou et al., 2019). Since the SM is associated with its temperature and water availability, a simple approach is proposed by interpreting the relationship between LST and NDLI. Within LST-NDLI Trapezoidal Space, a new temperature-soil moisture dryness index (TMDI, Table 7.1) to quantify surface SM is proposed (Le and Liou, 2022). It is defined as a function of land surface temperature variation and its relationship with surface water availability. The spatial pattern of TMDI has been analyzed over two seasons (dry and rainy) for the plain area of Tainan, Taiwan, using Landsat 8 OLI/TIRS image acquired on January 20 and October 19, 2014. The results indicated that TMDI exhibited a significantly negative correlation with the simulated evapotranspiration (ET) and positive correlation with in situ measured the near-surface air temperature (T_{air}) from seven stations ($r = -0.95$ and -0.9 for simulated ET and $r = 0.94$ and 0.78 for T_{air} corresponding to January 20 and October 19, respectively). It was further compared to the performance of the TMDI with the existing remotely sensed dryness assessment methods, including the temperature vegetation dryness index (TVDI) and surface energy balance algorithm for land (SEBAL) model. The advantages of the TMDI in reflecting surface SM, especially in the non-vegetation area, are clearly demonstrated. It is concluded that the TMDI is a reliable indicator for determining the surface SM status with a large degree of freedom for further applications because it does not require any other ground-based measurements.

Table 7.1 sums majority of soil moisture and drought indices frequently used in SM estimation, edited from Sadeghi et al. (2017). Of them, some are directly used in triangle-trapezoid methods and some in ML-based methods.

7.3.2 Triangle-Trapezoid-Based Methods

The triangle method is based on an interpretation of the image (pixel) distribution in surface radiant temperature (LST) to fractional vegetation cover (come from NDVI) space. A triangle emerges because the range of surface radiant temperatures decrease as the vegetation cover increases (Figure 7.2); a vertex emerges showing the narrow range of surface radiant temperature over the dense vegetation (Carlson, 2007). Vegetation subject to that water stress usually experiences an increase in its radiant temperature compared to vegetation that is well watered. There are, however, issues associated with the triangle method to be resolved. First, Moran et al. (1994) reported that the triangular space, in which the driest and wettest fully vegetated surfaces have the same LST, does not account for the impact of water stress on canopy transpiration and replaced the triangle with a trapezoid. In the trapezoidal space, the temperature of the extreme dry fully vegetated surface is higher than that of the

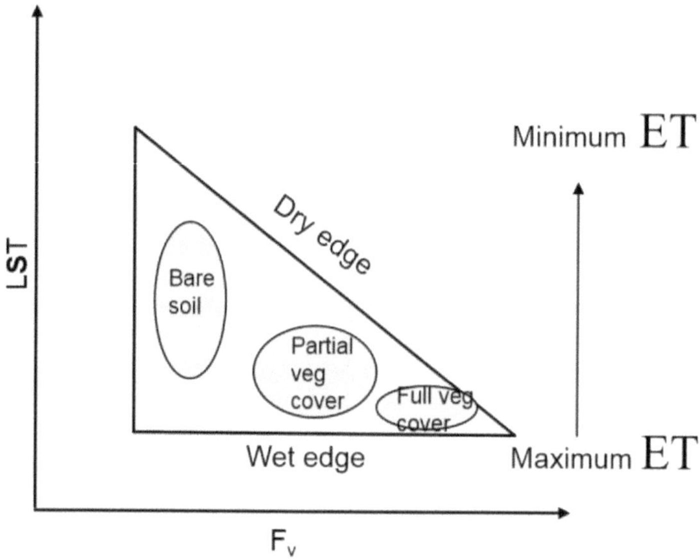

FIGURE 7.2 The idealized triangle space between LST and Fv.

wettest surface with full vegetation cover but is lower than extreme dry bare soil surface. Second, traditional approaches for determining the extreme boundaries of the triangle (or trapezoid) space are based on visual examination of the VI/LST data points, which implicitly requires that both conditions of extreme dry (lowest SM) and extreme wet points (highest SM) be present in a study scene to ensure the validity of the triangle (or trapezoid) method (Carlson, 2007). This is often not the case in reality. Moreover, substantial subjectivity results from the observed boundaries as well, because they can differ among users for the same study area. Furthermore, visual examination of extreme point boundaries would largely increase 'labor work' and therefore severely reduces the operability of the method. Third, one major assumption of the triangle (or trapezoid) method is that among-pixel variation in surface energy flux is primarily controlled by that in water availability but is not affected by differences in atmospheric conditions. As a result, the triangle (or trapezoid) method is applicable for areas with relatively homogeneous meteorology.

Compared with the triangular space, the trapezoidal space (Figure 7.3) does not require a large number of pixels. The advantages of the trapezoidal method are that (1) it has a robust physical basis, and Yang et al. (2015) demonstrated that the trapezoid is more theoretically sound than using other shapes and that (2) the space determined from the four vertices is the main limiting condition that is close to the true conditions of the land surface. In contrast, the disadvantages include that this method (1) requires more ground-based parameters to calculate the soil moisture index and that (2) the influences of water stress on vegetation cover have a time lag that implies uncertainties for SM estimations. The 'trapezoid' or 'triangle' model constitutes the most popular approach to remote sensing (RS) of surface SM based on coupled thermal (i.e., LST) and optical RS observations. The model, hereinafter

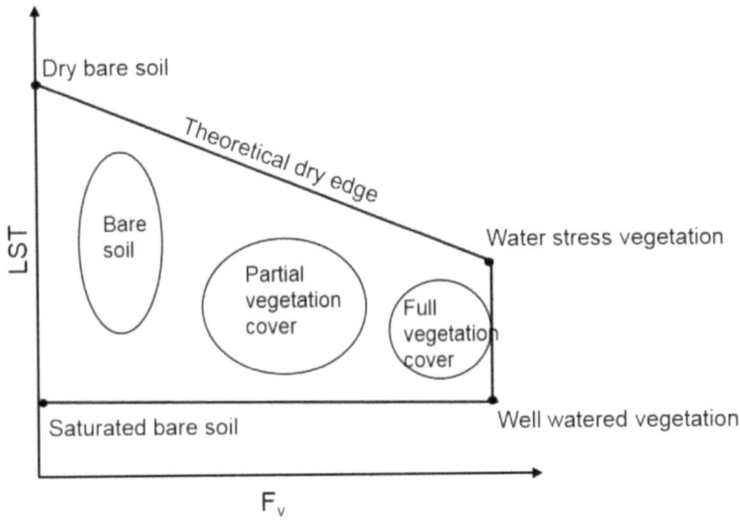

FIGURE 7.3 The simplified trapezoidal space between LST and Fv.

termed the thermal-optical trapezoid model (TOTRAM) (Sadeghi et al., 2017), is based on the interpretation of the pixel distribution within the LST-VI space, where LST is the land surface temperature and VI is a RS-based vegetation index. The most common vegetation index used in TOTRAM is the NDVI.

An inverse linear relationship between surface soil moisture (θ) and LST is then assumed:

$$SM = \frac{\theta - \theta_d}{\theta_w - \theta_d} = \frac{LST_d - LST}{LST_d - LST_w} \tag{7.5}$$

where SM is the soil moisture content normalized by the local minimum dry soil moisture content, θ_d, and the local maximum wet soil moisture content, θ_w; the LST_d and LST_w terms are the LSTs of the dry and wet soil, respectively, where LST_d and LST_w are obtained from the LST-NDVI trapezoid (Figure 7.4) for a specific location (satellite scene). The upper (dry) and lower (wet) edges of the trapezoid are used to solve for LST_d and LST_w at any given NDVI (i.e., fractional vegetation cover):

$$LST_d = i_d + s_s NDVI \tag{7.6}$$

$$LST_w = i_w + s_w NDVI \tag{7.7}$$

Combining Eqs. (7.5), (7.6), and (7.7), the SM for each pixel can be estimated as a function of LST and $NDVI$ as follows:

$$SM = \frac{i_d + s_d NDVI - LST}{i_d - i_w + (i_d - i_w) NDVI} \tag{7.8}$$

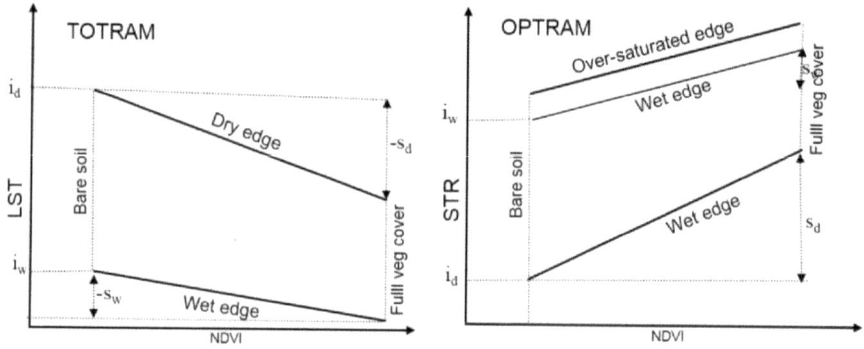

FIGURE 7.4 Sketch illustrating parameters of the traditional thermal-optical trapezoid model (Eq. (7.8), TOTRAM) and the new optical trapezoid model (Eq. (7.13), OPTRAM). TOTRAM and OPTRAM are parameterized are based on the pixel distributions within the LST-NDVI space and STR-NDVI space, respectively; NDVI is the normalized difference vegetation index; LST is the land surface temperature; and STR is the SWIR transformed reflectance (see Eq. (7.10)).

The new trapezoid model, OPTRAM, is based on the idea of replacing LST in TOTRAM with a measure for SM in the optical domain. Based on the Kubelka and Munk (1931) two-flux radiative transfer model, Sadeghi et al. (2015) developed a physical model exhibiting a linear relationship between surface moisture content and SWIR transformed reflectance as follows:

$$SM = \frac{\theta - \theta_d}{\theta_w - \theta_d} = \frac{STR - STR_d}{STR_w - STR_d} \tag{7.9}$$

where STR is the SWIR transformed reflectance and STR_d and STR_w are the STR at θ_d and θ_w, respectively. The STR is defined as follows:

$$STR = \frac{(1 - R_{SWIR})^2}{2R_{SWIR}} \tag{7.10}$$

The previously derived Eq. (7.9) has been tested for bare soils for two SWIR bands (i.e., 1,650 nm corresponding to band 6 of Landsat 8, and 2,210 nm corresponding to band 7 of Landsat 8), and it has been demonstrated that the model is highly accurate, especially at 2210 nm. Eq. (6) was also derived for vegetated soils based on the Kubelka and Munk (1931) radiative transfer model and holds for any fractional vegetation cover (i.e., any NDVI). For vegetated soils, θ in Eq. (7.9) is assumed to be correlated to root zone soil moisture through the vegetation response to soil moisture deficit in the root zone. Based on the assumption of a linear relationship between soil- and vegetation-water contents, that the STR-NDVI space forms a trapezoid is expected. Therefore, the parameters of Eq. (7.9) can be obtained for a specific location (e.g., satellite scene) from the dry and wet edges of the optical trapezoid, described in Figure 7.4:

$$STR_d = i_d + s_d NDVI \tag{7.11}$$

$$STR_w = i_{dw} + s_w NDV \tag{7.12}$$

Combining Eqs. (7.9), (11), and (12), the soil moisture for each pixel can be estimated as a function of STR and NDVI:

$$SM = \frac{i_d + s_d NDVI - STR}{i_d - i_w + (i_d - i_w) NDVI} \tag{7.13}$$

A comparison of Eqs. (7.8) and (7.13) results demonstrates that the new OPTRAM is comparable with OPTRAM only requiring observations in the optical electromagnetic frequency domain. In contrast to the LST-NDVI space, which varies over time due to variations of ambient atmospheric parameters, we expect the STR-NDVI space to remain nearly time invariant because reflectance is a function of only the surface properties and not the ambient atmospheric conditions. Therefore, it is expected that there is feasibility of a universal parameterization of OPTRAM that is valid for all observation dates at a specific location, which is a significant advancement that opens a new avenue for optical remote sensing of SM.

7.3.3 Thermal Inertial-Based Methods

As it was initially proposed by Price (1977), the thermal inertia (TI)-based concept exploits information on thermal conductivity, bulk density, and specific heat capacity of the target. However, the application of this method is not straightforward as the knowledge of the physical properties of the target is required. So, the apparent thermal inertia (ATI) was proposed to make the determination of TI easier using the diurnal gradient of surface thermal responses (physical temperature) and thus can be derived directly from multi-spectral remotely sensed imagery (Xue and Cracknell, 1995). The relationship between SM content and ATI has been investigated in numerous previous studies for both bare soil and vegetated areas. Xue and Cracknell (1995) applied the ATI approach to AVHRR data to determine land surface thermal inertia. Cai et al. (2007) used MODIS images over the northern Chinese Plain and obtained a mean difference between measured and estimated soil moisture values of 4.3% over bare and sparsely vegetated fields. To infer reliable SM estimates derived from ATI, the temperature difference should be larger than 10°C.

SM estimation is based on the solution of the equations linking soil thermal and optical properties, through its water content, by means of VIS-NIR and TIR images. These properties change with soil TI (J m^{-2} K^{-1} $s^{-1/2}$), expressing, for given material, the thermal fluctuation's quickness to energy flux exchanges as follows:

$$TI = \sqrt{\lambda \rho C} \tag{7.14}$$

where λ (W m^{-1} K^{-1}) is the soil thermal conductivity, ρ (kg m^{-3}) is its density, and C (J kg^{-1} K^{-1}) indicates the soil specific heat for unit mass. The mathematical relationship between TI and soil water content, SM (m^3 m^{-3}), is based on the λ and C univocal variation with SM. The higher SM the higher are both λ and C and, consequently, TI. For an operational use of Eq. (7.14), it needs to be written as explicit function of SM.

Some of these use the Apparent Thermal Inertia, ATI, given by Eq. (7.15) (Tramutoli et al., 2000):

$$ATI = \frac{1-\alpha}{\Delta T} \tag{7.15}$$

where α is the surface albedo and ΔT is maximum soil temperature variation usually calculated on a daily basis. Both α and ΔT can easily be mapped through multispectral images (VIS-NIR and TIR) to calculate ATI spatial distribution and therefore SM using an empirical relationship as follows:

$$SM = a * ATI + b \tag{7.16}$$

where parameters a and b are empirically derived (Price, 1980).

Recently Lu et al. (2009) proposed a procedure to directly estimate soil water content according to Murray and Verhoef (2007) thermal inertia model Eq. (7.17):

$$TI = TI_{dry} + (TI_{sat} - TI_{dry})K_{TI} \tag{7.17}$$

where, TI_{dry} and TI_{sat} represent the TI values of dry and saturated soils, respectively, whereas K_{TI} is the Kersten function (Groenevelt and Grant, 2004) given by:

$$K_{TI} = exp\left\{\varepsilon\left[1-\left(\frac{\theta}{\eta}\right)^{-\mu}\right]\right\} \tag{7.18}$$

where parameters ε and μ equal 0.6 and 0.71, respectively, in the cases of soils with fine texture; θ is the volumetric water content; and η is the soil porosity.

Based on this approach, given a known TI distribution, K_{TI} is obtainable through Eq. (7.18) and then θ by reordering the Kersten equation as follows:

$$\theta = \eta\left(1-\frac{lnK_{TI}}{\varepsilon}\right)^{1/-\mu} \tag{7.19}$$

The methodology has been applied on three kinds of soil, where water content was varied between residual and saturation values.

The subsequent application of Eqs. (7.17)–(7.19) allows calculating maps of volumetric SM water content, consecutively validated on the basis of thermo-gravimetric measures.

SM content controls soil density, heat capacity, thermal conductivity and other characteristics, and soil TI is closely related to soil moisture (Cheruy et al., 2017).

The ATI, which is simplified on the basis of TI, is simple to calculate and has been widely used to characterize SM (Lu et al., 2009). The calculation formula of the ATI model only involves surface albedo and a night–day surface temperature pair parameter, which can be directly derived from RS data. Therefore, the ATI model has become one of the most commonly used methods for retrieving SM in bare soil and low vegetation cover areas by TIR technology.

Among the methods of estimating SM by using TIR remote sensing, the ATI method is widely used in bare soil and low vegetation cover areas. However, high surface ET will cause ATI error, resulting in lower accuracy of SM estimation. The coupling of SM and surface evaporation, if the two are positively correlated, indicates that evaporation in this area is limited by water, and soil moisture plays a leading role in evaporation, that is, the greater the SM, the greater the evaporation. To overcome the problem that high surface ET will cause ATI error, the potential of ATI-ET space for estimating the SM of bare and vegetated areas is studied. ATI and ET data were utilized to construct triangle feature space, and six distance parameters are extracted from the positions of random pixels in the triangle (Figure 7.5). Some correlation estimates were made to derive those parameters that were useful for SM estimation, which were three in total. The SM estimation model consisting of these three parameters was built. Compared with the ATI model, the ATI-ET triangle model can not only be applied to areas with high ET but also has higher accuracy in estimating SM. The ATI-ET triangle model is more suitable for application in bare soil and low vegetation cover areas.

Luo et al. (2022) used MODIS data that can simultaneously provide characteristic information such as surface albedo, surface temperature, and ET to test the ATI-ET triangle space model (Figure 7.5) for estimating surface SM. The ATI-ET two-dimensional space fully considers the relationship between the ATI and the ET and overcomes the limitation that the ATI model cannot be applied in areas with high ET. According to the spatial characteristics of the triangle, the ATI-ET triangle model is constructed to estimate the soil moisture in the farmland area during the drought period. The calculation soil TI from TIR remote sensing is mainly to calculate ATI, which reflects the relative size and change trend of the real TI to a certain extent. The calculation formula of ATI in Eq. (7.15) is as follows:

$$\alpha = 0.160\rho 1 + 0.291\rho 2 + 0.243\rho 3 + 0.116\rho 4 + 0.112\rho 5 + 0.081\rho 7 - 0.0015 \quad (7.20)$$

where α is surface albedo. Using the method proposed by Liang (2001), ρ_i (i = 1, 2, 3, 4, 5, 7) represents the reflectance of the *i*th band of MODIS data. The ET and ATI data of the pixels were used to construct a two-dimensional spatial scatter map. In the triangular feature space, the surface type of the pixels located at the low ET and high ATI is high canopy. With the decrease in ATI, the surface type of pixels located at low ET and low ATI is dry soil. Along the soil line, with the increase in ET, the surface type of the pixels located at high ET and low ATI is wet soil (Figure 7.5a).

To build a new SM estimation model using this triangle, six parameters related to distance are extracted from the triangle space according to the position of a random pixel in the ATI-ET feature space (Figure 7.5b). Point P indicates the position of a

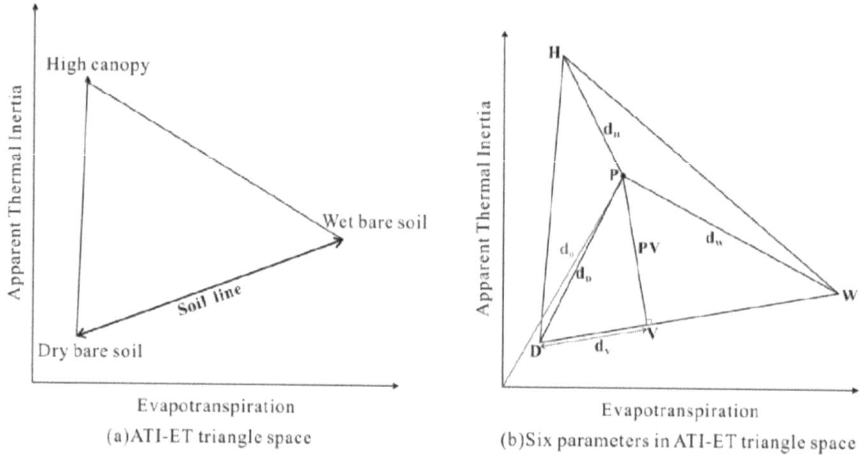

FIGURE 7.5 Six parameters used for estimating SM. (a) ATI-ET triangle space; (b) six parameters in ATI-ET triangle space.

Source: Luo et al., 2022

random pixel inside the triangle; points W and D represent wet soil and dry soil on the soil line; and point H represents high canopy. The PV is the distance from point P to the soil line; d_H is the distance from point P to point H. The larger the value of PV and d_H, the greater the vegetation cover at that pixel. d_W is the distance from point P to point W. We believe that larger values of PV, d_H, and d_W correspond to relatively high values of SM. The d_0 and d_D are the distances from point P to the axis origin and point D, respectively, and d_V is the distance from point V to point D. The larger the values of d_0, d_D, and d_V, the smaller the soil moisture value.

To find out the applicability of each parameter to SM estimation, the correlation between each of these parameters and in situ SM data at 10 cm depth was evaluated. The experimental result is presented in Table 7.2. The correlation coefficient between parameters d_D, d_W, and d_0 and in situ SM exceeds 0.3.

7.3.4 Machine Learning (ML)-Based Methods

Beyond the more traditional modeling approaches described to estimate SM, machine learning (ML) offers some alternative approaches. Due to the high complexity of physical models, the popularity of ML for the remote-sensing based estimation of biophysical parameters has grown significantly over the last decade. The flexibility of ML approaches is further highlighted by their potential to be used in various hydrometeorological applications, from the prediction of SM to precipitation, temperature, or wind (Ahn and Palmer, 2016; Tyralis et al., 2019). To address the complex nature of soil moisture, statistical learning theory tools such as Artificial Neural Networks (ANNs), Support Vector Machines (SVMs), and Relevance Vector Machines (RVMs), have been tested in recent years for their ability to spatially estimate distributed SM at the single-farm level. A key characteristic of these learning machine algorithms is

TABLE 7.2
Correlation Between Each of the Six Parameters and in Situ SM Data (Luo et al., 2022)

Parameter	a_D	d_W	d_0	d_V	d_H	PV
Correlation coeff.	0.41	0.39	0.35	0.29	0.21	0.13

their use of the underlying statistical characteristics of available information (predictors such as spectral/vegetation indices) without prior assumptions of their relationship. Examples of ML algorithms in spatial applications used in water resources and soil moisture applications can be found in (Gill et al., 2007; Andris et al., 2013; Bachour et al., 2014; Hassan-Esfahani et al., 2015). In the study by Liu et al. (2021), their results show that the SM estimation accuracies, for a farmland test area, by a combination of Sentinel-1 and Sentinel-2 and several different ML algorithms (SVR, deep neural networks, and generalized regression neural networks) were higher than those by traditional semi-empirical models based on either Sentinel-1 or Sentinel-2. Compared to more traditional approaches, ML-methods have two significant advantages: (1) they enable the construction of more objective, purely data-driven retrieval models, independent of necessary assumptions and (2) they allow the combination of data from different sources (like the combination of optical, thermal, and microwave remote sensing), exploiting their relationship with a target variable.

The following example is using a ML algorithm Gradient Boosted Regression Trees (GBRT) algorithm to map the high spatial resolution (50 m) mapping of SM based on the integration of Landsat-8 optical and thermal images, Copernicus Sentinel-1 C-Band SAR images, and modeled data, executable in the Google Earth Engine. GBRT belongs to the ensemble methods, which means several weak learners are built and combined into one powerful ensemble. Weak learners are combined sequentially, and each newly added model tries to correct the prediction bias of all previous models combined. GBRT, like Random Forest, belongs to the family of tree-based methods, which means that it is naturally compatible with different data types and ordinal scales. Further advantages are its insensitivity to differently scaled features and the low computational cost associated with algorithm training and target prediction. Moreover, it has proven to perform very well in similar applications. GBRT uses an additive model to combine weak learners:

$$F(x) = \sum_{m=1}^{M} \gamma_m h_m(x) \tag{7.21}$$

where, M is the total number of models; γ is a multiplication factor (the so-called step size); and $h(x)$ is the weak learner. An iterative process is building the model using decision trees of fixed size as weak learners. The loss L is minimized (with a least-square loss function, in this case) for each tree h_m given the previous ensemble F_m–1:

$$h_m = \underset{h}{argmin} \sum_{i=1}^{n} L\left(y_i, F_{m-1}(x_i) + h(x_i)\right) \tag{7.22}$$

where x_i and y_i are the feature and target values of the training dataset. GBRT solves the minimization problem numerically via steepest descent. The following equation determines the value of γ, based on line search (Armijo, 1966):

$$\gamma_m = argmin \sum_{i=1}^{n} L\left(y_i, F_{m-1}(x_i - \gamma \frac{\partial L\left(y_i, F_{m-1}\left(x_i\right)\right)}{\partial F_{m-1}\left(x_i\right)} \right). \tag{7.23}$$

For further detailed discussion on GBRT, the reader should refer to (Friedman, 2001). This study used the algorithm implementation of Scikit-Learn (Pedregosa et al., 2011). The substantial performing procedure is the following: (1) generation of training and test datasets: the test dataset consisted of 20% of the samples and was randomly selected, and the remaining 80% represented the training dataset; (2) tuning and training: cross-validation and a grid-search approach drove the optimization of hyperparameters (Kohavi, 1995); (3) feature selection: the training procedure was further nested in an automatic feature selection routine wherein the training was repeated iteratively with the least essential feature removed in each step; and (4) testing: the final assessment of the SM content estimation accuracy (the test score) was performed based on the independent test dataset extracted from the whole dataset before the training procedure. Based on the independent validation dataset, the resulting overall estimation accuracy, in terms of root-mean-squared-error and R^2, was 0.04 m^3/m^{-3} and 0.81, respectively.

7.4 ESTIMATING EVAPOTRANSPIRATION FROM TIR REMOTE SENSING DATA

7.4.1 BACKGROUND

As one of the most significant components, land surface evapotranspiration (ET—or latent heat flux, LE, in W/m^2) in the hydrological processes accounts for approximately 60% of the mean precipitation globally and controls the water and energy transfer in the interface of land surface and atmosphere. At present, measuring directly the actual ET remains an unresolved problem over large heterogeneous areas in practical applications. Traditional ET estimation methods (i.e., sap flow, weighing lysimeter, pan measurement, Bowen ratio system, eddy correlation system, and scintillometer) are generally of limited use because there are difficulties in spatial extrapolation due to the significant surface heterogeneity. To date, satellite remote sensing technology has been recognized as the most effective means to capture surface information at various spatial scales. The combination of physics-based models with satellite remote sensing data that characterize the surface heterogeneity is therefore an effective strategy to quantify the spatiotemporal variations in ET (Fisher et al., 2008). Since the advent of Earth observation from space in 1957, in particular the TIROS weather satellite series since 1960 and the Landsat multispectral and thermal sensors since 1972, there have been a large array of models developed to utilize satellite data for mapping ET across large spatial and temporal scales. Both empirical and physical models, which differ greatly in the model structure, input, and output, have been developed with varying degrees of complexity. Different combinations of the remote sensing spectral information from visible and NIR bands to mid and TIR bands can produce surface characteristic parameters that are indispensable to ET models. These parameters include NDVI, Leaf Area Index (LAI), fractional vegetation coverage (F_v), surface albedo, surface emissivity, and radiometric surface temperature.

Therefore, with the consideration of the characteristics of the various ET methods developed over the past decades and of the significance of land surface ET to hydrologists, water resources and irrigation engineers, climatologists, etc., how to calculate the ET over a regional scale or how to estimate ET precisely based on the remote sensing technology has become a critical question in various ET-related applications and studies. Satellite images in different spectral ranges contain different information contents useful for estimating ET from vegetated and non-vegetated surfaces. Thermal remote sensing provides the land surface temperature (LST), which is closely linked to the water status of the surface and therefore to ET. Many models have been developed to utilize thermal images for ET mapping. When information on the surface energy balance, such as net radiation (R_n) and ground heat flux (G), is available, LST can be effectively used to estimate ET as the residual of the surface energy balance after subtracting (R_n-G) by the sensible heat loss to the air, which is proportional to the difference between LST and air temperature (Bartholie et al., 1972). Shortwave remote sensing images contain information on vegetation type and structure, which can be used in various ways to help quantify ET.

To better understand the mechanisms of the water and energy interactions between the atmosphere and land surface, it is essential to summarize and compare various remote sensing-based ET approaches. This section builds upon and updates the review by Li et al. (2009) and Liou and Kar (2014), covering a range of remote sensing ET methods and models developed across different spatial and temporal scales. For each method or model, the underlying theory and assumptions are outlined, and their advantages, limitations, and potential applications are critically evaluated.

7.4.2 General Methodologies of Evapotranspiration Retrieval

Current remote sensing ET methods can be roughly categorized into the (semi) empirical method and the analytical method. The (semi)empirical method is developed by establishing empirical regression functions of remote sensing observations and minimum ground-based measurements, while the analytical physics-based method mainly considers the physical processes at the scale of interest and generally requires a greater variety of direct and indirect measurements.

7.4.2.1 Empirical Regression Methods

The main theory of the empirical regression method is to statistically fit the ET at different spatiotemporal scales to remote sensing or ground-based variables. For example, Jackson et al. (1977) related the daily ET from the irrigated wheat at Phoenix, Arizona to the difference between instantaneous surface temperature (T_s), and near-surface air temperature (T_a) measured near midday at local time about 13:30 to 14:00. Wang et al. (2007) developed a linear relationship to estimate ET using the vegetation index, daytime-averaged surface net radiation (R_n), and T_a data collected over the Southern Great Plains area of the USA from January 2002 to May 2005. The most general form of the empirical regression method can be expressed mathematically

$$ET = f_{regress}\left(R_n, T_s, T_a, VI, \ldots\right) \tag{7.24}$$

where the independent variable can be either R_n, T_s, T_a, VI (vegetation index) and wind speed, relative humidity, or a combination of several of these parameters. The regression method can be applied on a daily or monthly scale. The simplified regression method proposed by Jackson et al. (1977) and its refinements have attracted great attention in the subsequent operational applications of ET mapping. All contributions to the Jackson's simplified equation have shown that the error is on the magnitude of about 1 mm/day in the calculated daily ET, and the simplified equation can provide reliable information to the water availability in the practical applications. The main advantage of the empirical regression method is its simplicity, ease of use, and relatively high accuracy. However, the site-specific coefficients in the regression equation are limited to its applications at regional scales with variable heterogenous conditions.

7.4.2.2 Physics-Based Methods

This type of model makes use of land surface temperature (LST or T_s) observed by thermal sensors to estimate the sensible heat flux and to derive ET as a residual of the surface energy balance or to estimate the evaporative fraction and to derive ET from the available energy. Depending on whether land surface is treated uniformly as a big leaf or separately as a combination of soil and canopy components, they can be characterized as one-source and two-source models.

7.4.2.2.1 Single-Source Models

The one-dimensional form of surface energy budget equation at instantaneous time scale can be expressed mathematically as

$$ET = R_n - G - H \tag{7.25}$$

where R_n (W/m^2) is the net radiation including both shortwave and longwave radiation absorbed by the surface; G (W/m^2) is the ground heat flux; H (W/m^2) is the sensible heat flux. Eq. (7.25) indirectly estimates ET as the residual of the surface energy balance. This model and its simplified forms were applied to the Advanced Very High Resolution Radiometer (AVHRR) and Landsat thermal images to obtain instantaneous and daily ET of various surface types. All three components at the right-hand side of the energy budget equation can be estimated by combining remote sensing parameters and ancillary ground-based surface and meteorological variables (including air temperature, wind speed, and humidity).

Net radiation (R_n) can be obtained in terms of the radiation balance and considered as a balance between incoming and outgoing short-wave and long-wave radiation under steady atmospheric conditions.

Soil heat flux (G) represents the heat energy that is used to warm or cool substrate soil volume. It is proportional to thermal conductivity and vertical temperature gradient of the topsoil and is traditionally measured using sensors buried beneath the surface soil. Many studies have expressed G as a fraction of R_n (from 0.05 for full vegetation cover or wet bare soil to 0.5 for dry bare soil) (e.g., Price, 1982; Kustas and Norman, 1996; Li and Lyons, 1999) and related this fraction as a function of LAI, NDVI, T_s, and/or solar zenith angle based on field observations:

$$\frac{G}{R_n} = f\left(LAI, NDVI, T_s, \ldots\right) \tag{7.26}$$

Soil heat flux varies in both diurnal and yearly cycles over diverse surface conditions. Remote sensing satellite platforms, however, are not able to directly measure G under current circumstances.

Sensible heat flux (H) represents the heat energy that is driven to warm/cool the air above the surface. It can be calculated in analogy to Ohm's law by combining the difference between aerodynamic ($_{Ts}$) and air temperatures (T_a) with the aerodynamic resistance (r_a) in the single-source energy balance model:

$$H = \rho c_p \frac{T_s - T_a}{r_a} \tag{7.27}$$

where ρ is the air density (kg/m^3), and c_p is the specific heat at constant pressure (J/(kg·K); T_s is aerodynamic temperature; T_a is the air temperature; and r_a is the aerodynamic resistance depending on wind speed, surface roughness, and atmospheric stability.

A number of factors, including surface roughness (vegetation height, vegetation structure), wind speed, and atmospheric stability, are found to affect the aerodynamic resistance, and extremely elementary (a function of wind speed only) to quite rigorous methods (accounting for atmospheric stability, wind speed, surface 'aerodynamic' roughness, etc.) have been developed to estimate r_a. The one that has been most commonly applied is as follows (Norman and Becker, 1995):

$$r_a = \frac{ln\left[\frac{z_a - d}{z_{0m}} + \varnothing_m\right] ln\frac{z_a - d}{z_{0m}} + \varnothing_h}{k^2 \mu} + r_e \tag{7.28}$$

where k is the von Karman constant; μ is the wind speed at height z_a; and $\varnothing_m$ and $\varnothing_h$ are the stability functions for momentum and heat transfer, respectively, being zero for neutral stability; d is the zero-plane displacement height; z_{om} and z_{oh} are the momentum and heat transfer lengths, respectively. The excess resistance (r_e) is defined as:

$$r_e = \frac{\ln\left(\frac{z_{0m}}{z_{0h}}\right)}{k\mu^*} \tag{7.29}$$

where μ^* is the friction velocity.

There have been many different one-source surface energy balance (SEB) models for ET estimation. This model type is briefly introduced later. Table 7.3 summaries the list of the models for estimating ET from key assumptions, input parameters, and advantages/disadvantages of each model.

(1) *Surface Energy Balance Index (SEBI)*

SEBI, first proposed by Menenti and Choudhury (1993), along with its derivatives like Surface Energy Balance Algorithm for Land (SEBAL), S-SEBI (Simplified-SEBI), SEBS (Surface Energy Balance System), METRIC (Mapping EvapoTranspiration at high Resolution with Internalized Calibration), etc., are typically single-source energy balance models based on the contrast between dry and wet limits to derive pixel by pixel ET and EF from the relative evaporative fraction when combined with surface parameters derived from remote sensing data and a certain amount of ground-based variables measured at local and/or regional scale. Here under dry conditions, in the SEBI method, evaporation is assumed to be zero due to the limitation of water availability in soil for a particular set of boundary layer characteristics so that the sensible heat flux density takes its maximum value (maximum surface temperature). $T_{s,max}$ is inverted from the bulk transfer equation, which is expressed as (Van den Hurk, 2001):

$$T_{s,max} = T_{pbl} + r_{a,max} \frac{H}{\rho c_p} \tag{7.30}$$

where T_{pbl} represents the average planetary boundary layer temperature in K; $r_{a,max}$ is the maximum aerodynamic resistance to sensible heat transfer (s/m).

The minimum surface temperature ($T_{s,min}$) is evaluated for the wet region from Eq. (7.30) by computing the potential ET from the Penman-Monteith equation considering zero internal resistance:

$$T_{s,min} = T_{pbl} + \frac{r_{a,min} \dfrac{(R_n - G)}{\rho c_p} - \dfrac{(e_{sat} - e)}{\gamma}}{1 + \dfrac{\Delta}{\gamma}} \tag{7.31}$$

where $r_{a,min}$ represents the minimum aerodynamic resistance in s/m and e and e_{sat} stand for actual and saturation vapor pressure, respectively; Δ is the slope of saturated vapor pressure as a function of T_a (air temperature measured at a reference height) in k·Pa/°C; γ denotes the psychrometric constant in k·Pa/°C. Interpolating the observed surface temperature (T_s) with the maximum and minimum surface temperatures, the relative evaporative fraction can then be expressed as the equation given here (Van den Hurk, 2001):

$$\frac{LE}{LE_p} = 1 - \frac{\Delta T \times r_a^{-1} - \Delta T_{min} \times r_{a,min}^{-1}}{\Delta T_{max} \times r_{a,max}^{-1} - \Delta T_{min} \times r_{a,min}^{-1}} \tag{7.32}$$

TABLE 7.3
Comparisons of the Different Remote Sensing ET Methods Source: Li et al. 2009; Liou and kar, 2014

Method	Equations	Key input parameters	Key assumptions	Advantages	Disadvantages
Simplified equation	(7.25)	R_{nd}, T_s, T_a	(1) Daily soil heat flux is negligible; (2) instantaneous H at midday can express the influence of partitioning daily available energy into turbulent fluxes	Simplicity	Site-specific
SEBI	(7.32)	T_{pbl}, $h_{pbl,}$ u, T_s, $R_{n,}$ G	(1) Dry limit has a zero surface ET; (2) wet limit evaporates potentially	Directly relating the effects of T_s and r_a on LE.	Ground-based measurements are needed
SEBS	(7.36)	T_s, R_n, G	(1) Dry limit, latent heat flux is assumed to be zero; (2) at the wet limit, ET takes place at potential rate (LE_{wet})	(1) Consideration of the energy balance at the limiting cases; (2) characterizing actual turbulent heat fluxes without any prior knowledge; (3) representativeness of parameters associated with surface resistance.	(1) Too many required parameters and (2) relatively complex solution of the turbulent heat fluxes
S-SEBI	(7.37)	T_s, α_s, R_n, G	(1) EF varies linearly with Ts for a given surface albedo; (2) T_s,max corresponds to the minimum LE; (3) T_s, min corresponds to the maximum LE	Ground based measurements are not required	Extreme temperatures are location specific
SEBAL	(7.40)	u, z_a, T_s, VI, R_n, G	(1) Linear relationship between Ts and dT; (2) ET of the driest pixel is 0; (3) ET_{wet} is set to the surface available energy	(1) Minimum ground measurements; (2) automatic internal calibration (3); accurate atmospheric corrections are not needed	(1) Applied over flat surfaces (2); uncertainty in the determination of anchor pixels
METRIC	(7.40)	u, z_a, T_s, VI, R_n, G	(1) For the hot pixel, ET is equal to zero; (2) for the wet pixel, LE is set to $1.05ET_r$	Same as SEBAL but surface slope and aspect can be considered	Uncertainty in the determination of anchor pixels
TSM	Soil and canopy energy budgets	u, z_a T_a, T_s, T_c, F_r or LAI, Rn, G	(1) Fluxes of soil surfaces are in parallel or in series with fluxes of canopy leaves; (2) Priestly-Taylor Equation is employed to give the first guess of canopy transpiration	(1) Effects of view geometry are taken into account; (2) empirical corrections for the 'excess resistance' are not needed	(1) Many ground measurements are needed; (2) component temperatures of soil and vegetation are required

where $\Delta T = T_s - T_{pbl}$, $\Delta T_{min} = T_{s,min} - T_{pbl}$, and $\Delta T_{max} = T_{s,max} - T_{pbl}$; surface temperature T_s is determined by using image data from the thermal infrared region for each pixel, while T_{pbl} is the air potential temperature at higher elevation or at the top of the Planetary Boundary Layer (PBL). By modifying the CWSI, Choudhury and Menenti (1993) defined theoretically the pixel-wise ranges for LE and T_s to account for surface variability of actual evaporation caused by albedo and aerodynamic roughness. For a surface at particular surface albedo and roughness, pixel-wise maximum and minimum surface temperatures and redefined CWSI as a pixel-wise SEBI are computed to derive the regional ET from the relative evaporative fraction (Choudhury and Menenti,1993).

(2) *Surface Energy Balance System (SEBS)*

Another well-known model is the Surface Energy Balance System (SEBS). Su (2002) described a modified form of SEBI for the estimation of land surface energy balance using remotely sensed data, which has been named SEBS. SEBS estimates sensible and latent heat fluxes from satellite data and routinely available meteorological data. Computations of land surface physical parameters, calculation of roughness length for heat transfer, and estimation of the evaporative fraction based on energy balance at limiting cases are the main bases of SEBS. In SEBS, at the dry limit, latent heat flux is assumed to be zero due to the limitation of soil moisture, which means sensible heat flux reaches its maximum value (i.e., $H_{dry} = R_n - G$). At the wet limit, ET takes place at potential rate (LE_{wet}) (i.e., ET is limited only by the energy available under the given surface and atmospheric conditions), which can be calculated by a combination equation similar to the Penman-Monteith combination equation (Monteith, 1965); assuming that the bulk internal resistance is zero, the sensible heat flux reaches its minimum value, H_{wet}. The sensible heat flux at dry and wet limits can be expressed as:

$$H_{dry} = R_n - G \tag{7.33}$$

$$H_{wet} = \frac{(R_n - G)\gamma}{(\gamma + \Delta)} - \frac{\rho c_p (e_{sat} - e)}{r_a (\gamma + \Delta)} \tag{7.34}$$

where r_a is dependent on the Obukhov length, which in turn is a function of the friction velocity and sensible heat flux. The relative evaporative fraction (EFr) and evaporative fraction (EF) then can be expressed as:

$$EF_r = \frac{H_{dry} - H}{H_{dry} - H_{wet}} \tag{7.35}$$

$$EF = \frac{EF_r \times LE_{wet}}{R_n - G} \tag{7.36}$$

Remote sensing data-derived land parameters and ground-based meteorological measurements are used as inputs in SEBS. Using remote sensing data from ATSR and ground data from a Numerical Weather Prediction model, Jia et al. (2003) proposed a modified version of SEBS and validated the estimated sensible heat flux with large aperture scintillometers. Wood et al. (2003) used SEBS to the Southern Great Plains region of the United States and compared the latent heat fluxes with the measurements from the Energy Balance Bowen Ration (EBBR) sites. Their results indicate the potential usefulness of the SEBS approach in estimating surface heat flux from space for data assimilation purposes. Daily, monthly, and annual estimation of evaporation in a semi-arid environment have been done by SEBS. For an accuracy of ET value estimated from SEBS, the results could reach 85%–90% of that of in-situ measurements even when evaporative fraction ranged from 0.5 to 0.9 as shown by Su et al. (2005). Main advantages of the SEBS include: (1) consideration of the energy balance at the limiting cases, which minimizes the uncertainty involved in surface temperature or meteorological variables; (2) new formulation of the roughness height for heat transfer is developed in SEBS instead of using fixed values; (3) characterizing actual turbulent heat fluxes without any prior knowledge; and (4) representativeness of parameters associated with surface resistance. However, too many required parameters and relatively complex solution of the turbulent heat fluxes in SEBS can be the source of more or less inconvenience when data are not readily available.

(3) *Simplified-Surface Energy Balance Index (S-SEB*I)

A simplified new method derived from SEBI, called S-SEBI, has been developed to estimate the surface flux from remote sensing data. The main theory of S-SEBI is based on the contrast between a reflectance (albedo) dependent maximum surface temperature for dry limit and a reflectance (albedo) dependent minimum surface temperature for wet limit to partition available energy into sensible and latent heat fluxes. No additional meteorological data is needed if the surface extremes are available on the scene studied. By assuming steady global radiation and air temperature, a physical explanation to the observed surface reflectance and temperature in the S-SEBI method can be given when surface characteristics within the observed image change between dark/wet and dry/bright pixels. A theoretical explanation to S-SEBI, when a wide range of surface characteristics changing from dry/dark soil to wet/bright pixels exists, can be given: (1) at low reflectance (albedo), surface temperature stays almost unchangeable because of the sufficient water available under these conditions, such as over open water or irrigated lands; (2) at higher reflectance (albedo), surface temperature increases to a certain point with the increases of reflectance due to the decrease of ET resulting from less water availability, which is termed as 'evaporation controlled'; (3) after the inflexion, the surface temperature will decrease with the increases of surface reflectance (albedo), which is called 'radiation controlled' (see Figure 7.6). Here, the evaporative fraction (EF) is constrained by the dry and wet regions and formulated by interpolating the reflection-dependent surface temperature between the reflection-dependent maximum and minimum surface temperatures as shown in Eq. (7.37):

$$EF = \frac{T_H - T_s}{T_H - T_{LE}} \tag{7.37}$$

where T_H is the land surface temperature corresponding to dry condition and represents the minimum latent heat flux ($LE_{dry} = 0$) and maximum sensible heat flux ($H_{dry} = R_n - G$), and T_{LE} is the land surface temperature corresponding to the wet condition and represents the maximum latent heat flux ($LE_{wet} = (R_n - G)$) and minimum sensible heat flux ($H_{wet} = 0$) for a given surface reflectance as shown in Figure 7.6.

Using the following regression equations, T_H and T_{LE} can be, respectively, calculated:

$$T_H = c_{max} + d_{max}\alpha_s \tag{7.38}$$

$$T_{LE} = c_{min} + d_{min}\alpha_s \tag{7.39}$$

where the empirical coefficients c_{max}, d_{max}, c_{min}, and d_{min} are estimated from the scatter plot of T_s and α_s over the study area (Figure 7.6). Finally, the *EF* is calculated from Eq. (7.37) using Eqs. (7.38) and (7.39). The major advantages of S-SEBI are that (1) additional ground-based measurement is not needed to derive the *EF* except the surface temperature and reflectance (albedo) derived from remote sensing data if the surface extremes temperatures are present in the remotely sensed imagery and (2) extreme temperatures for the wet and dry conditions vary with changing reflectance (albedo) values, whereas other methods like SEBAL try to determine a fixed temperature for wet and dry conditions. However, it should be noted that atmospheric corrections to retrieve *Ts* and α_s from satellite data and determination of the extreme

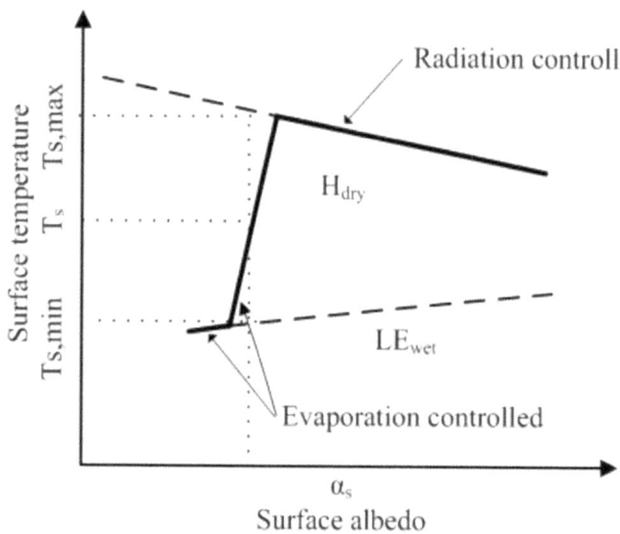

FIGURE 7.6 Theoretically schematic relationship between surface temperature and albedo in the S-SEBI.

Source: Li et al., 2009; Liou and Kar, 2014

temperatures for the wet and dry conditions are location specific when atmospheric conditions over larger areas are not constant any more.

(4) *Surface Energy Balance Algorithm for Land (SEBAL)*

SEBAL, an image-processing model for calculating evapotranspiration (ET) as a residual of the surface energy balance, was developed in the Netherlands by Bastiaanssen et al. (1998a, 1998b). Within the most promising methods currently available to estimate evapotranspiration, the SEBAL has been designed to calculate the energy balance components, at both local and regional scales with minimum ground data. This model is an intermediate approach using both empirical relationships and physical parameterization. It requires digital imagery data collected by any satellite sensor measuring visible, NIR, and TIR radiation, T_s, NDVI, and albedo maps. Latent heat flux (*LE*) is estimated as a residual of the energy balance equation on a pixel-by-pixel basis. Net radiation (R_n) is computed from the balance of short and longwave radiation. Soil heat flux (*G*) is calculated utilizing the equation proposed by Bastiaanssen, which is applicable to all sorts of vegetation cover and soil type (Bastiaanssen et al., 1998a, 1998b). Under several climatic conditions, this method has been verified at both field and catchment scales with typical accuracy at field scale being 85% and 95% at daily and seasonal scales, respectively, in more than 30 countries worldwide. The estimation of sensible heat flux (*H*) is required to obtain the parameters that will allow the computation of ET as a residual from the energy balance. To determine the value of *dT* for each pixel, SEBAL assumes the existence of a linear relationship between *dT* and the radiometric surface temperature T_s considering homogeneous metrological and surface conditions:

$$dT = d + cT_s \tag{7.40}$$

where *dT* is the near-surface air temperature difference; T_s is the radiometric surface temperature; and *c* and *d* are empirical coefficients obtained from the so-called 'anchor' pixels (Bastiaanssen, 1995) for a given satellite image. Generally, the anchor pixels within the image represent conditions of extreme evaporative behavior. Evaporation is assumed to be consuming most of the available energy ($R_n - G$) at a 'wet (cold)' pixel, which indicates that the sensible heat flux (H) and consequently *dT* are both assumed to be near zero ($dT_{wet} = 0$). At a 'dry (hot)' pixel where evaporation is near zero, all the available energy is modified essentially into sensible heat. Near surface air-temperature difference, *dT*, is then computed using aerodynamic theory for the two extreme conditions (Bastiaanssen, 1995) as:

$$dT_{dry} = \frac{H_{dry} \times r_{a,dry}}{\rho_{air,dry} c_p} \tag{7.41}$$

where H_{dry} (sensible heat flux at dry limit W·m^{-2}) is equal to ($R_n - G$). Once the surface-air temperature differences at both dry (hot) and wet (cold) points are determined, coefficients *c* and *d* in Eq. (7.40) can be easily estimated. The surface-air temperature difference *dT* at each pixel is calculated with T_s using this equation by using *c* and *d* coefficients. In SEBAL, the cold pixel is generally taken from a pixel located in deep water, and the hot pixel is taken from a pixel located in an area that shows high surface temperature. Finally, *H* is obtained iteratively with $r_{a,dry}$ corrected for stability using Eq. (7.27), where $dT = T_s - T_a$. An extrapolation of wind speed from ground level to a height of about 100 to 200 m is required in this procedure. Identifying the dry pixels is the most important aspect in SEBAL, while the wet pixels are frequently spotted at a location of well-watered areas or over a relatively large, calm water surface. The major advantages of SEBAL for the estimation of land surface fluxes from thermal remote sensing data are (1) minimum use of auxiliary ground-based data; (2) automatic internal correction, which prevents strict correction of atmospheric effects on surface temperature; and (3) internal calibration, which is done within each analyzed image. Major disadvantages of this method are that (1) subjective specifications of representative hot/dry and wet/cool pixels within the image are required to determine model parameters *c* and *d*. The resulting H flux and ET estimates from SEBAL can vary with differing extreme pixels selected by the operator, domain size, and spatial resolution of satellite sensors; (2) over mountainous regions, some adjustments are required based on a digital elevation model for T_s and *u* to account for the lapse rate; (3) estimated H is greatly affected by the errors in surface-air temperature differences or surface temperature measurements; and (4) ignoring the effect of radiometer viewing angle can cause variation in T_s by several degrees for some images.

(5) *Mapping Evapotranspiration at High Resolution and with Internalized Calibration (METRIC)*

To avoid the limitations of the SEBAL in mapping regional ET over more complicated surfaces, Allen et al. (2005, 2007) developed a similar SEBAL-based approach, named METRIC, to derive ET from remotely sensed data in the visible, NIR, and TIR spectral regions along with ground-based wind speed and near surface dew point temperature. In METRIC, an automatic internal calibration method similar to SEBAL (linearly relating T_s to the surface-air temperature difference) is used to calculate the sensible and latent heat fluxes. Gowda et al. (2008) have evaluated the performance of the METRIC model in the Texas High Plains using Landsat 5 TM data acquired on two different days in 2005 by comparison of resultant daily ET with measured values derived from the soil moisture budget. Santos et al. (2008) have found that combing a water balance model with ET estimated from the METRIC model could provide significant improvements in the irrigation schedules in Spain. Main distinctions between METRIC and SEBAL are: (1) METRIC does not assume $H_{wet} = 0$ or $LE_{wet} = R_n - G$ at the wet pixel, instead a daily surface soil water balance is run to confirm that for the hot pixel; *ET* is equal to zero, and for the wet pixel, *ET* is set to 1.05 *ETr*, where ET_r is the hourly (or shorter time interval) tall reference (like alfalfa) *ET* calculated using the standardized ASCE Penman-Monteith equation; (2) wet pixels in METRIC are selected in an agricultural setting where the cold pixels should have biophysical characteristics similar to the reference crop (alfalfa); (3) the interpolation (extrapolation) of instantaneous *ET* to daily value is based on the alfalfa ET_rF (defined as the ratio of instantaneous *ET* to the reference ET_r that is computed from meteorological station data at satellite overpass time) instead of the actual evaporative fraction, which can better account for the impacts of advection and changing wind and humidity conditions during the day.

7.4.2.2.2 Dual-Source Models (Also Called Two-Source Models)

Norman and Becker (1995) proposed a new model named two-source model, also known as dual-source model to improve the accuracy of LE (latent heat flux) estimates using satellite remote sensing data, especially over sparse surfaces. The basic principle of this model is partitioning the composite radiometric surface temperature into soil and vegetation components, and considered sensible and latent heat fluxes are transferred to the atmosphere from both soil and vegetation components. Dispensability of ground-based information or any prior calibration has made the applicability of the dual source model wider without resorting to any additional input data. In the dual-source model, satellite-derived surface temperature (T_s) is considered to be a composition of the soil (T_{soil}) and canopy temperatures (T_{veg}), and H and LE are also divided into soil and vegetation contributions, respectively. Canopy latent heat flux is computed using the Priestley-Taylor equation (Priestley and Taylor, 1972). An iterative method is used to obtain the T_{soil} and canopy T_{veg} from satellite-derived T_s setting an initial value of 1.3 for the Priestley-Taylor parameter α (Anderson et al., 2008; Kustas and Anderson, 2009). This nominal choice of α overestimates canopy latent heat flux under moisture-stressed conditions and yields negative soil evaporation (LE_{soil}) and is regarded as a nonphysical solution during the daytime. The α is therefore iteratively reduced until LE_{soil} approaches zero to obtain a final α as well as T_{soil} and T_{veg}. The LE and H are then calculated from these estimates. Both the one- and two-source models are sensitive to their use of the temperature differences to estimate H. Dispensability of precise atmospheric corrections, emissivity estimations, and high accuracy in sensor calibration are the main advantages of the dual source method. Coupling of the dual source models with PBL eliminates the need of ground-based measurement of T_a (Kustas and Daughtry, 1990) and, thus, is much better suited to applications over large-scale regions than other algorithms. Effects of view geometry are normally incorporated, while the empirical corrections for the 'excess resistance' are eliminated in the dual-source models.

Therefore, it is believed that dual-source models are an advance over single-source surface models and have a wider range of applicability (Anderson et al., 1997). In the dual-source models, the divergence of surface energy balance inside the canopy needs to be estimated and the clumping considered (affecting the wind speed profile, the radiation penetration, and the radiative surface temperature partitioning between soil and vegetation) (Kustas and Norman, 1997). The energy balance components of the total soil and vegetation system in the representative dual-source N95 model (Norman et al., 1995) are calculated from the following formulae: soil and canopy energy budgets:

$$R_{n,s} = H_s + LE_s + G \tag{7.42a}$$

$$R_{n,c} = H_c + LE_c \tag{7.42b}$$

Sensible heat flux:

$$H = H_s + H_C = \rho c_p \frac{T_0 - T_{AC}}{r_s} + \rho c_p \frac{T_c - T_{AC}}{r_X} \tag{7.42c}$$

Latent heat flux:

$$LE = LE_s + LE_c \tag{7.42d}$$

$$LE_c = \alpha f_g \frac{\Delta}{\Delta + \gamma} R_{n,c} \tag{7.42e}$$

Soil heat conduction flux:

$$G = \xi_s R_{n,s} \tag{7.42f}$$

where subscripts s and c represent the soil and canopy components, respectively; T_c and T_o are the vegetation canopy and soil component temperatures, respectively (K); r_X and r_S are the total boundary layer resistance of the complete canopy of leaves and resistance to heat transfer immediately above the soil surface, respectively (s/m); f_g is the fraction of the total LAI that is green; ξ_s is the fraction of the ratio of soil heat flux to surface net radiation at bare soil surface; α is the Priestley–Taylor coefficient for the potentially transpiring canopy with an initial approximation of 1.3; Δ is the slope of the saturated vapor pressure versus air temperature curve (kPa/°C); and γ is the psychrometric constant (kPa/°C).

Over the past years, temperature and energy components of both soil and vegetation over various landscapes have been investigated by a series of studies through a set of applications, validations, and modifications to the dual-source energy balance models. For example, Norman et al. (1995) developed a TSM (Two-Source (soil+canopy) Model) to accommodate the difference between radiometric surface and aerodynamic temperatures to partition surface energy balance into energy components of both soil and vegetation using data either from a single view angle or from multiple view angles. On the basis of N95, the ALEXI (Atmosphere-Land Exchange Inverse (Mecikalski et al., 1999)) removes the need for the measurements of near-surface air temperature and is relatively insensitive to uncertainties in surface emissivity and atmospheric corrections on the remotely sensed surface temperatures. By combining low- and high-resolution remotely sensed data, a two-step disaggregated ALEXI (DISALEXI) was proposed without a need for local observations (Norman et al., 2000; Kustas et al., 2003). More details of these Two-source models are found in Li et al. (2009), while the revision and recent advancements of these Two-source models are found in the literature (Mauser and Schädlich, 1998; Kustas and Norman, 1999; Li et al., 2005; Wang and Dickinson, 2012).

7.4.3 Penman–Monteith Method

The Penman–Monteith (PM) evapotranspiration (ET) equation predicts the rate of total evaporation and transpiration from the earth's surface using commonly remote sensing data and meteorological data (solar radiation, air temperature, vapor content, and wind speed). The PM equation follows a single-layer or 'big leaf' approach, where single-surface resistance and single aerodynamic resistance terms represent the transport properties of the cropped surface. The PM method is useful in providing direction in determining changes in crop coefficients under new cropping conditions. Determination of crop coefficients is an expensive and time-consuming process, as it requires additional lysimeter or micrometeorological studies. The PM equation has the form:

$$ET = \frac{\Delta(R_n - G) + \rho c_p (e_s - e_a) / r_a}{\Delta + \gamma\left(1 + \frac{r_s}{r_a}\right)} \tag{7.43}$$

where c_p is the specific heat of dry air; e_s is the saturation vapor pressure of the air at some height above the surface; e_a is the actual vapor pressure of the air; r_a is aerodynamic resistance to turbulent heat and/or vapor transfer from the surface to some height z above the surface; r_s is a bulk surface resistance that describes the resistance to flow of water vapor from inside the leaf, vegetation canopy, or soil to outside the surface.

The PM equation combines both the energy driver and atmospheric demand of evaporation and is a more robust and theoretically sound approach. Evapotranspiration from the PM equation is not overly sensitive to any of the inputs. When surface conductance is high (e.g., the energy limited condition), ET is sensitive to wind speed (a determinant of aerodynamic resistance) and the sensitivity depends upon the value of surface conductance. The key and difficulty of the use of the PM equation is to properly determine the surface bulk conductance, which needs to account for the effect of soil water availability, vapor pressure deficit, carbon dioxide concentration, atmospheric temperature, and canopy physiology through a coupled biophysical model or an empirical discount function.

Kang et al. (2009) applied Advanced Spaceborne Thermal Emission and Reflection Radiometer (ASTER) remote sensing data and estimated bare soil evaporation with the PM model, the Priestley-Taylor model, and the aerodynamics model. Evaporation estimated by each of the three models was compared with actual evaporation, and error sources of the three models were analyzed. The mean absolute relative error was 9% for the PM model, 14% for the Priestley-Taylor model, and 32% for the aerodynamics model; the PM model was the best of these three models for estimating bare soil evaporation. The error source of the PM model is the neglect of the advection estimation. With remotely sensed leaf area indices from the MODIS and the PM equation, gridded meteorology, and a two-parameter biophysical model for surface conductance (G_s), Zhang et al. (2008) estimated the eight-day average evaporation at 1-km resolution. Their results show that the evaporation model can be easily applied to estimate steady state evaporation and runoff and that evaporation can be used with rainfall-runoff models to improve accuracy of estimated runoff in ungauged catchments. Ciraolo et al. (2012) reports an application of the well-known PM approach, applied in a distributed way, using high resolution remote sensing data to map the potential evapotranspiration and demonstrated the feasibility of the use of recent and new satellite typical pixel size to assess ET spatial distribution directly related with agro-hydrological purposes and for precision farming applications.

7.4.4 Modified Priestley–Taylor Method

Although the PM equation can provide in theory accurate ET estimates, it also has difficulty in preparing the input variables for the parameterization of aerodynamic resistance and stomatal resistance, especially when applied globally. To overcome the problems encountered in the PM equation, Fisher et al. (2008) adapted the Priestley–Taylor (PT) equation (Priestley and Taylor, 1972) to soil water stress conditions by exerting plant physiological limitations and soil drought constraint. Inputs to Fisher's model include surface net radiation (R_n), NDVI, soil adjusted vegetation index, maximum air temperature (T_{max}), and actual vapor pressure (e_a). The Fisher model has been used in many sites that represent a wide range of plant functional types. Total LE in Fisher et al. (2008) is composed of canopy transpiration (LE_c), wet and moisture soil evaporation (LE_s), and canopy interception evaporation (LE_i). Each of the three components is estimated from the adjusted PT equation as follows:

$$LE = LE_s + : LE_c + LE_i \tag{7.44}$$

$$LE_c = \left(1 - f_{wet}\right) f_g f_T f_M \alpha \frac{\Delta}{\Delta + \gamma} R_{n,c} \tag{7.45}$$

$$LE_s = \left[f_{wet} + f_{SM}\left(1 - f_{wet}\right)\right] \alpha \frac{\Delta}{\Delta + \gamma}\left(R_{n,s} - G\right) \tag{7.46}$$

$$LE_i = f_{wet} \alpha \frac{\Delta}{\Delta + \gamma} R_{n,c} \tag{7.47}$$

where f_{wet} is the relative surface wetness, which can be expressed as the fourth power atmospheric relative humidity (RH^4); f_g is the green canopy fraction, expressed as the ratio of f_{APAR} (fraction of PAR absorbed by green vegetation cover) to f_{IPAR} (fraction of PAR intercepted by total vegetation cover); f_T is the plant temperature constraint factor; f_M is the plant moisture constraint factor; f_{SM} is the soil moisture constraint factor; $R_{n,c}$ is the canopy net radiation (W/m^2); and $R_{n,s}$ is the soil net radiation (W/m^2).

This modified PT equation has been used by Fisher et al. (2008) to produce global LE at a monthly temporal scale and a 1-degree spatial scale during the years 1986–1993 using datasets from ISLSCP-II and AVHRR. Although Fisher et al. (2008) have claimed that their model can be potentially applied based solely on remote sensing inputs at finer spatial and temporal resolutions, to overcome the difficulty of the satellite-based estimation of relative humidity and vapor pressure deficit, Yao et al. (2013) subsequently excluded the plant soil moisture constraint on the canopy transpiration and introduced the temperature diurnal range to parameterize the soil moisture constraint. The original inputs are also changed to only include surface net radiation, NDVI, air temperature, and diurnal temperature range. Validation results at 16 sites located in China revealed that Yao et al. (2013) indicate that they improved the monthly *LE* estimates from Fisher et al. (2008) when Yao et al. (2013)'s models were applied to MODIS data at a 0.05-degree spatial resolution. More importantly, the modified PT equation is also satisfactory in reproducing the inter-annual variability at flux tower sites with at least five years of data. The R^2 between measured and predicted annual *LE* anomalies is 0.42 ($p = 0.02$). The PT equation is then applied to detect the variations of long-term terrestrial *LE* over Three-North Shelter Forest Region of China and to monitor global land surface drought. It is described here to demonstrate the ability to map regional terrestrial LE and identify global soil moisture stress, without requiring precipitation information (Yao et al., 2014).

Literature conducted over the past several years has shown that there appears to be a trend in the scientific community to estimate continental/global ET in the context of climate change using the modified PM equation and PT equation, although the efficacy of these two equations still needs to be further improved. Using the retrieved vegetation biophysical parameters other than surface temperature from remote sensing data as part of the inputs has made the modified PM equation and PT equation good alternatives to the conventional energy balance models for the global ET estimation.

7.5 MAPPING ICE/SNOW AREAS WITH TIR REMOTE SENSING IMAGERY

Ice/snow cover can be detected and monitored with a variety of remote sensing technologies. The greatest number of applications have been found in the VNIR region and TIR range of the electromagnetic spectrum. Many studies with optical and TIR remote sensing derived spectral/vegetation indices and LST conducted projects on mapping ice/snow areas via vegetation index thresholding, supervised classification, object-based image analysis, and LST and derived apparent thermal inertia approaches. The following is a brief review of such studies.

In deriving a daily snow cover map of East Asia, Lee and Choi (2022) proposed a daily snow cover mapping algorithm based on Geostationary-Korean Multi-Purpose Satellite-2A multitemporal images. After processes for cloud masking and misclassification removal are conducted, a threshold with consideration for the characteristics of a geostationary satellite is applied to the normalized difference snow index (NDSI), and a daily snow cover map is generated utilizing a stacking process. The steps of the algorithm involve (1) cloud masking if considerable changes in TOA reflectance and NDSI exist, (2) masking the remaining cloud pixels using the Clouds proposed in this study and by using the cirrus channel and BTD (3.7 μm–10:5 μm), and (3) extracting NDSI over 0.1 and accumulating the NDSI images observed at 10-min intervals to create daily snow cover maps. For the quantitative validation, the over accuracy was 0.92 for Landsat-8 and 0.97 for the interactive multisensor snow and ice mapping system (IMS). In addition, when the snow cover extent calculated from the IMS was compared, the correlation reached 0.91 from December 2020 to January 2021.

Detailed and accurate information about the size and spatial extent of debris-covered glaciers is vital for climate change and other glaciological applications such as observing glacier changes, conducting mass balance studies, predicting glacial lake outburst floods, and much more. Recently, several remote sensing-based studies were carried out for glacier cover mapping using various pixel-based classification approaches. However, object-based image analysis (OBIA), which works on the groups of homogeneous pixels called image objects, emerged as the alternative for the accurate extraction of fine details from high spatial resolution (HSR) images. It has significant advantages over traditional pixel-based classification approaches in utilizing spatial information such as image texture, contextual information, pixel proximity, and geometric characteristics from HSR imagery. Applications of OBIA in glaciology using remote sensing data have received much attention in the past decade. Recently, a few studies reported the mapping of the Indian Himalayan glaciers using HSR data, which include Mitkari et al. (2017) and Jawak et al. (2018). The brief review suggests that researchers tried to evaluate the potential of OBIA for glacier cover/glacier facies mapping using the spectral properties and that the true power of OBIA to map distinct glacier cover classes of debris-covered glaciers using spatial information is yet be investigated. It is also prudent to mention that when glaciers are covered with varying amounts of debris, the spectral similarities between certain glacier cover classes such as supraglacial further add to the inaccurate delineation of glacier boundaries. However, the errors in the glacier boundary delineation can be reduced by using multiple data sources such as temperature and slope with the spectral data.

By considering mapping of the actual glacier area, in which the most significant parameter differentiating the debris covers is the surface temperature, Karimi et al. (2012) proposed an approach to distinguish the supraglacial and periglacial debris and outline the debris covered glacier boundary by means of optical (Worldview-2 satellite imagery) and thermal (Landsat-TM sensor) remote sensing data. After collecting GCPs from different members of glaciers in field operation, Neural Network (NN) classification has been used to delineate the actual area of glaciers. The basic approach is comprised of the following four steps: (1) reprocessing of satellite data; (2) extraction of morphometric parameters of glacier; (3) evaluating temperature and analysis of different temperatures among components of glacier; (4) applying supervised multispectral classification (NN); and (5) determining the actual glacier boundary. The accuracy of such a thermal–optical based method was analyzed by comparison of its result to the geomorphometric-based techniques and manual delineation. These analyses shows that the result of thermal–optical approach has a significant difference with geomorphological approach especial in the ablation area and this can be related to the recession of glacier in recent years. Overall accuracy and Kappa coefficient with values of 92.9% and 0.84 respectively demonstrated the high efficiency of thermal–optical approach to delineate the actual area of this alpine glacier.

Thermal remote sensing provides a set of variables that are useful for characterizing thermal and thermophysical surface characteristics of the natural and built environment such as mapping glaciers. In mapping rock glaciers and debris-covered glaciers, Brenning et al. (2012) explored the utility of remotely-sensed land surface temperature (LST) and derived apparent thermal inertia (ATI) for thermally characterizing and discriminating periglacial mountain environments, particularly rock glaciers and debris-covered glaciers. In their case study, they demonstrated this approach is useful by utilizing ASTER daytime and nighttime data in a study area in the Andes of Central Chile, where in-situ near-surface ground temperature (NSGT) measurements are available for comparison with remotely sensed LST. Spatial patterns of LST and NSGT were mostly consistent with each other both at daytime and at nighttime. Daytime LST over ice-debris landforms was decreased and ATI consequently increased compared to other debris surfaces under otherwise equal conditions. ASTER-derived ATI of rock glaciers and debris-covered glaciers differed from the inertia of other vegetation-free debris surfaces in the study area, and ice-debris landforms exhibited reduced daytime LST under otherwise equal conditions. While their results demonstrated the utility of thermal imaging and ATI mapping in a mountain cryospheric context, further research is needed for a better interpretation of ATI patterns in complex thermophysical conditions.

7.6 SUMMARY

Remote sensing can provide a means of observing hydrological state variables over large areas, including retrieving LST, estimating surface soil moisture and evapotranspiration, and mapping ice/snow areas. Therefore, in this chapter, the reasons why we need to perform studies of TIR remote sensing in hydrology were first introduced, then remote sensing of hydrometeorological states variables was introduced, including land surface temperature (LST), surface soil moisture, and ice/snow cover areas. Next, introduced and discussed were various methods and techniques of estimating soil moisture from TIR remote sensing data, including (1) spectral/vegetation-based methods, (2) triangle-trapezoid-based methods, (3) thermal inertial-based methods, and (4) machine learning (ML)-based methods. Next, introduced and reviewed/detailed were relatively extensive methods and models of estimating evapotranspiration from TIR remote sensing data from general methodologies of evapotranspiration retrieval (including empirical regression methods and physics-based methods). Per physics-based methods, depending on whether land surface is treated uniformly as a big leaf or separately as a combination of soil and canopy components, they can be characterized as one-source and two-source models. For the one-source model, we reviewed and discussed in detail (1) Surface Energy Balance Index (SEBI), (2) Surface Energy Balance System (SEBS), (3) Simplified-Surface Energy Balance Index (S-SEBI), (4) Surface Energy Balance Algorithm for Land (SEBAL), and (5) Mapping Evapotranspiration at High Resolution and with Internalized Calibration (METRIC) models. We also introduced and reviewed the physics-based Penman–Monteith method and Modified Priestley–Taylor method. Finally, many studies with optical and TIR remote sensing derived spectral/vegetation indices and LST, conducted projects on mapping ice/snow areas via vegetation index thresholding, supervised classification, object-based image analysis, and LST and derived apparent thermal inertia approaches were reviewed and discussed briefly.

7.7 APPENDIX: NOMENCLATURE USED IN THIS CHAPTER

c_p	Specific heat of air at constant pressure	J/(m·K).
C_{rad}	Correction coefficient used in sloping terrain	
d	Zero plane displacement height	m
dT	Surface-air temperature difference	K
dT_{dry}	Surface-air temperature difference at dry pixel	K
$dTwet$	Surface-air temperature difference at wet pixel	K
dTs	Surface temperature difference of two times in the morning	K
D	Day of year	day
EF	Evaporative fraction	
EF_r	Relative evaporative fraction	
ET	Evapotranspiration	mm/h
ET_i	Instantaneous ET	mm/h
ET_r	Reference ET (over the standardized 0.5 m tall alfalfa)	mm/h
ET_d	Cumulative daily ET	mm/d
$ET_{r,d}$	Cumulative daily reference ET	mm/d
ET_rF	Reference ET fraction	
F_r	Fractional vegetation cover	
$f(\theta)$	Fraction of canopy in the field of view of the radiometer	
g	Acceleration due to gravity of the earth	m/s^2
G	Soil heat flux density	W/m^2
h_{pbl}	Height of the PBL	m
H	Sensible heat flux	W/m^2
H_{dry}	Sensible heat flux at dry limit	W/m^2
H_{wet}	Sensible heat flux at wet limit	W/m^2
k	Von Karman's constant	
L	Latent heat of vaporization	J/kg
LE	Latent heat flux density	W/m2
LE_d	Daily ET	W/m2
LE_p	Potential ET	W/m2
LE_{wet}	Latent heat flux at wet limit	W/m2
r_a	Stability-corrected aerodynamic resistance to heat transfer between surface and reference height	s/m
$r_{a,max}$	Maximum aerodynamic resistance to sensible heat transfer	s/m
$r_{a,min}$	Minimum aerodynamic resistance	s/m
r_e	Excess resistance	s/m
r_{rc}	Effective radiometric-convective resistance	s/m
r_s	Soil-surface resistance	s/m
R_s	Incoming shortwave solar radiation	W/m2
t	Duration time starting at sunrise	h
T_a	Air temperature measured at a reference height	K
T_c	Vegetation canopy temperature	K
T_0	Soil surface temperature	K
T_{pbl}	Average planetary boundary layer temperature	K
$T_{RAD(?)}$	Directional radiometric surface temperature	K
$TB(\theta)$	Brightness temperature	K

(Continued)

(*Continued*)

$T_{s,max}$	Maximum surface temperature	K
$T_{s,min}$	Minimum surface temperature	K
u	Wind speed	m/s
$u*$	Friction velocity	m/s
z_a	Measurement height of wind speed and air temperature	m
z_{oh}	Surface roughness length for heat transfer	m
z_{om}	Surface roughness length for momentum transfer	m
α_s	Surface shortwave albedo	
γ	Psychrometric constant kPa/°C	kPa/°C
Δ	Slope of saturated vapor pressure	kPa/°C
e_a	Atmospheric emissivity	
e_s	Surface emissivity	
l	Geographical latitude (expressed in decimal degrees) in ET estimation	degrees
ρ	Density of a certain entity	kg/m^3
ρ_w	Density of water	kg/m^3
σ	Stefan-Boltzmann constant (5.67×10^{-8})	W/(m^{2}K^4)
θ	View zenith angle	degrees

8 Thermal Infrared Remote Sensing Applied in Biomass Burning

Biomass burning profoundly influences atmospheric chemistry, composition, climate, vegetation ecosystems, and, of course, human activities. Remote sensing offers essential tools for observing biomass burning patterns over vast spatial and temporal scales. This chapter provides a concise and precise review of thermal infrared (TIR) remote sensing studies related to biomass burning, focusing on active fire detection, burned area mapping, fire radiation power retrieval, and smoke plume monitoring. To address diverse research and application needs, this chapter systematically explores the necessity (Section 8.1), methodologies for active fire detection (Section 8.2), techniques for mapping burned areas using TIR data (Section 8.3), principles of fire radiation retrieval (Section 8.4), and algorithms for satellite-based smoke plume monitoring (Section 8.5).

8.1 INTRODUCTION

Biomass burning is a major natural disturbance that has tremendous impact on the environment, humans and wildlife, ecosystem, weather, and climate. There appears to be an increasing trend of natural fire activity. Satellite remote sensing has played a growing role in fire detection and monitoring over the past three decades. Whereas fire scars generally contrast strongly with unburned surfaces in the visible (0.4–0.7 μm), near infrared (0.7–1.5 μm), and middle infrared (1.5–4 μm, MIR) portions of the electromagnetic spectrum, researchers often use thermal detectors operating from about 4 to 15 μm to identify active fires. Since vegetation fires occur over wide areas, are sporadic and rapidly changing in nature, are international in scope, and often occur in isolated regions, remote sensing has become a key tool in their identification and study. Thermal remote sensing is used widely to map fire extents, examine fire regimes, characterize fire impacts, and estimate and characterize the chemical composition of fire emissions.

Biomass burning is an important factor in ecosystem management, land cover change, and wildlife habitat studies (Levine, 1991). On one hand, biomass burning is a natural part of several ecosystems for maintaining their health and diversity in numerous ways, such as regulating plant succession and fuel accumulations; structure and species composition of vegetation; affecting insect and disease populations; influencing nutrient cycles and energy flows; regulating biotic productivity, diversity, and stability; and determining habitats for wildlife. On the other hand, biomass burning causes huge economic losses to forestry resources and a wide range of global

DOI: 10.1201/9781003276173-8

environmental impacts, such as the loss of animal habitat and biodiversity (Lovejoy, 1994), modifying vegetation successional patterns (Christensen, 1993), and altering biological nutrient cycling (Menaut et al., 1993). Meanwhile, vegetation fires emit substantial amounts of trace gases (e.g., CO_2, CO, CH4, NOx) and particulates into the atmosphere; smoke from such burning, if not properly managed, can have significant health impacts on exposed populations. Therefore, accurately monitoring the effects of wildfire events (i.e., time, location, and severity) is one of the central questions in ecology and natural resource management. Remote sensing from space is especially suitable for forest fire research. The wide area coverage repetitively provided by satellite sensors, as well as their information on non-visible spectral regions, makes them a very valuable tool for the prevention, detection, and mapping of wildfires.

Detaining information on biomass burning impacts on Earth's land and atmospheric properties is included in reviews such as Akagi et al. (2011). Amongst the most recent burned area estimates are those of Giglio et al. (2010), who used datasets derived from (mainly optical) satellite remote sensing to map that between 1997 and 2008 global vegetation fires cumulatively burned 44.5 million km^2, equivalent to the combined area of North and South America, or ~40 % of Earth's total terrestrially vegetated area. Some of this burning is planned and under human control; other areas are ignited and left to spread largely unhindered by man; still others are ignited accidentally or by natural phenomena (primarily lightning). Indeed, such burning is in part responsible for maintaining the structure and function of savannah ecosystems, which in total constitute ~20 % of Earth's land surface area (Bond and van Wilgen, 1996). The maps of Giglio et al. (2010) indicate that, on average, ~3.4 % of the Earth's terrestrially vegetated area burns annually, with resulting in large-scale effects on surface properties and land cover, landscape heterogeneity, and ecology. Obviously, the importance of a wide-scale disturbance phenomenon like biomass burning is highly significant within the Earth's system. This includes impacts from large-scale slash-and-burn practices and severe forest fires, particularly in disturbed areas, that contribute significantly to tropical deforestation and forest degradation.

Wildfire detection and monitoring should be able to achieve the following outcomes: (1) recorded time and location of hotspots, (2) a good approximation to the actual burn area and extent, (3) real-time mapping of fire movement and spread of fires, (4) the impacts of the fire on vegetation regrowth, erosion, and other environmental characteristics (associated with burn severity) and (5) accurate and proper indicators that can reflect the changes of the vegetation for post-fire, etc. Thus, to provide a theoretical basis and data support for wildfire monitoring, prevention, and management, Landsat series, NOAA/AVHRR polar orbit environmental satellite (POES) series, and other satellite data are used in wildfire monitoring research. The earliest satellite platforms used in forest fire monitoring were NOAA/AVHRR series satellites and the Geostationary Orbiting Environmental Satellite (GOES) series. Up to now, these two series of satellites are still the main remote sensing monitoring platforms, playing an important role in the monitoring of forest fires, especially the AVHRR. NOAA/AVHRR data has been widely and deeply applied to various types of fire research on a global scale. In 1981, Dozier (1981) used AVHRR data to propose a theoretical model for studying sub-pixel temperature, which became the theoretical basis for future fire monitoring application models. Since then, scientists have continuously applied NOAA/AVHRR data to different regions and different ecosystems and then developed a series of fire monitoring models based on NOAA/AVHRR data into a global fire monitoring system. The GOES satellites are almost synchronized with the NOAA/POES series satellites, and the main application is the Visible Atmosphere Sounder (VAS) sensor mounted on the satellite. Compared with NOAA/AVHRR series satellites, GOES satellites have a higher temporal resolution. The GOES-8 satellite launched in 1994 scans every 15 minutes in North America and every half an hour in other regions. Although the NOAA/AVHRR series satellites only pass twice a day, NOAA satellites have better spatial resolution and large coverage globally (Franc and Cracknell, 1994). In order to improve the spatial and spectral resolutions of remote sensing data, NASA launched the Earth Observation System (EOS) satellite series. The medium-resolution imaging spectrometer (MODIS) has been greatly improved in terms of resolution and accuracy (MODIS has 36 channels of data). The accuracy and precision of wildfire monitoring have been greatly improved. At the same time, more detailed information has been obtained for the earth observations. It provides a new data source for local, regional, and global fire mapping (Key and Benson, 2003). Especially its higher resolution and extensive coverage, free-to-use functions are an excellent data source for the research of scientists from various countries, so MODIS remote sensing data has received special attention. Compared with AVHRR, MODIS has the following advantages: the sensitivity and quantization accuracy of MODIS sensor is much higher than that of AVHRR, so MODIS has increased the ability of detecting and measuring fire; MODIS has many channels for fire detection, which not only avoids the saturation problem of AVHRR but also analyzes the nature of wildfire qualitatively and quantitatively; MODIS has a precise positioning function from the very beginning and does not require users to perform complicated operations, which greatly improves the accuracy of fire positioning and the convenience of users; MODIS also provides a ground surface background with a resolution of 250 m, which is greatly conducive to intelligent wildfire prediction and high-speed judgment of disaster-affected areas. With the increase in the number and types of satellites, the temporal and spatial resolutions of data have gradually improved. Satellite imagery offers the advantages of extensive regional coverage, zero disturbances of the area to be viewed, as well as a method for acquiring data in less accessible areas on a regular and cost-effective basis. Therefore, the most commonly used remote sensing systems for wildfire monitoring research and applications are space-based, multispectral sensors, such as NOAA/VHRR and NASA/MODIS.

The focus of this chapter is primarily to explore the background and techniques related to the thermal remote sensing of the biomass burning itself. Hence, the theory and methods of both fire detection and mapping and fire characterization as well as smoke pluming monitoring from airborne and spaceborne platforms are briefly reviewed and discussed.

8.2 ACTIVE FIRE DETECTION

During the las several decades, researchers have developed many active fire detection algorithms. There are generally two types of active fire detection algorithms: fixed threshold algorithms and background contextual algorithms.

8.2.1 Fixed Threshold Algorithms

Commonly, mid-infrared (MIR) and thermal (TIR) bands are used for this purpose. Based on a review of fire properties by Matson and Dozier (1981) flaming temperature can be anywhere between 800 K and 1200 K and as hot as 1,800 K. Figure 8.1 shows the radiation emitted in the different bands of the electromagnetic spectrum by fires burning at several temperatures to indicate the significance of MIR band (AVHR channel 3) in active fire detection. The algorithm and sensitivity studies were based, therefore, on the assumption that the flaming temperature is 1000 K ± 200 K and the smoldering temperature is 600 K ± 100 K. These temperatures are easily detectable in the MIR part of the spectrum. Fires also radiate in the thermal part of the spectrum, i.e., between 8 μm and 12 μm; however, the peak radiation at these wavelengths corresponds to a normal environmental temperature of 300 K (Robinson, 1991). Flannigan and Haar (1986) were among the first studies to the fixed threshold method for active fire detection. Their method of hotspot extraction based on thermal infrared brightness temperature has laid a foundation for subsequent remote sensing in wildfire hotspot detection. They tested the ability of AVHRR to monitor a severe fire in north-central Alberta, Canada in 1982. Bands 3 (3.5–4 μm) and 4 (10.5–11.5 μm) were used in combination, and if the temperature for both channel 3 and channel 4 exceeded the mean background temperature, and if channel 3 was greater than channel 4 by a critical threshold, the fire was assumed to exist. They successfully detect 80% of the wildfires listed by the Alberta Forest Service. Kaufman et al. (1990) used a fixed threshold-based fire detection method with NOAA AVHRR satellite sensor data in an assessment of trace gases and particulate emissions from fires in Brazil. They noted that lower threshold values resulted

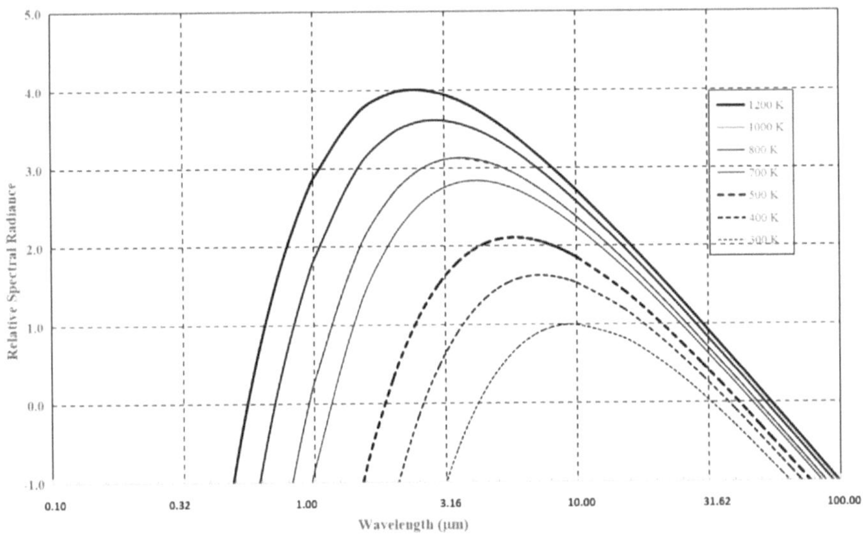

FIGURE 8.1 Radiation emitted in the different bands of the electromagnetic spectrum by fires of several burning temperatures.

in an increased number of false alarms caused by warm surface areas, such as hot desert pixels or areas that were overheated during summertime. To minimize the rate of false alarms, Kennedy et al. (1994) modified the threshold set by Kaufman et al. (1990) to detect the savanna fires in West Africa and improved fire detection effectiveness. Later, a comprehensive description of thresholding algorithms and their application can be found in Arino et al. (1993). Li et al. (2000) employed fixed multiple thresholds applied to multiple channels to detect active fires in the Canadian boreal forests. Wang et al. (2007) also adjusted the threshold set by Kaufman et al. (1990), and they found the temperature threshold of 320 K is too high and suggested the lowering of this threshold to 293 K, which they found significantly increased the ability of the fire detection algorithm to detect smaller wildfires. These studies show that a fixed threshold for wildfire detection may not be appropriate, and the algorithm thresholds may need adjustment for spatial and temporal variation if the lack of these adjustments may produce unacceptable rates of error.

One example of these fixed threshold algorithms is that proposed by Kaufman et al. (1990) for NOAA AVHRR satellite sensor data. According to this algorithm and area was classified as a fire when it met a set of constraints:

T3 (Temperature measured in the MIR channel) $\geq$ 320 K
T3 $\geq$ T4 (Temperature measured in the Thermal channel 4) + 10 K
T4 > 250 K

Therefore, fires can be detected as a local or absolute maximum in the MIR and TIR spectra. Fixed threshold algorithms are based on measuring the emitted energy for a pixel in either a single channel or over multiple channels and applying an empirical value (an absolute (or regional) maximum), and any pixel above a given threshold temperature is defined as containing a fire. Figure 8.2 presents a flowchart of the fire-detection algorithm with multiple channel thresholds' process-chain to produce an active fire (Li et al., 2000).

Fixed threshold algorithms are simple to apply and use, but differences in fire characteristics among regions in the world lead to problems of false alarms and/or missed fires using this method. It limited in use to local and regional scales and only to the particular season under study. Thresholds that have been developed and optimized for one location, such as woodland or savannah, and time, such as summer or dry/wet season, when applied to other operating environments can lead to significant errors of omission or commission.

8.2.2 Background Contextual Algorithms

Fixed threshold approaches can work well for individual scenes or time periods but often provide poor performance during multi-regional and/or multi-seasonal analyses and are not really appropriate for use in studies where the sensor used may change over time. To overcome the limitations in using fixed threshold algorithms, modern active fire detection algorithms use contextual methods, which are also known as relative or adaptive algorithms. Lee and Tag (1990) proposed one of the first contextual fire detection algorithms in their analysis of gas waste flares from AVHRR nocturnal

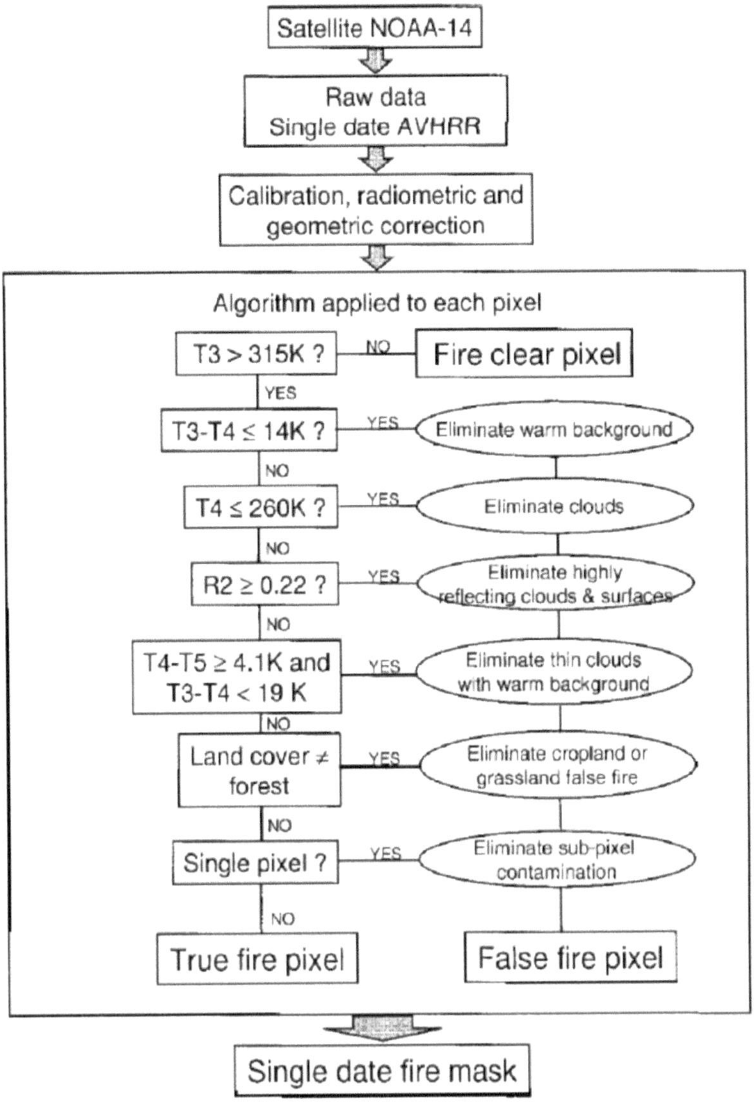

FIGURE 8.2 The flowchart of the hotspot detection algorithm for use with NOAA-14 AVHRR data. Presenting a process-chain to produce an active fire.

Source: Li et al., 2000.

data. Flasse and Ceccato (1996) proposed a contextual algorithm over central Africa to eliminate these false alarms. According to this algorithm, a hotspot is considered a fire only if its temperature is above the temperature of its surrounding areas. Ravail and San-Miguel-Ayanz (2002) evaluated the performance of Flasse and Ceccato (1996)'s contextual algorithm in Spain in 1997 and 1998 on AVHRR images. They found that over 90% of the detected hotspot pixel were false fire detections. These 'false alarms' were caused by the warm surface areas in Central Spain and by the systematic error on land-water interfaces.

Contextual algorithms use adaptive thresholds for fire detection. Multispectral criteria are aimed at detecting the difference between a fire pixel (active fire) and the background temperature (environmental temperature in the proximity of the fire pixel (Flasse and Ceccato, 1996; Giglio et al., 2003)). This is accomplished via statistical investigation of the background characteristics of the local pixels. These approaches generally have two pathways by which pixels containing actively burning fires can be identified. The first 'fixed threshold' pathway may use a single algorithm stage and is designed to detect pixels unambiguously containing large and/or intensely burning fires. The stage generally consists of thresholding tests in Figure 8.2 with fixed thresholds set sufficiently high such that in theory only pixels certain to contain fires pass the tests. The second 'contextual' pathway typically consists of two or more stages, whereby potential fire pixels (PFPs) are first identified using a fixed threshold algorithms based on relatively low thresholds (thus selecting many non-fire pixels as well as true fires) and then testing each PFP against the statistical properties of its immediately surrounding 'ambient background' pixels in order to confirm whether or not it is a 'true' fire pixel. The entire algorithm, incorporating the 'two pathway' procedure, is generally termed a 'contextual active fire detection algorithm.'

For instance, Gong et al. (2006) proposed an integrated algorithm; they combined the strengths of a fixed multi-channel threshold algorithm and an adaptive-threshold contextual algorithm and modified the fire detection algorithm developed by the Canada Center for Remote Sensing for fire detection in boreal forest ecosystems, to effectively eliminate false alarms caused by highly reflective clouds and surfaces and by warm backgrounds.

These contextual algorithms work much in the same way that the human eye will identify a fire visually by the contrast between a hotspot and the background. The relative nature of the algorithms means that contextual algorithms are more suited to the task of global fire detection in comparison to the fixed absolute threshold algorithms, by automatically adjusting the threshold levels for different regional and temporal conditions (Flasse and Ceccato, 1996). Therefore, most current methods for identifying actively burning fires include some form of contextual approach, including the algorithms used to generate the GEOS Advanced Biomass Burning Algorithm (ABBA) fire products, the MODIS Active Fire and Thermal Anomaly products (Giglio et al., 2003), the BIRD Hotspot Recognition Sensor (HSRS) fire products, and the Meteosat SEVIRI fire products.

Contextual algorithms have the advantage of making the detection process independent from the season and the zone analyzed since the thresholds are obtained by means of a statistical analysis of the environment. However, they have a serious drawback when they are applied to images in which the clouds have not been filtered since cloud edges cause false alarms. For the contextual algorithms, the selection of initial thresholds for identifying potential wildfire pixels is sensitive. An overly high setting leads to large omission errors, as the confirmation tests are only limited to potential fire pixels. On the other hand, an overly low setting introduces too much noise in determining the statistics for the background. Considering the different functions and consequences of threshold settings, it is recommended that more liberal thresholds be used to identify potential wildfire pixels and more conservative thresholds be used to calculate non-fire background statistics. Since the severity of wildfire varies considerably with fuel type, fuel amount, and weather conditions, the thresholds should also be contingent upon these variables. As a first approximation, one may use different thresholds over different land cover types in different seasons or set the threshold as a function of a vegetation index.

The following is a brief summary to introduce the active fire detection algorithm, as an example, Giglio et al. (2016) employed the contextual algorithm to perform active detection using the MODIS data to produce MODIS active fire product, Collection 6.

The detection algorithm uses native (i.e., unprojected swath) 4-, 11-, and 12-μm brightness temperatures derived from the corresponding 1-km MODIS channels, denoted by T_4, T_{11}, and T_{12}, respectively, and, for daytime observations, 0.65-, 0.86-, and 2.1-μm reflectance (denoted by $\rho_{0.65}$, $\rho_{0.86}$, and $\rho_{2.1}$, respectively), aggregated to 1-km spatial resolution. Full details to describe the algorithm may be found in the product User's Guide. A flowchart of the detection algorithm is shown in Figure 8.3 and the statistical quantity terms computed for contextual window is listed in Table 8.1. Pixels lacking valid data (e.g., during a calibration maneuver) are summarily assigned to a class of missing data and excluded from further consideration. According to functions in Figure 8.3, the following are brief descriptions to them.

8.2.2.1 LAND AND WATER MASKING

Water, land, and coastal pixels are identified using the updated 1-km land/sea mask contained in the MODIS Collection 6 MOD03/MYD03 geolocation product. Water and coastal pixels are not processed further by the detection algorithm.

8.2.2.2 CLOUD MASKING

The fixed thresholds used to mask clouds have been adjusted slightly for Collection 6 compared to Collection 5 algorithm. Daytime pixels—if the following combination of conditions is satisfied: $(((\rho_{.65} + \rho_{0.86}) > 1.2)$ or $(T_{12} < 265$ K) or $((\rho0.65 + \rho0{:}86) > 0.7)))$ and $((T_{12} < 285$ K) or (water pixel)) and $(\rho_{0.86} > 0.25)$ and $(T_{12} < 300$ K). Nighttime pixels are flagged as cloud if the single condition $T_{12} < 265$ K is satisfied.

8.2.2.3 IDENTIFICATION OF POTENTIAL FIRE PIXELS

As with previous versions of the algorithm, a preliminary classification is used to eliminate obvious non-fire pixels. Those pixels not eliminated during this stage are considered to be potential fire pixels and undergo further processing.

A daytime pixel is identified as a potential fire pixel if $T_4 > T_4^*$, $\Delta T > \Delta T^*$ (where $\Delta T \equiv T_4 - T_{11}$), and $\rho_{0.86} < 0.35$ (for nighttime pixels this last condition is dropped). For Collection 6, the thresholds T_4^* and ΔT^* are set dynamically for each of the 1,354 sample positions of the current MODIS scan (a MODIS scan is composed of ten lines, each containing 1,354 1-km samples, acquired in parallel). This along-scan adjustment helps compensate for local variations in the land surface as well as the systematic increase in atmospheric path length (and hence atmospheric absorption) with scan angle.

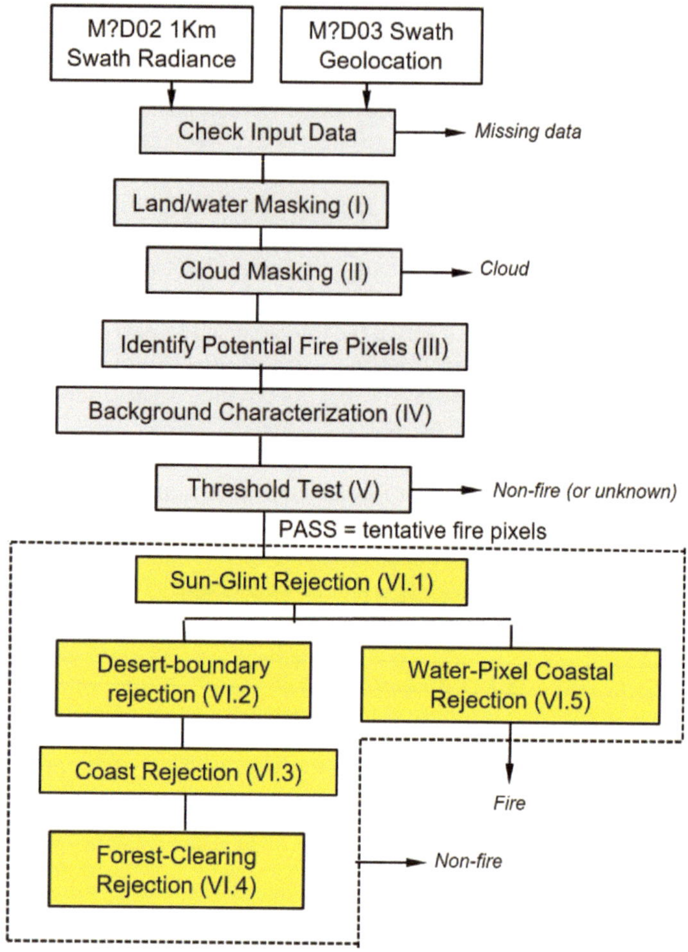

FIGURE 8.3 Flowchart of Collection 6 detection algorithm with possible output classes signified by italicized text. Grey boxes indicate algorithm steps performed for both daytime and nighttime pixels, while yellow boxes show steps restricted to daytime pixels. The individual rejection tests described in Section VI are shown inside the dashed box; those tentative fire pixels failing any rejection test are summarily classified as non-fire. Failure of the threshold tests (center box) applied to potential fire pixels will produce an output class of non-fire or, in those very rare cases when background characterization was not possible, unknown.

Source: modified from Giglio et al., 2016.

As a crude means of preventing intense fires from nontrivially biasing the broad-scale averages, pixels having $T_4 > 360$ K (320 K at night) are ignored during averaging. A potential fire threshold is then independently calculated for each sample position by applying a 5 K offset to the large-window averages. Specifically, $T_4^* = \text{mean}(T_4)_j + 5$ K, with the result constrained so that 300 K $\leq T_4^* \leq$ 330 K, and $\Delta T^* = \text{mean}(\Delta T)_j + 5$ K, with the result similarly constrained so that 10 K $\leq \Delta T^* \leq$ 35 K. Under this scheme, the two averages required for each sample position will be computed from up to 9,030 pixels. Averages based on fewer than 2,000 pixels are deemed unreliable, and for such cases the Collection 5 fixed potential fire thresholds are used as a conservative fallback since no reliable information about the large-scale local background is available to the algorithm. The dynamic thresholds calculated in this step are applied only to land pixels.

8.2.2.4 BACKGROUND CHARACTERIZATION

During the subsequent background characterization phase of the algorithm, the radiometric signal of each potential fire pixel in the absence of fire is estimated by computing statistical summaries of neighboring pixels within a small spatial window surrounding the potential fire pixel. Remains identical for Collection 6, with the important exception that valid background pixels are now restricted to either land or water pixels (but never both) to match the land/water state of the potential fire pixel. In lieu of a detailed description of the local statistics computed for the background window, Table 8.1 provides a summary, and the reader is referred to Giglio et al. (2003) for additional information. New quantities introduced in Collection 6 include the mean 4-μm radiance, $\overline{\rho_{0.68}}$; the mean 0.86-μm reflectance (daytime pixels only), N_c; the number of coast pixels within the background; and N_L, the number of land pixels excluded from the background when the potential fire pixel lies over water.

8.2.2.5 TENTATIVE FIRE DETECTION

At this phase, the algorithm tentatively identifies pixels containing active fires. For nighttime land pixels, this will in fact be a final identification. For water and daytime land pixels, a subsequent series of rejection tests will be performed, as described in the following subsection. The absolute threshold criterion of Kaufman et al. (1998) remains unchanged for Collection 6:

$$T_4 > 360\text{ K (320 K at night).} \tag{8.1}$$

Being deliberately conservative, test (8.1) provides a direct route by which large and/or particularly intense active fires can be identified. A suite of contextual tests, which offer much greater sensitivity to the much larger proportion of smaller and/or cooler fires, are consequently used to supplement the fixed threshold test. In addition to radiometric information about the potential fire pixel itself, the contextual tests make use of information extracted from the immediate neighborhood—or 'background'—as described earlier. These tests are as follows (a detailed description of each test may be found in Giglio et al. (2003)):

$$\Delta T\ \overline{\Delta T} > +3.5\delta_{\Delta T} \tag{8.2}$$

$$\Delta T\ \overline{\Delta T} > +6\text{ K} \tag{8.3}$$

$$T_4 \overline{T_4} > +3\delta_4 \tag{8.4}$$

$$T11\ \overline{T_{11}} > +\delta_{11} - 4\text{ K} \tag{8.5}$$

$$\delta_4^{'} > 5\text{ K} \tag{8.6}$$

TABLE 8.1
Statistical Quantities Computed for Contextual Window. The Acronym MAD Denotes the Mean Absolute Deviation. More Detailed Descriptions Can Be Found in Giglio et al. (2003).

Variables	Description
$\overline{L_4}$	Mean 4-µm radiance
$\overline{T_4}$	Mean 4-µm brightness temperature
$\overline{T_{11}}$	Mean 11-µm brightness temperature
$\overline{T_\Delta}$	Mean brightness temperature difference (≡ T4 – T11)
δ_4	4-µm brightness temperature MAD
δ_{11}	11-µm brightness temperature MAD
$\delta_{\Delta T}$	Brightness temperature difference MAD
N_v	Number of valid pixels within contextual window
$\overline{\rho 0.68}$	Mean 0.86-µm reflectance (daytime pixels only)
$\overline{T'}4$	Mean 4-µm brightness temperature of pixels rejected as background fires
δ_4'	4-µm brightness temperature MAD of pixels rejected as background fires
N_f	Number of pixels within contextual window rejected as background fires
N_C	Number of coast pixels (always excluded) within the contextual window
N_W	Number of excluded water pixels within the contextual window
N_L	Number of excluded land pixels within the contextual window

A daytime potential fire pixel is tentatively classified as a fire pixel if either (1) test (8.1) is satisfied or (2) tests (8.2) through (8.4) are satisfied and either test (8.5) or test (8.6) is satisfied, otherwise it is classified as a non-fire pixel. A nighttime potential fire pixel is definitively classified as a fire pixel if either (1) test (8.1) is satisfied or (2) tests (8.2) through (8.4) are satisfied, otherwise it is classified as a non-fire pixel. For potential fire pixels in which background characterization was not possible (because too few valid background pixels were available), only test (1) is applied. If satisfied, the potential fire pixel is tentatively classified as a fire pixel, otherwise it is assigned a final class of unknown.

8.2.2.6 REJECTION TESTS

In the final phase of the detection algorithm, pixels tentatively flagged as containing active fires undergo a series of false-alarm rejection tests. These include a sunglint rejection test, applied to both land and water pixels, followed by three tests applied only to tentative land fire pixels: a desert boundary rejection test, a land-pixel coastal rejection test, and a new forest clearing rejection test, supplemented with a new water-pixel coastal rejection test reserved for tentative fire pixels detected over water. Once all applicable rejection tests have been carried out, those tentative fire pixels that remain are definitively classified as fire, while those that were at any point rejected are classified as non-fire.

8.2.2.6.1 SUN GLINT REJECTION

The sun glint rejection test remains basically the same as Collection 5 for Collection 6, though two angular thresholds were increased to make the test slightly more aggressive in rejecting false alarms. The tentative fire pixel is rejected if one or more of the following conditions are satisfied:

$$\theta_g < 2^o \tag{8.7}$$

$$(\theta_g < 10^o) \text{ and } (\rho_{0.65} > 0.1) \text{ and } (\rho_{0.86} > 0.2) \text{ and } (\rho_{2.1} > 0.12) \tag{8.8}$$

$$(\theta_g < 15^\circ) \text{ and } (N_{aw} + N_w) > 0) \tag{8.9}$$

where., N_{aw} is the number of water pixels within the eight pixels surrounding the tentative fire pixel, and θ_g is the angle between vectors pointing in the surface-to-satellite and specular reflection directions (Giglio et al., 2003). Note that test (8.9) effectively renders the detection of offshore gas flares impossible in the presence of sun glint.

8.2.2.6.2 DESERT BOUNDARY REJECTION

The specific thresholds used in the MODIS algorithm largely restrict this type of false alarm to the perimeter of hot, arid, and sparsely vegetated or barren regions most commonly associated with deserts. The desert boundary rejection test, which was introduced in Collection 4 to help eliminate such false alarms, remains unchanged for Collection 6. A detailed description of the test can be found in Section 2.2.7 of Giglio et al. (2003).

8.2.2.6.3 LAND-PIXEL COASTAL REJECTION

When processing a potential fire pixel over land, unmasked water pixels within the background window can depress both $\overline{T_4}$ and $\overline{\Delta T}$, potentially generating a false alarm (Giglio et al., 2003). The land-pixel coastal rejection test, which employs the 0.86- and 2.1-μm reflectance and Normalized Difference Vegetation Index (NDVI) of valid background pixels, is designed to prevent such false alarms. A complete description of the test can be found in Section 2.2.8 of Giglio et al. (2003).

8.2.2.6.4 FOREST CLEARING REJECTION

For Collection 6, a new rejection test is introduced to reduce the frequency of the tropical forest clearings as fires of commission error. A daytime fire pixel (land only) is rejected as a false alarm if $T_{11} > \overline{T_{11}} + 3.7\delta_{11}$, $\rho_{0.86} > 0.28$, and $T_4 < 325$ K are satisfied. The first of these three conditions exploit the fact that forest clearings tend to be significantly warmer than the surrounding intact forest, while the second helps restrict the test to larger patches of tropical forest (which are bright in the near infrared) where a ~1-km clearing might be encountered. The third condition limits the test to fire pixels having a comparatively weak fire signature, which is a consistent property of this type of false alarm.

8.2.2.6.5 WATER-PIXEL COASTAL REJECTION

Whereas contamination of the background window with unmasked water pixels can produce a false alarm over land (Subsection 8.2.2.6.3), a similar class of coastal false alarm can occur over water when the potential fire pixel itself is an unmasked land pixel. As a misidentified land pixel, both T_4 and ΔT will be elevated relative to the (generally) cooler water background, to the point that the unmasked land pixel may resemble a small and/or low intensity fire. This scenario is most likely to arise in the vicinity of coastline, where the distinction between land and water pixels is inherently ambiguous. The presence of land or coast pixels in the background window is an obvious indicator for precisely this situation, thus we reject any tentative fire pixel detected over water for which $N_L+N_c > 0$ and test (8.1) are satisfied.

8.3 BURNING SCAR MAPPING

Burned area (BA) estimates are of critical importance for land managers, climate scientists, and policymakers. BA estimates provide accurate spatial representations of wildfire extents and perimeters. Detailed and current information concerning the location and extent of the BA are important for assessing economic losses and ecological effects, monitoring land use and land cover changes, and modeling atmospheric and climatic impacts of biomass burning. Biomass burning produces a significant change in the structure and the reflectance of vegetation and the soil properties within the BA that are noticeable differences in the microwave, visible, and especially the infrared part of the electromagnetic spectrum compared to these for non-burned areas.

There appears to be a reasonable degree of agreement concerning some basic aspects of BA detection and mapping. For example, the NIR (Figure 8.4, channel 2 of AVHRR sensor) is considered the best spectral region for the detection of burned surfaces by a large number of authors working on a broad range of ecosystems (e.g., Razafimpanilo et al., 1995; Pereira et al., 1997). Remote sensing data (e.g., Roy et al., 2002; Li et al., 2004; Loboda et al., 2007; Veraverbeke et al., 2011) has suggested that spectral separability between burnt and unburnt areas is strongest in the NIR and SWIR range of the reflectance spectrum (Figure 8.4). Since healthy vegetation tends to be very reflective in the NIR, while charred surfaces are strong absorbers of radiation in this spectral range, the contrast between the two cover types tends to be very clear (Figure 8.4). When vegetation is burned, there is a drastic reduction in visible-to-NIR reflectance and an accompanying increase in the SWIR surface reflectance of most satellite sensors. Burned patches are relatively easy to discriminate visually for this reason. NIR detectability of the scar signal will be higher if the soil background of bare or sparsely vegetated patches is at least moderately dark. However, due to the wide range of spectral signatures and spatial heterogeneity caused by fire regimes, the type of vegetation burned, and the environmental conditions, it is difficult to detect the BA automatically.

The commonly used mapping methods for burned areas are generally classified into four categories:

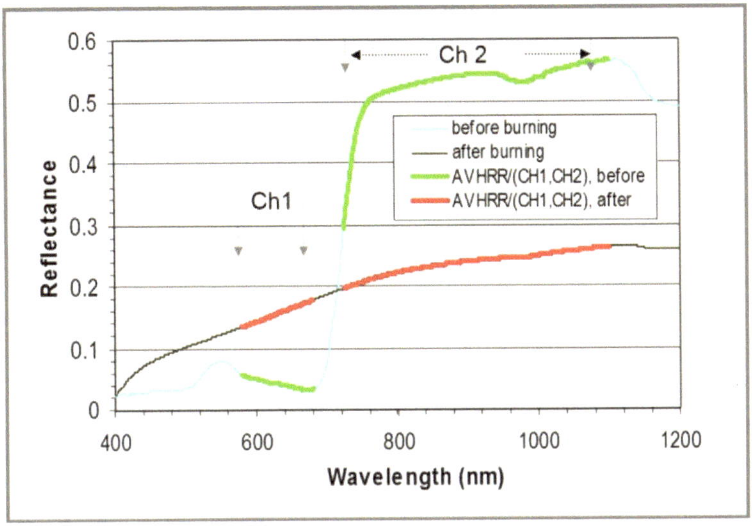

FIGURE 8.4 This diagram shows you the NDVI difference between before and after burning. In this diagram, green line is normal vegetation area and red line is after burning. It is easy to see NDVI decreasing after burning. So the NDVI difference image can be used for wildfire mapping.

(1) Image classification (e.g., Pereira et al., 1999).
(2) Image processing (e.g., Koutsias et al., 2000).
(3) Spectral index (e.g., Quintano and Fernández-Manso, 2011).
(4) Integrated approach (e.g., Mouillot et al., 2014).

Image classification is based on the spectral difference of images before and after burning, using image classification techniques, such as supervised classification, regression tree classification, object-oriented classification, etc., to distinguish burned and non-burned areas. Recently machine leaning (ML) classifiers have been of increasing interest to many researchers to map burned area. Otón et al. (2019) reported that a synthetic burned area (BA) index was created to improve the detection of the BA signal, which included red and near infrared reflectance, surface temperature, two spectral indices, and their temporal differences; and monthly models were generated using the random forest classifier, using the 12-monthly composites of each year as the predictors. The final product (named FireCCILT10) estimated BA in 0.05-degree cells for the 1982–2017 period. This product is the longest global BA currently available, extending almost 20 years back from the existing NASA and ESA BA products. BA estimations from the FireCCILT10 product were compared with those from the NASA MCD64A1 product for continental regions, obtaining high correlation values ($r^2 > 0.9$) with better agreement in tropical regions rather than boreal regions. After analyzing the contribution of different features derived from optical, thermal, and Synthetic Aperture Radar images to extract BA information from the Turkish red pine (*Pinus brutia* Ten.) forest in a Mediterranean ecosystem,

Abdikan et al., (2022) employed the pixel-based random forest image classification method to classify the BA detection in 24 scenarios created using these features. Their results show that the L-band data provided a better contribution than C-band data and the combination of features created from Landsat LST, normalized burn ratio index, and coherence of L-band ALOS-2 achieved the highest accuracy, with an overall accuracy of 96% and a Kappa coefficient of 92.62%. The principle of image processing is to use principal component analysis, histogram transformation, spectral mixture analysis, and other image processing technologies to obtain more targeted information from the images for burned and non-burned areas, so as to improve the mapping and monitoring accuracy of the burned area. The spectral index method is to extract the burned area according to the change of vegetation index or burned area index. The commonly used indexes are NDVI, global environment monitoring index, burned area index, normalized burn ratio, and disturbance index. The spectral indices are very popular tools for the BA mapping, both in single date and in multitemporal frameworks because of their clear concept and high calculation efficiency. The integrated method is to use multi-source data information, such as fire point data, ground truth data, and remote sensing images with different resolutions, to integrate the advantages of different data sources, so as to improve the detection accuracy of BA. However, the spectral properties of burned surfaces are prone to confusion with various land cover types, affecting the accuracy of BA estimates derived from remotely sensed data. The development of methodologies capable of producing more accurate BA estimates from remotely sensed data is, therefore, an active topic of research at geographic scales ranging from local to global.

Hotspots have been typically used as seeds or indicators of burn scars in various mapping approaches taking advantage of region growing algorithms. Active fire detection can also be used to locate training samples to define spectral characteristics for BAs. Active fire data have also directly been used as the basis of a global BA estimation method correlating active fire detection with BA mapping results (Giglio et al., 2006). It must be noted, however, that Giglio et al. (2006) pointed out that this approach faced serious problems, particularly in persistently cloudy humid tropics like insular Southeast Asia. While the focus of the spaceborne and airborne thermal remote sensing methods discussed thus far has been on the detection of actively burning fires, such measurement capabilities also have some relevance to the identification and mapping of BAs. This has often been simply using active fire detections to gauge the ultimate size of the fire-affected area, generally via some form of empirical relationship linking the number of active fire detections to the size of the area burned (e.g., Giglio et al., 2006). Since fire behavior varies greatly between environments, such associations require careful testing and calibration. Reported slopes of the linear relationships linking the two metrics (i.e., active fire count and its BA) ranged in the global study by Giglio et al. (2006) between 0.29 km^2 of burned area per active fire pixel in southern-hemisphere South America to 6.6 km^2 of burned area per active fire pixel in Central Asia (Table 8.2). Once such relationships are established for the areas of interest, the types of long term active fire detection records available from such sources as MODIS, Tropical Rainfall Measuring Mission, Visible and Infrared Scanner (TRMM VIRS) and Advanced Along Track Scanning Radiometer (AATSR) can be used to estimate BA trends,

TABLE 8.2

Correlation (r) Between Predicted and Observed Burned Area Within Each Region for Linear Regression and Regression Tree Approaches. The Slope of the Linear Regression (α) and the Total Number of Non-zero Observations (N) Are Also Shown. All Correlations Are Highly Significant with a Probability $p < 0.001$ (Giglio et al., 2006)

Region	The slope of linear regression (km^2/pixel)	r	N	Tree (r)
Boreal North America	1.4	0.69	1,018	0.85
Temperate North America	0.8	0.94	982	0.98
Central America	0.4	0.73	301	0.85
NH South America	1.0	0.78	352	0.00
SH South America	0.3	0.35	4,034	0.56
Europe	3.1	0.91	225	0.95
Middle East	0.4	0.34	215	0.78
NH Africa	5.2	0.86	910	0.90
SH Africa	2.9	0.60	1,670	0.73
Boreal Asia	1.3	0.90	2,104	0.94
Central Asia	6.6	0.85	282	0.92
South Asia	2.9	0.75	531	0.83
Equatorial Asia	0.5	0.71	192	0.78
Australia	3.4	0.82	5,563	0.89

Note: NH—Northern Hemisphere, SH—South Hemisphere

provided of course that the active fire products are appropriately inter-calibrated for differences in sensor spatial resolution (and thus minimum active fire size), satellite overpass time, and image repetition frequency.

In terms of approaches to exploiting thermal band data in BA mapping algorithms, probably the most successful has been the inclusion of active fire detections into BA data processing chains. Specifically, the locations of detected active fire pixels (made using the types of thermally based algorithms covered in Section 8.2) can very usefully act as 'seed locations' for optical waveband change detection methods aimed at identifying new BAs from optical wavelength data. Among the earliest examples is the Hotspot And NDVI Differencing Synergy (HANDS) algorithm, used to map BAs across the Canadian boreal forest (Fraser et al., 2000). HANDS synergistically combines two methods commonly used for burn assessment: hotspot detection and multitemporal NDVI differencing. Hotspots are used to dynamically train an NDVI difference threshold for separating scars. NDVI differencing is used to remove falsely detected hotspots and to fill in patches within individual burns that were not detected as hotspots. The general strategy is to first confirm hotspot pixels using an NDVI difference image. Confirmed hotspots are then used to derive coarse, regional-level NDVI difference thresholds, followed by more restrictive local thresholds. The local thresholds are derived for individual burn patches, which are isolated using GIS clumping techniques. The overall result is a close approximation to actual burn boundaries, within the limits of the input data. Using the technique, burn mapping may be performed in a consistent and automated manner over large regions. Table 8.3 summarizes steps of the modified HANDS algorithm applied in a case of mapping burned scars in California, US (Gong et al., 2006). More recently Giglio et al. (2009) utilized the technique to produce a global, multi-year burned area product from the 10+ year MODIS data record, a product that is now used as the primary burned area dataset within version 3 of the GFED.

8.4 FIRE RADIATION POWER RETRIEVAL

The widespread, highly dynamic nature of biomass burning means that earth observation (EO) from space is considered key to characterize its magnitude, impact, variability, and trend for use in earth system models. Polar orbiting instruments such as AVHRR and more latterly the MODIS have long been used to investigate

TABLE 8.3
A Summary of the Modified HANDS Algorithm (Gong et al., 2006)

Step	Description
Step 0	AVHRR data preparation: two NDVI composites of an interesting interval, one corresponding hotspot composite (fire mask)
Step 1	Normalize NDVIpost to NDVIpre: means of NDVIs calculated only with interested land covers $Ratio.C = \frac{mean\,of\,NDVIpre}{mean\,of\,NDVIpost}$ normalized NDVIpost = Ratio.C * NDVIpost, not hotspot pixels used for calculation of NDVI mean.
Step 2	Calculate NDVI difference: normalized NDVIpost – NDVIpre, burn scar expected to have a negative value in the NDVI difference image.
Step 3	Confirm hotspot pixels (CBP) using NDVI difference: a CBP is assumed to have a negative NDVI difference, otherwise the hotspot pixel is considered as a false fire.
Step 4	Calculate NDVI difference statistics (mean, standard deviation, SD) of NDVI decrease for all CBPs for each land cover type, different land covers expected to have a different mean and SD.
Step 5	Select potential burn scar pixels (BSPs): NDVI difference of a potential BSP < (mean + 0.5SD), the potential BSPs selected for different land cover types with the mean and SD from Step 4.
Step 6	Apply a sieve filter to the selected BSPs: filter out a burnt patch of < 3 pixels, a burn patch less than three pixels is considered to be caused by noise.
Step 7	Confirm a BSP with a neighbor CBP and later on up to four neighbor confirmed BSPs to create connected burn patches, first iteration with a neighbor CBP, second iteration with a neighbor CBP or a conformed BSP (CBSP), third to fifth iteration with two to four CBSPs, after fifth iteration using four CBSPs only.
Step 8	Filter out a BSP patch of < 3 pixels: the patch of < 3 CBSPs is considered to be a false burnt patch, caused by noise.
Step 9	Output a burnt area mask (in TIFF format).

the regional and global and seasonal patterns of biomass burning via the thermally based detection of actively burning landscape fires and more recently the fire radiative power (FRP) has been assessed via EO in order to provide information relatable to fuel combustion and smoke emission rate (Ichoku and Kaufman, 2005; Wooster et al., 2005; Giglio et al., 2006; Ichoku et al., 2008). Further advances have come through use of higher spatial resolution polar orbiting satellite sensors such as the Bi-spectral Infrared Detection (BIRD) satellite, which are more suited to measuring the FRP of even rather small/low intensity fire events (Zhukov et al., 2006).

Biomass burning emits thermal radiation over a very wide range of wavelengths, peaking in the middle infrared (MIR) spectral region. Planck's Radiation Law indicates that at MIR wavelengths (Figure 8.1), the radiative emission from a flaming vegetation fire may be 10,000× more intense than that from the surrounding ambient background, whilst at TIR wavelengths the intensity is orders of magnitude lower than in the MIR (Robinson, 1991). This differential emission signature enables even small active fires (e.g., those covering « 1% of the pixel area) to raise pixel integrated MIR brightness temperatures (BTs) significantly above those of the surrounding background. Such properties are the foundation of most active fire detection algorithms.

Once detection of a fire pixel has been unambiguously confirmed, the fire activity contained within it can be characterized in terms of its FRP, a measure of the rate of radiant energy emission from the fire integrated over all wavelengths and directions. Wooster et al. (2003, 2005), Ichoku and Kaufman (2005), Freeborn et al. (2008), and Ichoku et al. (2008) have provided a full description of how this measurement can be derived from EO imagery and how it relates to rates of combustion and smoke emission.

There are two approaches that are generally used for measuring FRP, (1) the bi-spectral method coupled to the Stefan-Boltzmann Law, which employs the differential fire signal in two or more co-registered spectral channels or (2) single band approaches that use the fire signal in the MIR channel only. Bi-spectral or multi-spectral methods appear most appropriate for application to higher spatial resolution data (Wooster et al., 2003; Zhukov et al., 2006; Dennison et al., 2006), and their use with coarser spatial resolution imagery can be affected by difficulties in separating the fire signal from the background signal in channels and by interchannel spatial misregistration effects (Giglio and Kendall, 2001; Giglio and Justice, 2003; Wooster et al., 2003). On the contrary, the single band approach or MIR radiance method of FRP estimation relies on quantifying the fire signal in the MIR only, where its strength is maximized (Wooster et al., 2003, 2005):

$$FRP_{MIR} = \frac{A_{samp}\sigma\varepsilon}{a\varepsilon_{MIR}} L_{MIR} \tag{8.10}$$

where A_{samp} is the pixel sampling area (m^2); σ is the Stefan-Boltzmann constant (5.67×10^{-8} J s^{-1} m^{-2} K^{-4}); ε is the fire broadband emissivity and ε_{MIR} the spectral emissivity in the MIR (fires are assumed to be grey bodies so that $\varepsilon = \varepsilon_{MIR}$); a is the power-law scaling constant whose derivation is described in Wooster et al. (2005) (mW m^{-2} sr^{-1} $(cm^{-1})^{-1}$ K-4) and L_{MIR} (mW m^{-2} $sr^{-1}(cm^{-1})^{-1}$) is the contribution of the fire to the total MIR spectral radiance of the fire pixel (i.e., with the contribution from the ambient background part of the pixel removed, usually via differencing the fire pixel from mean surrounding background pixel signals). Integrating the FRP over the lifetime of a fire pixel provides a measure of the total fire radiative energy (FRE) emitted during the combustion process, which in turn can be used to estimate how much fuel must have been burned to produce this radiative energy and via the application of the appropriate emissions factor the amount of trace gas or aerosol released (Freeborn et al., 2008; Ichoku et al., 2008; Wooster et al., 2005).

Since the late 1990s, active fire (AF) detections have been increasingly accompanied by efforts at fire characterization, mostly in terms of retrievals of fire radiative power (FRP; usually expressed in MW). An empirically derived algorithm for direct estimation of FRP was first proposed and demonstrated with MODIS airborne simulator data by Kaufman et al. (1998)—Eq. (8.11). An underlying assumption is that, because heat yields are relatively constant among vegetation types (Stott, 2000), remotely sensed FRP retrievals provide data useful for estimating rates of fuel consumption and smoke emission, as first demonstrated by Kaufman et al. (1998), Wooster et al. (2005), Freeborn et al. (2008), and Ichoku et al. (2008), respectively. Eq. (8.11) was used to retrieve FRP in the early (Collections 1–4) MODIS AF Products (Giglio et al., 2003), in units of emitted power per unit area of the pixel:

$$\mathrm{FRP} = 4.34\times10^{-9}\sum\left(T^{8}_{MIR,fire} - T^{8}_{MIR,bg}\right) \tag{8.11}$$

where $T_{MIR,fire}$ and $T_{MIR,bg}$ are the MIR BT (K) of the AF pixel and the mean of the surrounding 'background' pixels, respectively.

Whilst Eq. (8.11) performs well for MODIS, its empirical nature means it is specific to data from that sensor. It was found that it underperformed when applied to finer spatial resolution data that record higher BTs due to fires covering a greater proportion of their pixel area (Section 3) (Wooster et al., 2003). Wooster et al. (2003, 2005) derived a more physically based approach to FRP retrieval, based on a power-law approximation to the Planck function that linearly related FRP (MW) to the AF pixels' excess MIR spectral radiance above the background:

$$\mathrm{FRP} = \frac{A_{samp}\sigma\varepsilon}{a\varepsilon_{MIR}}\left(L_{MIR,fire} - L_{MIR,bg}\right) \tag{8.12}$$

where L_{MIR} is the MIR spectral radiance of the AF pixel (W m^{-2} sr^{-1} μm^{-1}), and $L_{MIR,bg}$ is the estimate of what the AF pixel spectral radiance would be if it did not have a fire within it (typically taken as the mean or median MIR spectral radiance of the surrounding background pixels). For other terms' meaning, see Eq. (8.10).

Moving beyond the single-band FRP retrieval methods discussed earlier, another approach to FRP estimation is to exploit outputs of the 'bi-spectral' method introduced earlier, namely the effective fire temperature (T_f, K) and sub-pixel proportion (p_f) of the Matson and Dozier (1981) and Matson and Holben (1986) approach:

$$L_{MIR} = \tau_{MIR} p_f B_{MIR}(T_f) + (1-p_f)L_{MIR,bg} \tag{8.13}$$

$$L_{LWIR} = \tau_{LWIR} B_{LWIR}(T_f) + (1-p_f)L_{LWIR,bg} \tag{8.14}$$

$$FRP = \sigma\left(T_f^4 - T_{bg}^4\right)p_f A_f \tag{8.15}$$

where L_x is the AF pixel spectral radiance (W m^{-2} sr^{-1} µm^{-1}) observed in the denoted spectral band x; $B_x(T)$ is the Planck function (W m^{-2} sr^{-1} µm^{-1}); τ is the atmospheric transmittance; $L_{X,bg}$ is the ambient background spectral radiance (i.e., non-fire, W m^{-2} sr^{-1} µm^{-1}); σ is the Stefan-Boltzmann constant (5.67 × 10^{-8} J s^{-1} m^{-2} K_4) and $T_{x,bg}$ is the brightness temperature (K) of the ambient background in band x.

8.5 SMOKING PLUME MONITORING

Biomass burning emits a large amount of greenhouse gases and aerosol particles into the atmosphere. Approximate estimation showed that the annual amount of CO released into the atmosphere due to biomass burning is about 114 Tg in the tropics and 62.3 Tg in boreal zone. Trace gases and aerosol particles produced by fires play important roles in atmospheric chemistry, and fire impact on weather is mainly due to the attenuation of sunlight by smoke particles, which is usually short lived (Li et al., 2001). Smoke particles are composed of ~50–60% organic carbon and ~5–10% black carbon. Biomass smoke particles effectively scatter and absorb solar radiation. To a lesser extent, smoke can have an impact that extends far beyond the region of fire activity. Smoke plumes may travel over hundreds or even thousands of kilometers horizontally and reach up to the stratosphere under certain atmospheric circulation conditions (Fromm et al., 2000) (Figure 8.5).

The impact of smoke is twofold: cooling due to smoke particles and warming due to greenhouse gases. Smoke particles scatter and absorb incoming solar radiation, thus having a cooling effect at the surface but a warming effect on the atmosphere. Since the magnitude of the scattering effect outweighs that of absorption, smoke has a net cooling effect at the top of the atmosphere-surface system. Smoke can also modify the short-wave reflective properties of clouds by acting as cloud condensation nuclei. Under a limited supply of water vapor, an increased number of nuclei result in smaller cloud droplets that have higher reflectivity than larger cloud droplets. The cooling due both to the direct and indirect effects of smoke could potentially offset

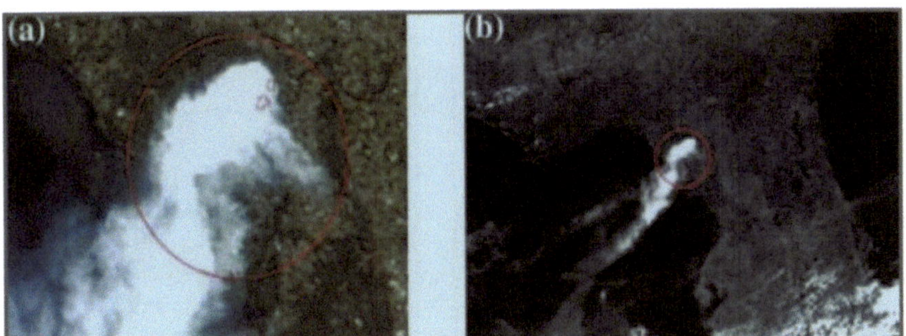

FIGURE 8.5 Observation of fire smoking in Florida (USA) from the MODIS (a) and the AVHRR (b) sensor.

Source: courtesy of the NASA Earth Observatory.

the warming effect of increasing CO content, but they act to vary spatiotemporally. Understanding such numerous and complex effects of smoke on weather and climate requires a good knowledge on the spatial and temporal variation of smoke and its optical properties, which is solely feasible by means of satellite observation. Discrimination of smoke on satellite imagery is a prerequisite to study and retrieve physical, chemical, and optical properties of smoke.

The determination of key parameters for estimating atmospheric effects of biomass burning is not straightforward. Smoke properties vary between fires depending on fuel type and moisture, combustion phase, wind conditions, and several other variables. Also, as the physical, chemical, and optical properties of biomass-burning aerosols can change rapidly as they disperse, it is difficult to relate the properties of individual fires to the ensemble smoky hazes that influence the atmosphere's radiative balance. Determining the impacts of these hazes on the meteorology of a region is hampered by high uncertainty in both the measurement methodologies and in the models themselves. As demonstrated later, there is a large overlap in the spectral signature of satellite measurements between smoke and other scene types such as clouds and background surfaces. So far, very few investigations have focused on the identification of smoke except for some studies that used somewhat ad hoc approaches to identify smoke for pursuing other research themes. The commonly used methods may be classified three categories: RGB false color composite image, multi-threshold, and machine-learning (ML) methods. The following is a brief introduction to the three categories of methods.

8.5.1 RGB False Color Composite Image

The most common identifying smoke is to assign different colors to different channels or channel combinations. The resulting false-color images can provide visual separation of smoke from other objects. This method achieves an appropriate color display of ground objects for visual interpretation by regulating the band combinations. Although fast and straightforward, this method has low accuracy. For example, Kaufman et al. (1990) assigned AVHRR channel 1 to red, channel 2 to green, and inverse channel 4 to blue, generating a composite image showing smoke plumes. Such an approach can hardly be used for automatic processing masses of satellite imageries. To fully exploit the geographic and spectral information provided by airborne or space-borne sensors, multi-threshold and ML methods have been proposed to extract smoke plumes.

8.5.2 Muti-Channel Threshold

To use more spectral information, multi-thresholds concerning the linear relationship among multiple bands classify smoke features (Li et al., 2001; Xie et al., 2007; Wang et al., 2011).

For example, Li et al. (2001) used the multi-threshold approach to identify smoke from their underlying surfaces based on differences in the reflectance of AVHRR channels 1 and 2 and in the brightness temperature of channel 4. They first converted AVHRR data gray level to [0,1] relative values:

$$R = (V - V_{min})/(V_{max} - V_{min}) \tag{8.16}$$

where V_{min} and V_{max} denote the minimum and maximum VHRR channel observation values. Then the multi-threshold approach involves two major steps: marking potential smoke pixels and removing false pixels. Both steps are accomplished using threshold tests. The first employs thresholds of the ratio of AVHRR channel 1 and channel 2 reflectance (R1 and R2) and a brightness temperature of AVHRR channel 4 (BT4):

$$0.9 \leq R2/R1 \leq 0.5 \text{ and } BT4 \leq 298 \text{ K} \tag{8.17}$$

Pixels passing this test are considered to be either smoke or cloud. Otherwise, they are deemed land pixels. The second test employs channel 4 to further separate smoke and cloudy pixels:

$$BT4 \leq 280 \text{ K} \tag{8.18}$$

Pixels passing this test are regarded as clouds as they are usually higher and colder than smoke. The third test is introduced to eliminate warm non-bright clouds from 'smoke' pixels:

$$BT4 \leq 284 \text{ K and } R1 \geq 0.35 \tag{8.19}$$

Xie et al. (2007) proposed a multi-threshold method for detecting smoke plumes with eight MODIS spectral bands based on the analysis of spectral characteristics of different cover types. A series of tests are applied to all pixels to filter out non-smoke pixels step by step with water masking. At each step, specific thresholds are utilized. The results have been validated with true color images for a number of cases from different areas and time, showing that the algorithm works well except for a few missing or incorrect identified smoke pixels. Figure 8.6 presents a flowchart of the multi-threshold method for smoke detection (Xie et al., 2007). Lin et al. (2010) developed a lookup table to detect smoke plumes with MODIS data over land. They used biomass burning cases that are from two separate areas—the California forest fire in October 2003 and the Indochina case in March 2004. The MODIS dataset of the former event was used for the analysis of the radiometric characteristics of smoke plumes. Accordingly, the MODIS reflectance measurements can be applied in identifying areas covered by smoke plumes via the constructed lookup table (Table 8.4). If the reflectivity of a MODIS land (ocean) pixel satisfies all of the criteria in Table 8.4, the pixel will be defined as a smoke covered pixel. For the validation of the constructed lookup table, the satellite-observed dataset of the severe biomass burning event (March 24–29, 2004) over the Indo-China Peninsula was examined. The results of the smoke plume detection are further compared with the TOMS observed aerosol index. The strong agreement of the smoke covered areas is obvious. Although the training area and testing area are different, both results of the smoke plume detection are acceptable. Meanwhile, the number of biomass burning cases discussed in this research is still insufficient to conclude a representative result. In order to increase

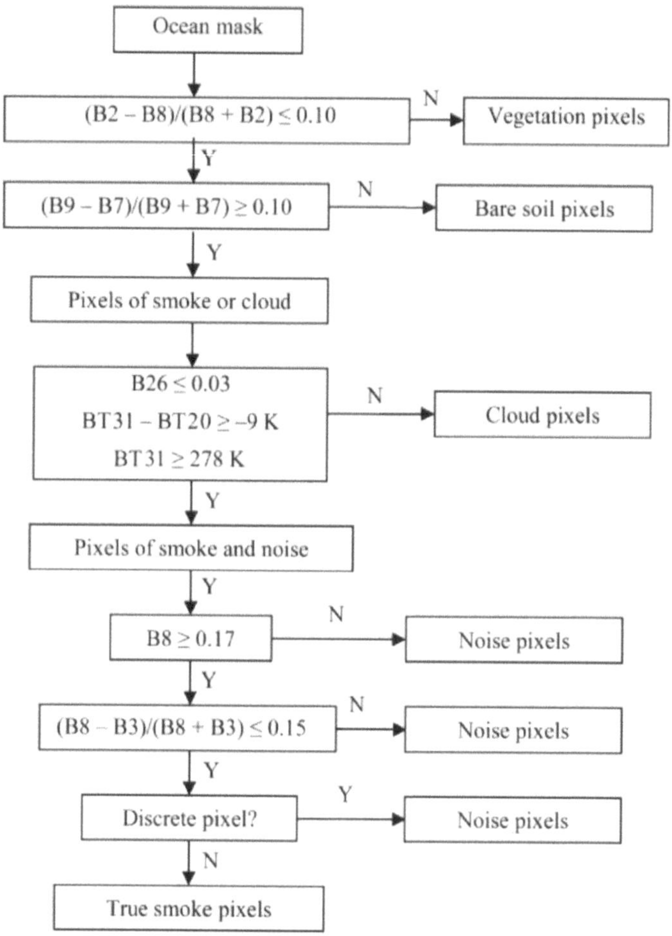

FIGURE 8.6 Flowchart of the multi-threshold method for smoke detection.

Source: Xie et al., 2007.

the applicability for different regions and improve the accuracy of the smoke plumes detection model, more biomass burning events should be collected and analyzed to render the lookup table to be more complete in the future, although the occurrence of severe biomass burning event seldom occurs. Based on the KMeans algorithm and multiple thresholds, Wang et al. (2011) developed a method to distinguish smoke in Heilongjiang (China) that could remove underlaying surface pixels using KMeans to improve the accuracy. Despite balancing computational efficiency and accuracy, these threshold methods require prior knowledge of specific bands and the thresholds among regions for smoke detection. Therefore, these methods are difficult to apply to other satellite datasets. Researchers have begun to explore simplified and adaptable approaches to detect smoke plumes. Studies of machine learning (ML) algorithms are among them.

TABLE 8.4
The Lookup Table Used in the Detection of Smoke Plumes Over Land and Ocean for MODIS Channels

Land or Ocean	Characteristic Spectral Band	Criterion	Flag
Land	Bands 1, 2	Band2–Band1 < 0.1	Smoke
	Band 3	Band3 > 0.1	Smoke
	Bands 4, 5	Band4–Band5 < 0.1	Smoke
	Band 8	Band8 > 0.15	Smoke
	Band 9	Band8–Band9 < 0.025	Smoke
Ocean	Band 3	Band3 > 0.075	Smoke
	Bands 4, 5	Band4–Band 5 > 0.03	Smoke
	Bands 5, 6	Band5–Band6 > 0.0025	Smoke
	Bands 6, 7	Band6–Band7 > 0.001	Smoke
	Bands 7, 8	Band8–Band7 > 0.11	Smoke
	Band 8	Band8 > 0.1	Smoke
	Bands 8, 9	Band8–Band9 < 0.04	Smoke
	Bands 17, 18	Band17–Band18 > 0.0025	Smoke
	Bands 18, 19	Band19–Band18 > 0.0015	Smoke

Note: Band#—MODIS Channel number
Source: Lin et al., 2010

8.5.3 Machine-Learning Methods

Machine learning algorithms are designed to fit the complicated linear or non-linear relationships between input and output and find the optimal function to detect smoke (e.g., Li et al., 2001, 2015). The ML algorithms have the capability to learn patterns whose complexity makes them difficult to analyze using other conventional approaches.

For example, Li et al. (2001) employed the multilayer perceptron neural network (MLPNN) to classify smoke plume pixels. They selected the training data from AVHRR images containing active forest fires. Input parameters to the MLPNN include reflectance from channels 1 and 2 and brightness temperatures from channels 3, 4, and 5 without considering any of their combinations. Training pixels were obtained from representative polygons containing smoke, clouds, land cover, and water. Three outputs were generated by the MLPNN corresponding to the three types of classified objects (smoke, cloud, and land). To compare the capability between MLPN and multithreshold approaches, both MLPNN and multithreshold approaches were explored. The MLPNN approach has the capability of learning from training datasets and handling complex relationships between various channels in linear or nonlinear forms. Moreover, it provides quantitative and continuous indices of smoke as well as other objects. The smoke index provides a measure of both the concentration of smoke and the mixing with other scene types (i.e., smoke/cloud, smoke/land, and smoke/land/cloud). The main disadvantage of the MPLNN approach is that it is time consuming to process large number of images like those covering all of Canada (it should be noted that training is most time consuming. But once trained, the network can classify scenes relatively quickly). A multithreshold approach was introduced that has certain advantages (quick processing) and disadvantages (categorized output) relative to the BLPNN. Both approaches suffer from limitations inherent in the characteristics of the input satellite data. They also have greater difficulty in identifying thin dispersed smoke compared to fresh dense smoke. For dealing with large data volume, an effective approach is to apply the threshold method to identify dense smoke and then to apply the MLPNN to deal with thin smoke.

In order to get satisfactory smoke plume detection results, Li et al. (2015) proposed a smoke detection algorithm where a multi-threshold method is used for extracting training sample sets to train back-propagation neural network (BPNN) classification for smoke plume detection. The algorithm contains the following four steps: (1) pre-processing the MODIS data and then transferring bands information to reflectance or brightness temperature; (2) extracting seasonal training samples of smoke, cloud, and underlying surface by using the multi-threshold approach; (3) analysis of spectral characteristics of smoke and other cover types to select the input feature vectors of BPNN; (4) training BPNN to separate smoke plumes from other types in different seasons based on the seasonal training sample sets and eliminating interference noise. Then the BPNN can be used to identify smoke. The performance of the algorithm was evaluated using the error matrix to calculate the overall accuracy and Kappa coefficient in conjunction with visual inspection of the true-color composition RGB images. The overall accuracy and Kappa coefficient are 97.63% and 96.29%, respectively. The results of all cases are relatively satisfactory when compared to true-color composition images.

Ismanto et al. (2020) proposed classification tree analysis (CTA) for automatic smoke plume detection using Himawari-8 Satellite data over the Maritime Continent of Sumatera and Borneo Islands in Indonesia. The CTA models (Entropy-09, Gini-09, Entropy-10, and Gini-10) were spatially, temporally, and statistically designed and analyzed. The CTA model classification was supervised using a collection of 21 single and transformation bands from the satellite data. The region of interest (ROI) sampling including cumulonimbus (Cb) top, low-middle cloud, smoke, bare soil, cirrus cloud, vegetation, and water classes, were used as the input to determine the best CTA models. The CTA models were validated using METAR data (weather airport observation), with results showing that Entropy-10 have the highest overall accuracy value of 0.79, and the lowest false alarm rate value of 0.11. For those smoke pixels undetected, it is possibly because there is no smoke revealed in said pixel, cloud covers the smoke pixel, or the smoke is transparent (the pixel value is embedded with land/underlying surface).

8.6 SUMMARY

Remote sensing can provide a means of observations for active fire detecting and burned area mapping from biomass burning at regional or global scales. Therefore, in this chapter, the reasons why we need to perform studies of TIR remote sensing in biomass burning were first introduced, then introduced methods and algorithms of active fire detection and burned area mapping, including fixed threshold algorithms and background contextual algorithms for active fire detection and image classification, image processing, spectral index, integrated approach, and active fire as 'fire seed' for burned area mapping and monitoring. Next, we introduced and discussed principles and methods of fire radiation power, including (1) the bi-spectral method coupled to the Stefan-Boltzmann Law, which employs the differential fire signal in two or more co-registered spectral channels and (2) single band approaches that use the fire signal in the MIR channel only. Finally, many studies with optical and TIR remote sensing to monitor smoke plume with three categories of methods: false color composite image, muti-channel threshold, and machine-learning methods, were reviewed and discussed briefly.

9 Thermal Infrared Remote Sensing Applied in Urban Thermal Environments

Urbanization and global climate change have significantly altered urban thermal environments. Thermal remote sensing offers essential tools to observe spatiotemporal properties of urban thermal environments, such as retrieving land surface temperature (LST) and urban biophysical parameters, crucial for evaluating urban heat islands (UHIs). To meet diverse research and application needs, a concise and accurate review of thermal infrared (TIR) remote sensing applications in urban thermal environments is essential. This chapter provides an overview of the rationale for studying TIR remote sensing in urban thermal environments (Section 9.1), techniques for mapping and analyzing urban heat islands (UHIs) (Section 9.2), and current challenges and future directions in urban thermal environment research (Section 9.3).

9.1 INTRODUCTION

Urbanization and industrialization improve our material lives and comfort; however, they also induce many problems to human beings, such as global warming, industrial waste, and air pollution. Besides the adverse global impacts, they impact regional urban areas more seriously and obviously where industrial activities and heavy use of synthetic construction material are commonly observed. As a result, the natural environment and ecology are tremendously affected and have lost the necessary balance between. Two major phenomena were observed in large cities as compared to its surroundings: a higher temperature or heat content called the urban heat island (UHI), which temperature in the urban central area is much higher than that in surrounding rural areas. Higher urban heat is mainly caused due to the anthropogenic heat released from vehicles, power plants, air conditioners, and other heat sources and due to the heat stored and re-radiated by massive and complex urban structures. Huge amounts of solar radiations are mainly stored and re-radiated in urban areas due to massive urban construction material and decreased sky view factor. The urban areas also possess less vegetation relative to their typical land use. High roughness structure is another problem of urban areas, which reduces the convective heat removal horizontally. It has been suggested to reduce anthropogenic heat release and make proper urban design changes such as the use of high albedo, cooler roofs, suitable building material, and proper building design to reduce high heat intensity due to the aforementioned problems.

DOI: 10.1201/9781003276173-9

The UHI—a phenomenon in which temperature tends to be higher in urban zones than surrounding non-urban areas (Figure 9.1)—represents a major anthropogenic alteration to the Earth's environments (Oke, 1982). The UHI phenomenon has been observed worldwide (Peng et al., 2012; Li et al., 2017). It has been widely attributed to the change in the energy exchange between urban land surfaces and the urban atmosphere, which is primarily caused by increased impervious surface cover replacing evaporative vegetation surfaces and by anthropogenic heat releases (Oke, 1982; Rizwan et al., 2008). The UHI has contributed to a suite of environmental changes such as regional climate, vegetation growth, and water and air quality. These factors, in turn, affect human health and well-being dramatically and may potentially lead to increases in morbidity and mortality, energy consumption, and even violent incidents in urban areas. The UHI effect and the associated consequences are expected to be more severe under a warming climate and a rapidly urbanizing world. UHI, therefore, has gained considerable research interest and has been the subject of active investigation, especially in the last decade.

UHI intensive studies have been conducted to understand UHI phenomena, which, in general, can be classified into two broad categories based on the ways and heights that they are formed (Oke, 1082): 'air' (or 'atmospheric') and 'surface' UHIs. Air UHI refers to UHI effects in the canopy (CLHI) or boundary (BLHI)

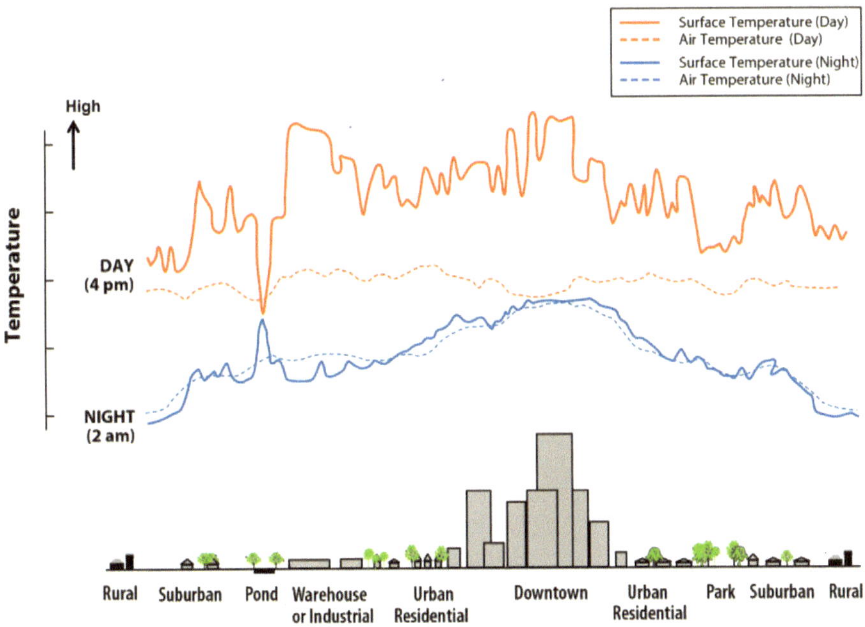

FIGURE 9.1 The temperature profile in different areas, with the formation of the UHI (adapted from USGS, 2023). During the morning period, the SUHI presents great amplitude and variable behavior, especially in urbanized areas. In these same areas, the UHI presents lower temperatures when compared to those with the presence of vegetation, due to the amount of mass per unit area for heating, i.e., places with fewer horizontal and vertical constructions will need less time to heat.

layer. The CLHI is usually measured by in situ sensors mounted on fixed meteorological stations or traverses of vehicles, while more special platforms such as tall towers, radiosondes, and aircraft are needed to measure the BLHI. The development and installation of these measurement devices are usually very time-consuming and expensive, and, thus are only available in a few large cities worldwide to date (Mirzaei and Haghighat, 2010). Due to limited monitoring stations, the measured air UHIs usually fail to provide sufficient spatial details for urban land use planning and climate change research (Wang et al., 2017c). In contrast, the surface UHI (SUHI), which represents the radiative temperature difference between urban and non-urban surfaces, is primarily measured by satellite thermal remote sensing data (Voogt and Oke, 2003), which mainly refers to UHI because it is easily measured in land surface temperature (LST) in this chapter unless in case indicted in special cases. Remote sensing can provide consistent and repeatable observations of the Earth's surface, which offers the ability to study the urban thermal environment at various spatial (from local to global scales) and temporal (diurnal, seasonal, and inter-annual) scales (Weng, 2009; Deilami et al., 2018). It can also be used to predict air UHI in a spatially explicit manner. Rao (1972) reported in 1972 the first satellite-based observations of the SUHI phenomenon in the Eastern United States. Since then, advancements in TIR remote sensing and spatial science have considerably boosted the number and quality of SUHI studies, which currently form the major body of the UHI literature. A thorough knowledge of past literature on SUHI can not only provide a basis for future research but also is crucial for formulating effective UHI mitigation and adaptation strategies.

Previous reviews of SUHI research have concentrated mostly on summarizing the methods, applications, and limitations of thermal remote sensing techniques based on representative literature and/or a more general topic that only partially covered SUHI. For example, Gallo et al. (1995) reviewed several satellite-based SUHI studies between 1972 and 1993 and stressed the application of Normalized Difference Vegetation Index (NDVI) data in SUHI research. Arnfield (2003) compared different types of UHI briefly while synthesizing progress in urban climatology during the 1980s and 1990s. Voogt and Oke (2003) focused on the principles of thermal remote sensing, the relationship between surface and air UHIs, and the potential use of thermal remote sensing data in urban climate models. Rizwan et al. (2008) provided a systematic review of the generation as well as determination and mitigation of UHI based mainly on air UHI studies. Weng (2009) presented a review on the methods and applications of thermal remote sensing but concentrated on land surface temperature (LST) retrieval, LST-vegetation relationships, and some typical applications of thermal infrared remote sensing data in SUHI studies. Huang and Lu (2018) examined the literature trends of UHI research from 1991 to 2015 using bibliometric methods, regardless of UHI types. The most recent work by Deilami et al. (2018) focused solely on SUHI but reviewed only the spatial and temporal factors affecting the SUHI based on the literature after 2007. Xia et al. (2019) proposed a weighted combination of kernel-driven and fusion-based methods (CKFM) to further improve the spatiotemporal resolutions of time series LSTs. The CKFM is more accurate and robust compared to the kernel-driven method with an improvement of 0.1–0.6 K, and it may reconstruct more spatial details than those by a fusion-based method when it was used to downscale daily MODIS 1 km LSTs to 30 m resolution LSTs with Landsat 8 data for UHI studies.

Traditionally, UHI was evaluated by the air temperature difference, measured by permanent meteorological stations or moving observations between urbanized and rural areas. Recently, more studies used land surface temperature (LST) derived from satellite thermal infrared (TIR) imagery (Li et al. 2011). Apparently, traditional observation data cannot provide a synoptic view of temperature over a city, while remotely sensed data can offer a continuous view of the whole city, which is important to investigate the spatial pattern of urban surface temperature (Weng, 2012). Extensive research attentions have been paid to deriving LST from moderate resolution imageries (TM, ETM+, and ASTER). The LST with a 60–120 m resolution significantly facilitated the studies on the spatial pattern of UHI and its relationship with LULC and surface biophysical parameters (Nichol, 1996; Weng et al., 2004, 2006; Yuan and Bauer, 2007). Therefore, it is of great scientific significance to investigate how satellite-derived LSTs can be utilized to characterize UHI phenomena.

Urban optical and TIR remote sensing data can be used to explore the interaction between urban environmental processes and human activities. Various methods have been employed to analyze the mechanism of UHI, such as the comparison of statistical index, correlation, and regression methods. Most studies focused on biophysical and meteorological factors, such as vegetation coverage (Weng et al., 2004), impervious surface area (Weng et al., 2006; Yuan and Baue, 2007), land cover characteristics (Hart and Sailor, 2009; Weng et al., 2007), and ecological setting (Imhoff et al., 2010). Social-economic factors have also been emphasized, such as pavement area, building area (Weng et al., 2008), building and population density (Chen et al., 2006; Xiao et al., 2008), and building energy use (Zhou et al., 2012), etc.

Satellite remotely sensed data provide more comprehensive and uniform spatial sets of observations and wider coverage to the UHI. However, satellite imagery is also described by sensor characteristics and image resolution. Image resolution is usually considered through three aspects, which include spectral, spatial, and radiometric resolutions. The temporal resolution, which is an indication of the revisiting frequency of a sensor over a particular place, is also an important guide for satellite imagery. Remote sensing is extremely useful for understanding the spatiotemporal land cover change in relation to the basic physical properties of materials used on urban structures. These properties related with the surface radiance and emissivity data. The electro-magnetic wavelength region of 3–35 μm is called thermal-infrared (TIR) region. The useful spectral bands in this spectral range are limited by the intensity of radiation emitted and the atmospheric windows. An excellent atmospheric window lies between 8 and 14 μm, in which most remote sensors are set up to detect the thermal radiative properties of the ground materials.

Satellite-derived surface temperature data (LST) have been utilized for urban climate and thermal environment analyses in several studies. Currently available satellite TIR sensors provide different spatial resolution and temporal coverage data that can be used to estimate LST (see Table 6.1 for a list of TIR sensors mostly used for retrieving LST) and thus to evaluate UHI phenomenon. The Geostationary Operational Environmental Satellite (GOES) has a 4-km resolution in the TIR, while NOAA, Advanced Very High-Resolution Radiometer (AVHRR), and Terra and

Aqua Moderate Resolution Imaging Spectroradiometer (MODIS) have 1-km spatial resolutions. Significantly, better-resolution data come from the Terra-Advanced Space borne Thermal Emission and Reflection Radiometer (ASTER), which has a 90-m spatial resolution, meanwhile the Landsat 5 Thematic Mapper (TM) has a 120-m resolution, Landsat 7 ETM+ has a 60-m resolution, and Landsat 8 TIRS has a 100 m resolution in the TIR bands. Better resolution and less frequent TIR observations from Landsat might be used to understand the spatial variation within coarser-resolution observations made by MODIS and AVHRR, which provide more frequent measurements. Recent satellite systems, e.g., MODIS and ASTER, include features to allow easier calibration and provide LST as standard products.

9.2 MAPPING AND ANALYZING URBAN HEAT ISLANDS (UHIS)

9.2.1 Generation of UHI

Generation of UHI is the mutual response of many factors which could broadly be categorized as controllable and uncontrollable factors. The controllable and uncontrollable factors could further be categorized as the temporary effect variables, such as air speed and cloud cover; permanent effect variables such as green areas, building material, and sky view factor; and cyclic effect variables such as solar radiations and anthropogenic heat sources (Rizwan et al., 2008). The heat generated and stored in a specific urban area comes from the sun in the form of solar radiations and from anthropogenic heat sources including power plants, automobiles, air-conditioners, and other sources. Almost all anthropogenic heat enters into the environment instantly and directly. On the other hand, only part of solar radiations heat up the environment directly, and the rest of them are absorbed by the complicated urban built structures and heat up the urban environment indirectly. The basic heat transfer and energy conservation processes, such as conduction, convection, and radiation, play their characteristic roles in this heat exchange (Rizwan et al., 2008). The urban structures on ground level, such as walls and roof facets, irrigated gardens, non-irrigated green spaces, lawns and paved areas, etc. capture solar radiation to store to different extents. These natural and man-made structures continuously absorb and store this radiation in the form of heat energy from sunrise till late afternoon. Afterwards, the sun starts setting and the environment starts cooling down. The heat energy stored in the urban structures is then released to the urban environment. The method and quantity of heat released by the urban structures, however, depends on other controllable factors such as the sky view factor and building material (Rizwan et al., 2008). It is believed that the low ability of heat release by long-wave radiation in cities is due to decreased sky view, which results in high heat storage in building structures. It is also believed that the albedo, the reflected light in comparison to the irradiance energy, is also very low in cities due to typical street canyon configurations, and thus is one of the main reasons for high air temperatures. The values of albedo and sky view factor are, therefore, reported as two important factors in creating UHI. The lack of greenspace cities also exhibits little latent heat of vaporization. It was reported that evapotranspiration in Tokyo has been reduced by 38% from 1972 to 1995 (Kondoh and Nishiyama, 1999). In urban areas, high roughness of structures reduces the amount of convective heat removal and transfer by wind. Typical thermal properties of building materials used in urban areas, such as low admittance, are other potential contributors. It is believed that air pollutants, in particular aerosols that are abundant over polluted urban areas, can absorb and re-radiate long wave radiation and block the corresponding radiative surface cooling producing a pseudo-greenhouse effect, which also is responsible for causing UHI.

UHI has also been reported to be affected by temporary effect variables in certain ways. Many studies have reported the influence of wind speed and cloud cover on UHI; the results show that the UHI is negatively correlated with wind speed and cloud cover (e.g., Oke, 1982; Kim and Baik, 2005). On the other hand, the UHI has been reported to be positively correlated with the city population. Tran et al. (2006) studied UHIs in 12 Asian mega cities and reported that the magnitude and extent of the UHI was positively correlated with city population. Tran et al. (2006) observed maximum UHI of 8°C in Bangkok with a population of 11 million and maximum UHI of 7°C with a population density of 12.55 million in Shanghai. They have also reported maximum UHI of 7°C in Manila with a population density of 15,617 (persons/km^2) and maximum UHI of 12°C with a population density of only 6,218 (persons/km^2) in Tokyo. The population could have twofold effects on heat generation, a direct effect as more people means more metabolisms and an indirect effect as number of buildings, vehicles, factories, etc. will likely be increased with the increasing population. However, other variables such as sky view factor, anthropogenic heat, building design and material, etc. do play an important part in increasing heat contents. These factors are not population dependent and may or may not favor increasing the heat contents with increased population. Therefore population (population density might be a better term) could—but would not necessarily—increase the heat contents of an area (Rizwan et al., 2008). There has been some research on quantifying the significance of temporary effect variables; an example is the study conducted by Kim and Baik (2002) who reported that a decrease in maximum UHI could be visible with wind speed greater than 0.8 m/s while with critical wind speed of 7.0 m/s maximum UHI of 0.3°C or less could be reduced. In another study Klysik and Fortuniak (1999) presented that a UHI of higher than 1°C could still be observed when the city average wind speed is 4 m/s during nighttime and 2 m/s during daytime. Morris et al. (2001) reported that the UHII is approximately the fourth root of both the wind speed and cloud cover. Yet, more work is required in understanding the degree of influence of various temporary and permanent factors on UHI.

9.2.2 Factors and Methods of Measuring of UHI Intensity

9.2.2.1 UHI Measurement Methods

The UHI phenomenon has been studied in different parts of the world using various approaches to measure its magnitude and possibly its extent. The phenomenon has long been studied by ground-based observations taken from fixed thermometer networks or by traverses with thermometers mounted on vehicles from which empirical data are generated and analyzed (Ngie et al., 2014). For the sake of spatial mapping over larger areas of UHI, satellite remote sensing has recently been used (Mirzaei and Haghighat, 2010). This has provided new avenues for the observation of UHI and the study of its causation through the combination of thermal remote sensing and urban micrometeorology (Voogt and Oke, 2003). UHIs may be identified by measuring surface (LST) or air temperatures and other factors. LSTs have an indirect but significant influence on air temperatures. For example, parks and vegetated areas, which typically have cooler surface temperatures, contribute to cooler air temperatures. Dense built-up areas, on the other hand, typically lead to warmer LST and thus air temperatures. Because air mixes within the atmosphere, though, the relationship between surface and air temperatures is not constant. Therefore, the measurement of UHI is done at various levels and by different tools. According to Voogt (2002), UHI is best observed on clear and calm nights since maximal differences in radiative cooling between central urban areas and their surroundings occur. The following are list of factors affecting UHI intensity to be measured.

9.2.2.2 LST Retrieval

Satellite TIR sensors measure top of the atmosphere (TOA) radiances, from which brightness temperatures (also known as blackbody temperatures) can be derived using Planck's Law (Dash et al., 2002). The TOA radiances are the mixing result of three parts of energy: (1) upwelling radiance emitted from the ground, (2) upwelling radiance from the atmosphere, and (3) downwelling radiance emitted by the atmosphere and reflected from the ground. The difference between the TOA and land surface brightness temperatures ranges generally from 1 K to 5 K in the 10–12 μm spectral region, subject to the influence of the atmospheric conditions (Prata et al., 1995). Therefore, TOA radiances are then converted to land surface temperature (LST) by correcting for three main effects; atmospheric attenuation, angular effects, and spectral emissivity values at the surface. The emissivity accounts for the roughness and heterogenous properties of the land surface, the amount and nature of vegetation cover, and the thermal properties and moisture content of the soil (Friedl, 2002).

Over the past decades, the scientific community has successfully developed a variety of new theories and influential LST retrieval algorithms including the single-channel algorithm (SCA) (e.g., Hook et al., 1992), the day/night algorithms (e.g., Wan and Li, 1997), the mono-window algorithms (MWA) (e.g., Qin et al., 2001b), the generalized split-window (SWA) algorithm (e.g., Wan and dozier, 1996), and physics-based methods (Fu et al., 2019). Many studies have been carried out, and different approaches have been proposed to derive LSTs from satellite TIR data, using the variety of methods to deal with the emissivity and atmospheric effects (e.g., Hook et al., 1992; Wan and Dozier, 1996; Wan and Li, 1997; Gillespie et al., 1998; Qin et al., 2001b; Jiménez-Muñoz and Sobrino, 2003). The advantage of the SCA algorithm is its simplicity because it only needs two parameters for LST retrieval: the ground emissivity (ε) and the water vapor content (w) (Yu et al., 2014). However, SCA and MWA algorithms show a significant overestimation of LSTs (Zhou et al., 2015). The third one can be carried out by utilizing two thermal bands, which is also known as the SWA algorithm (Wan and Dozier, 1996). Generally, the first two algorithms depend on a single TIR channel, whereas the third algorithm requires two adjacent TIR channels. A major disadvantage of the third one is that the coefficients are only valid for the datasets used to derive those (Dash et al., 2002). In other words, a set of thermal responses for a specific landscape phenomenon or process measured using a particular TIR sensor cannot be extrapolated to predict the same TIR measurements either from other sensors or from images sensed at different times using the same sensor (Quattrochi and Goel, 1995). The most significant sources of error are associated with atmospheric effects, which may introduce an error on the estimation ranging from 0.2 K to 0.7 K and with land surface emissivity, which can produce an error up to 0.2 K to 0.4 K (Jiménez-Muñoz and Sobrino, 2006). The aforementioned algorithms have not been frequently referenced and applied in urban climate and environmental studies as is deserved. The majority of urban studies are interested in relative LST measurements, i.e., in mapping the spatial variations of LST and/or environmental modeling with relative LSTs (Ngie et al., 2014; Balew and Semaw, 2022). Most urban studies found this straightforward approximation sufficient, which replaced the sensor response function with a delta function at the sensor's central wavelength to invert LSTs with the assumption of uniform emissivity (Lu and Weng, 2006; Weng et al., 2006).

9.2.2.3 Emissivity

The effect of land surface emissivity (LSE) on satellite measurements can be generalized into three categories: (1) emissivity causes a reduction of surface-emitted radiance; (2) non-black surfaces reflect radiance; and (3) the anisotropy of reflectivity and emissivity may reduce or increase the total radiance from the surface (Prata, 1993). Therefore, retrieval of LST from multispectral TIR sensor data requires an accurate measurement of emissivity values of the surface. The emissivity of a surface is controlled by such factors as water content, chemical composition, structure, and roughness (Snyder et al., 1998). For vegetated surfaces, emissivity can vary significantly with plant species, stem/areal density, and growth stage (Snyder et al., 1998). In the meantime, emissivity is a function of wavelength, commonly referred to as spectral emissivity (Dash et al., 2002). Emissivity for ground objects from passive sensor data has been estimated using different estimate techniques. Among those techniques, depending on the spectral channels of the sensor and the available temporal sampling, are (1) classification-based emissivity method; (2) NDVI-based emissivity method; (3) day/night temperature independent spectral-index; and (4) temperature emissivity separation method (TES). See detailed descriptions for the four methods in Chapter 5. The previous techniques are applicable to separate temperatures from emissivity, so that the effect of emissivity on estimated LST can be determined. However, lack of knowledge of emissivity can introduce an error ranging from 0.2 K to 1.2 K for mid-latitude summers and from 0.8 K to 1.4 K for the winter conditions for an emissivity of 0.98 and at the ground height of 0 km, when a single-channel method of LST estimation is used (Dash et al., 2002).

Hulley (2013) examined a water vapor scaling model (WVS) to improve to the accuracy of LST and emissivity retrievals using the TES algorithm for ASTER and MODIS observations over the Salton Sea on July 15, 2000. The emissivity of water is high (~0.98) and spectrally flat, and so this provides a good test for assessing the accuracy of the emissivity retrieval and consequently the LST retrieval. The LST and emissivity was retrieved in the same manner for both ASTER and MODIS data using the methods outlined in Hulley and Hook (2011). The results show that emissivity spectra from the standard atmospheric correction (STD) method are too low for both ASTER and MODIS, especially in the 8–10 μm region when compared to the lab spectra from the ASTER spectral library. Using the WVS coefficients results in a dramatic improvement in emissivity accuracy in both magnitude (up to 0.06 for ASTER band 11 and 0.09 for MODIS band 29) and spectral shape.

9.2.2.4 Socioeconomic Factors

Socioeconomic development is vital for the urban thermal environment because intense human activities (e.g., power plants, automobiles, air-conditioners, and other sources) can cause large amounts of anthropogenic heat to be released into a city, which warms the urban surface and air temperature. Previous studies have identified some potential socioeconomic factors, including population, GDP, and nightlight satellite data. The population and GDP data can be obtained from the statistical departments of the local cities and nightlight data can be directly downloaded from a NASA website. The factors account for a huge replacement of soil and vegetation with pavements, building structures, and dark surfaces with urban material (concrete, asphalt, and metal), leading to the growth in thermal radiation which alters surface energy balance, with a consequent increase in LST (Voogt, 2002). This leads to the increase in sensible heat flux at the expense of latent heat flux and eventually air temperature (Wang et al., 2007).

9.2.2.5 Climate/Topographic Factors

Climate and topography, which are in part determined by a city's geographic location, influence UHI formation. For example, large bodies of water moderate temperatures and can generate winds that convect heat away from cities. Nearby mountain ranges can either block wind from reaching a city or create wind patterns that pass through a city. Local terrain has a greater significance for heat island formation when larger scale effects, such as prevailing wind patterns, are relatively weak. Proximity to large water bodies and mountainous terrain can influence local wind patterns and UHI formation. The effects of climatic factors (e.g., precipitation, temperature, wind speed, and solar irradiance) are highly variable; nevertheless, some recent studies have indicated that climatic background might be an ultimate determinant for the variability of UHIs over large areas (Cao et al., 2016; Manoli et al., 2019).

9.2.2.6 Vegetation Abundance

Vegetation abundance describes the amount and nature of vegetation cover and density and modulates the proportions of vegetation vs. non-vegetated ground (e.g., bare soil) visible to a sensor (Weng, 2009). The differences in radiative temperature between the vegetation canopy and the ground affect the measurement of LST (Sandholt et al., 2002). For non-vegetated areas, LST measurements typically represent the radiometric temperatures of sunlit non-vegetated surfaces, such as bare soil and road surface. As the vegetation abundance increases, the radiative temperature sensed by a sensor approximates more closely the temperatures of green vegetation and the canopy temperature at spectral vegetation maximum or complete canopy cover (Goward et al., 2002). It is of significance to scrutinize the temperatures of each part of the vegetation-ground system (such as shaded ground, sunny ground, shade vegetation, and sunny vegetation) and to examine the effects of different canopy structures (Casells et al., 1992a, 1992b). In general, for image pixels that are not completely occupied by a single homogeneous vegetation or bare soil, called mixture pixels, LST measurements actually reflect a mixture of ground and vegetation canopy temperatures, resulting from mixed spectral signature. The observed portion of ground and vegetation cover can vary with the viewing angle, and thus the amount of vegetation (ground) alters as the observation angle increases (Casells et al., 1992a). Moreover, LST measurement will also be subject to the influences of the lower atmosphere and the temperature difference between the vegetation canopy and the soil background (Friedl, 2002). Thermal responses for vegetation can be highly varied as a function of the biophysical properties of the vegetation itself and vegetation landscape indices (Table 9.1).

TABLE 9.1
Description of Landscape Indices Used in This Study Source: Cardille and Turner, 2017

Category	Metric	Abbreviation	Description
Aggregation metric	Number of patches	NP	Reflecting the spatial pattern of the landscape, the value is positively correlated with landscape fragmentation
	Aggregation index	PD	The density of corresponding patches within an analysis unit
	Aggregation index	AI	Degree of aggregation of the corresponding patches within an analysis unit
	Contagion	CONTAG	Reflecting different patch types and clustering or extension trends in the landscape, small values indicate high landscape fragmentation
	Interspersion and juxtaposition index	IJI	Reflecting the spatial pattern of the landscape, larger values indicate the proximity of patch types to each other and high dispersion
Shape metric	Shape index distribution	SHAPE_MN	Average shape index of the corresponding patches within an analysis unit
	Perimeter-area ratio distribution	PARA_MN	Reflecting the complexity of landscape patch shapes and the extent to which land use is influenced by human activities
Area and edge metric	Percentage of landscape	PLAND	Landscape percentage of the corresponding patch
	Largest patch index	LPI	The percentage of the landscape occupied by the largest patch
	Patch area distribution	AREA_MN	The average size of the patches
Diversity metric	Shannon's evenness index	SHEI	Uniformity of distribution of landscape types
	Shannon's diversity index	SHDI	Reflecting the abundance and complexity of landscape types

In measuring vegetation abundance for UHI, the NDVI and SAVI (soil adjusted vegetation index) are important indexes and frequently used for controlling the urban thermal environment because they can determine the abundance and growth status of vegetation, such as forest and grass, which are highly related to the cooling rate in reducing the urban thermal environment via evapotranspiration. Based on the multi-spectral satellite data, NDVI and SAVI can be calculated from the red and NIR bands.

9.2.2.7 Soil Properties

For any surface material, certain internal properties, such as heat capacity, thermal conductivity, and inertia, play important roles in controlling the temperature of a body at equilibrium with its surroundings. These thermal properties vary with soil type and its moisture content (Sandholt et al., 2002). Dry, bare, and low-density soils, for example, have been linked to high LST as a result of relatively low thermal inertia (Carnahan and Larson, 1990). The emissivity of soils is a function of soil moisture conditions and soil density (Larson and Carnahan, 1997). Therefore, for the areas characterized by partial vegetation cover, surface thermal properties can largely influence the measurement of LST through the thermal processes of conduction, convection, and radiation.

For thermal infrared data, the parameters related to the soil thermal properties are mainly related to soil moisture. This method includes two categories, (1) the thermal inertia method and (2) the temperature index method. Thermal inertia is the intrinsic property of materials and describes their impedance to variations of temperature. When soil moisture increases, the thermal inertia of soil increases, and there is a lower diurnal temperature fluctuation. And the thermal inertia method is routinely used to estimate soil moisture in bare soil and sparsely vegetated regions (e.g., Verhoef, 2004). Combining visible and thermal infrared remotely sensed data can provide more information for estimating soil moisture than the single one. Soil moisture content controls soil density, heat capacity, thermal conductivity and other characteristics, and thus soil thermal inertia is closely related to soil moisture.

9.2.2.8 Geometry of Urban Structures

Urban geometry is also another leading factor of UHI with canyon geometry as a result of narrow streets and tall buildings, which intercept air movement. The geometry also contributes to this influence through wind flow. Two primary weather characteristics that affect UHI development are wind and cloud cover. In general, UHIs form during periods of calm winds and clear skies, because these conditions maximize the amount of solar energy reaching urban surfaces and minimize the amount of heat that can be convected away. Conversely, strong winds and cloud cover suppress UHIs by convecting the warm air away or reflecting it, respectively (Kim and Baik, 2005).

9.2.2.9 LULC Types

The changes in land use/cover (LULC) pattern in relation to urbanization also impact on the intensity and spatial pattern of the UHI effect (Chen et al., 2006). This has been affirmed with different land uses influencing UHI differently as a result of the surface properties (Tran et al., 2006). Huang et al. (2008) used four types of land covers in Nanjing (China) namely urban bare concrete cover, urban woods or the shade of trees, urban water areas, and urban lawn with their microclimates to be examined, and the UHI was analyzed using air temperature data measured at four fixed observation spots. The results revealed among others that the microclimate of these four types of land covers had significantly different impacts on UHI in the order of bare concrete cover > lawn > water areas > woods or the shade of trees during daytime with reversed order during nighttime. A similar study has been done recently in Delhi (India), which confirmed the difference in influence of land covers on the UHI phenomenon. It is vital in studying LULC change to understand the UHI intensity effect of the study area. When Balew and Semaw (2022) analyzed urban LULC change and its impact on the seasonal, spatial, and temporal SUHI intensity of Addis Ababa city and its surrounding from 1987 to 2019 using Landsat images, the result indicated that the impervious surface of Addis Ababa city and the surroundings have expanded from 81.49 km^2 in 1987 to 591.85 km^2 in 2019 with a 6.2% rate of change. On the other hand, vegetation cover, which has a high thermal cooling effect, has been degraded from 217.66 km^2 in 1987 to 157.8 km^2 in 2019 (Figure 9.2). The spatial pattern of LST increased from northern highlands toward southern lowlands. The mean temperatures of LST for January and February 1987 were 26.22°C and 27.76°C, respectively, whereas the mean temperatures of LST on January 25 and February 26, 2002 were 28.25°C and 31.26°C, respectively (Figure 9.3).

9.2.3 Methods and Techniques of Evaluating UHI Intensity

Generally, the methods for evaluating UHI can be summarized into four basic types: (1) historical weather station data, (2) field observation, (3) computer simulation, and (4) remote sensing technology. The limitations of the first three methods have been well documented (Rizwan et al., 2008; Ngie et al., 2014; Hu and Brunsell, 2015; Zhao and Li, 2015). In this section, we more focus on the methods that use remote sensing technology. A number of algorithms (or methods) have been developed to estimate UHI from remotely sensed data, including simple empirical approaches (e.g., linear regression) to complex methods (e.g., machine learning, ML) based on remotely sensed data using various models (Shi et al., 2021). The structure of the UHII pattern centroid in three dimensions indicates the overall variation of the intensity and distribution of the UHI in space and time. The simplified relationship of thermal sensor data and UHI has been applied from a local spatial scale using airborne very high-resolution images to a broad scale with AVHRR, MODIS, ASTER, and Landsat data at regional and continental levels. Processing procedures of UHI often require remotely sensed data over different spectral domains to retrieve input parameters that characterize surface properties such as thermal properties, LST, emissivity, albedo, NDVI, and other indices. Detailed knowledge of UHI, especially latent and sensible heat flux components, is important for monitoring the climate change of the land surface. The main methods classically used to measure UHI are appropriate to field observations, but these do not allow for an estimation of UHI at large spatial scales. For operational applications to ecological conservation and city planning, managers and researchers need accurate estimates of land surface temperature (LST) and UHI

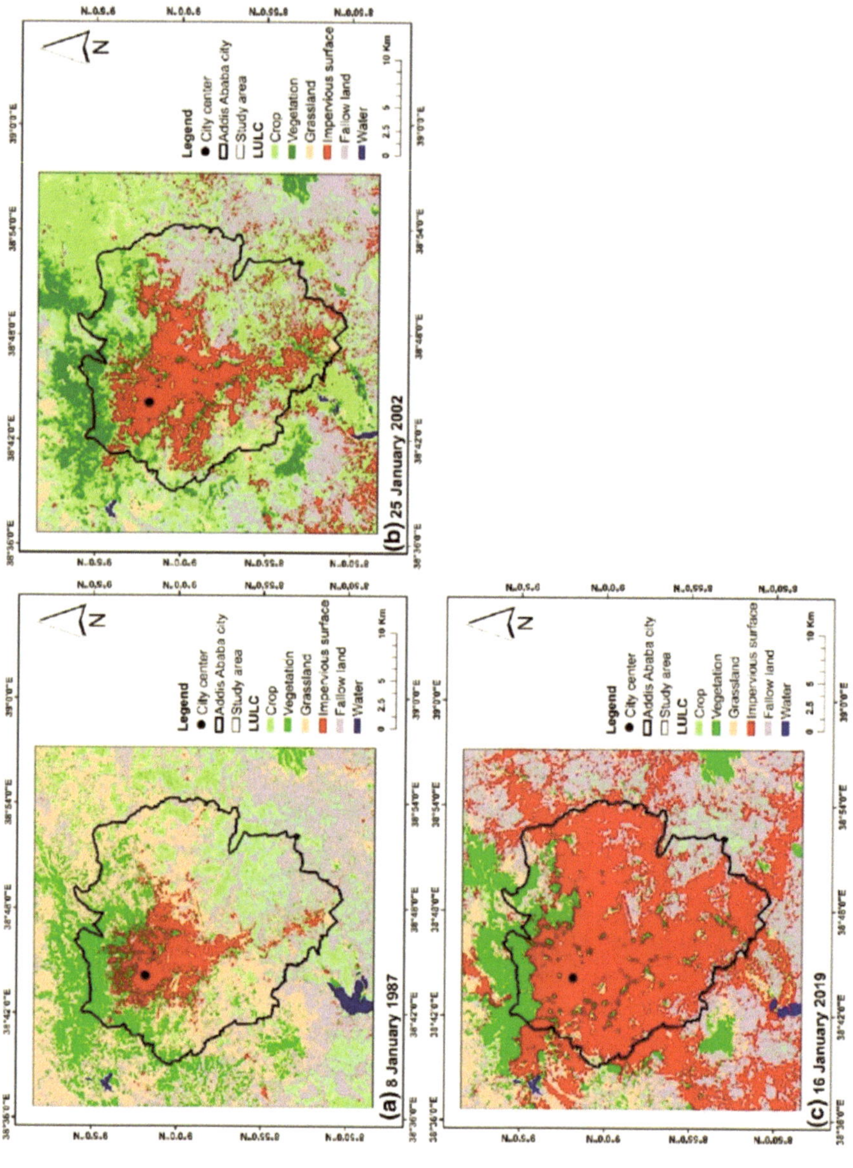

FIGURE 9.2 LULC map of Addis Ababa city and its surroundings in 1987, 2002, and 2019.

Source: Balew and Semaw, 2022.

at broad spatial scales. New algorithms based on remotely sensed data have been developed to use the TIR imagery of various spatial resolutions and temporal frequency to evaluate UHI (Fu et al., 2019; Hulley et al., 2019). It is frequently difficult to classify these methods because their complexity depends on the balance between the empirical and physical-based modules used. Nevertheless, we summarize some algorithm (model/method) categories in the following Subsections.

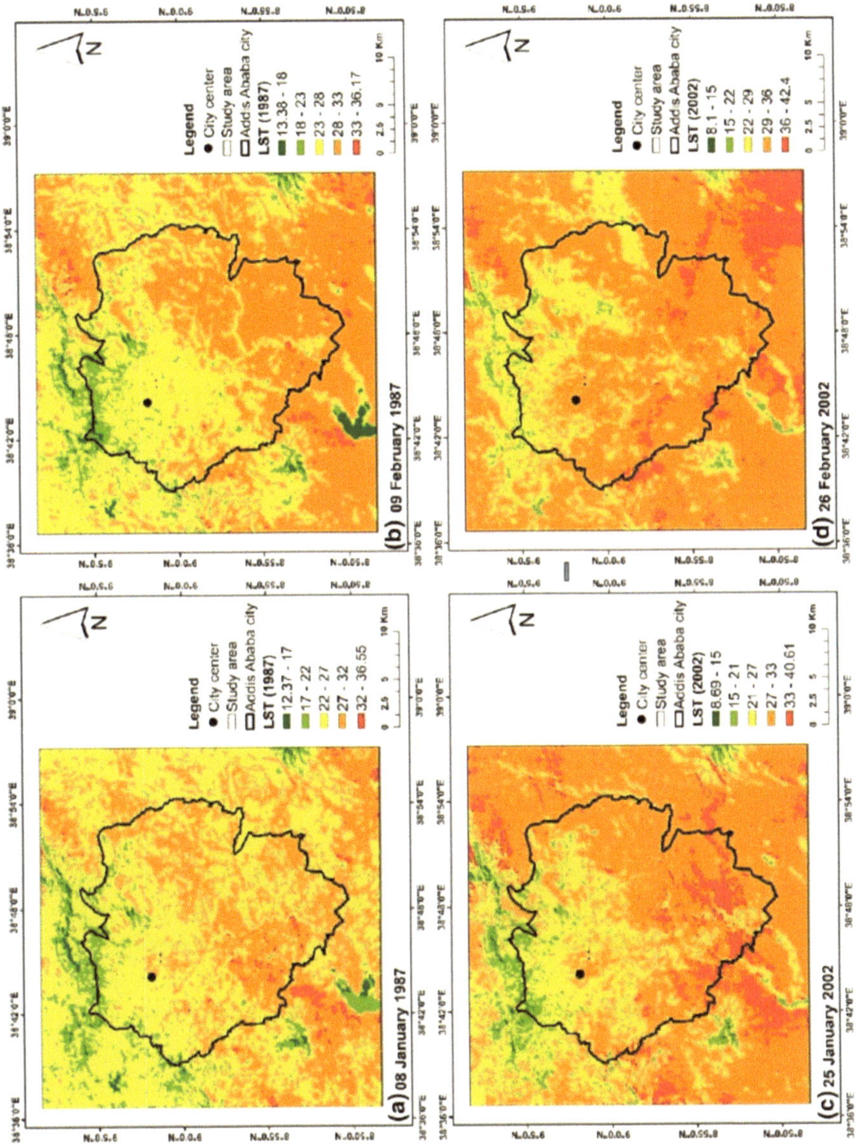

FIGURE 9.3 The spatial pattern of minimum and maximum LST for Addis Ababa and its surrounding in 1987 and 2002. The minimum and maximum temperatures of the whole study area on January 8, 1987 were 12.37 and 36.55°C, respectively, whereas on February 9, 1987 the minimum temperature increased to 13.38°C and the maximum temperature declined to 36.17°C. The mean temperatures for January and February 1987 were 26.22 and 27.76°C, respectively.

Source: Balew and Semaw, 2022).

9.2.3.1 LST and UHI Intensity Calculation

LST calculation, including empirical direct methods where TIR remotely sensed data are introduced directly in semi-empirical models to estimate LST, is the simplified relationship between TIR remotely sensed and meteorological data. This method allows for the characterization of UHI intensity both at the local scale, using ground measurements—and over large areas, using satellite data, by calculating LST differences (Li et al., 2013). Most current operational models (Li et al., 2021a, 2021b) use remote sensing data directly to estimate the input parameters such as landscape factors and LST (Table 9.2). Seasonal information captures the annual profile of LST and its trend over long time periods and is essential to the study of UHI. Therefore, remote sensing has been used to accurately monitor and compare the LST difference in the same season in different years and trends over long time periods. In the last 20 years, thermal sensor technology has been rapidly developing. Five types of methods have been developed to estimate LST with TIR remotely sensed data: single-channel algorithm (SCA), the day/night algorithms, the mono-window algorithms (MWA), the generalized split-window (SWA) algorithm, and physics-base methods. Recently, Peng et al. (2020) proposed a wavelet coherence approach to exploring spatial heterogeneity and the scale-dependence of the relationship between LST and multiple influencing factors across scales and locations, in order to identify the priority research scales and influencing factors. The results showed that around 1 km could be used as the prioritizing research scale for simultaneously exploring multiple LST influencing factors, where more influencing factors had significant influence on LST.

9.2.3.2 Comparing the Difference Between UHI Area and Non-urban Area

Many studies have documented the use of LST data to examine meso-scale temperature differences between urban and rural areas in cities worldwide. The land surface temperature (LST) of UHI is generally higher than the surrounding rural areas and has a strong correlation with land cover. UHII analysis is the most common method to compute the magnitude and extent of UHI by evaluating the LST difference between urban and surrounding non-urban areas (Gallo and Owen, 1998). These analyses are often supported with land surface information, such as land cover and impervious surface area (ISA). Deterministic models are generally based on more complex models that compute the intensity of UHI in space and time. Remotely sensed data are used at different modeling levels, either as the extracted input parameters to characterize the different land surface covers or in assimilation procedures, which finally retrieve adequate parameters for the LST estimation. UHI intensity was typically quantified in two steps in these studies (Zhou et al., 2019). First, urban and non-urban areas were defined and delineated from land cover or ISA maps. Urban areas are usually defined as land with a relatively higher proportion of ISA (Chen et al., 2006), whereas non-urban areas have various definitions in different studies but generally include non-urban land cover classes, such as vegetation areas in surrounding areas. Different sized rural and suburban zones have been used as reference areas.

TABLE 9.2
Summary of the Urban Landscape Factors Used in Li et al. (2021)'s Study

Catalog	Factor	FII name	Description	Unit	Formulae
Surface biophysical parameter	MNDWI	Modified Normalized Difference Water Index	Representing the general water status in urban region	None	$MNDWI = \frac{band_{green} - band_{MIR}}{band_{green} + band_{MIR}}$
	NDBI	Normalized Difference Building Index	Representing the general building status in urban region	None	$NDBI = \frac{band_{MIR} - band_{NIR}}{band_{MIR} + band_{NIR}}$
	NDVI	Normalized Difference Vegetation Index	Representing the general urban green space status in urban region	None	$NDVI = \frac{band_{NIR} - band_{red}}{band_{NIR} + band_{red}}$
Building form	BN	Building number	The ratio of the number of buildings to the total area of UFZ	$/km^2$	BN = sum of number/area of UFZ
	BD	Building density	The ratio of building area to the total area of UFZ	km^2/km^2	BD = building area/area of UFZ
	BH	Building height	The average height of the building within the UFZ	m	BH=average height
Landscape pattern metric	PLAND	Percent landscape	Patches area as a percentage of the entire landscape (at the class level)	%	$PLAND = p_i(\sum_j^n \frac{a_{ij}}{A}(100)$
	ED	Edge density	Shape complexity for a patch type or the landscape (at both the class and landscape level)	m/ha	$ED = \frac{E}{A}$
	LSI	Landscape shape index	The total length of edge divided by the shortest possible edge length for the area of a patch (at both the class and landscape level)	None	$LSI = \frac{0.25E}{A}$
	DIVISION	Landscape division index	Degree of separation of patch distribution in a landscape type (at both the class and landscape level)	%	$\text{DIVISION} = 1 - \sum_{j=1}^{n} (\frac{a_{ij}}{A})^2$
	SHDI	Shannon's diversity index	To determine the extent to which one or more types of landscape structures dominate the landscape (at the class landscape level)	None	$\text{SHDI} = \sum_{i=1}^{n} (p_i ln p_i)$

Note: P_i means the proportion of landscape patch type i; a_{ij} means area (m^2) of land cover patch ij; A means the area of UFZ block (m^2); E means total length of patch boundary; UFZ means urban function zone.

Other land covers, such as water bodies, cropland, forest, and low-intensity ISA, have also been used as references in the studies (Martin-Vide et al., 2015). Second, the mean urban-reference LST differences were calculated to reflect the UHI intensity or magnitude. A positive value of UHI intensity indicated an urban heating effect, while a negative value represented a cooling effect. The urban-reference difference method facilitates a comparative analysis of UHIs among cities and urban agglomerations in regions—and across the globe—but the validity of such comparisons can be limited by the large uncertainties associated with urban and reference definitions (Wu et al., 2013). Recent research (Xian et al., 2021) performed a comprehensive and consistent analysis of surface UHI using Landsat LST ARD time series and dynamic land cover datasets in the Sioux Falls, SD, area. It shows that the use of time series of LST and land change dynamic data provided a consistent and quantitative analysis for the distribution and change of UHI intensity.

9.2.3.3 UHI Evaluation by Using Urban Ecological Indices

Many studies have compared UHI to ecological indices, vegetation fraction, and percent ISA, finding strong correlations with mean LST. Landscape metrics (Table 9.1) indicate that urban landscape configuration also influences the surface UHI (Li et al., 2011). Different papers use different these approaches in the various journals, and these approaches use land cover (Zhang et al., 2010), LST pattern (Schwarz et al., 2011), and a combination of land cover and LST pattern (Xu et al., 2010; Pan, 2016; Wang et al., 2019) as monitoring indicators of UHI. Urban ecological status is closely related to the quality of human activities and the development of urban economies. A timely and objective understanding of urban ecological status, particularly in urban and non-urban interface areas, has become increasingly important (De Noronha Vaz et al., 2012). Researchers have been developing a remote sensing-based ecological index for the measure of urban ecological status under UHI (Chen et al., 2019). This urban ecological status index (UESI) integrates four important ecological indicators that are frequently used in evaluating urban ecology. The four indicators include greenness, wetness, dryness, and heat and can be represented by four remote sensing indices or components: NDVI, normalized difference built-up and soil index (NDBSI), wetness component of the tasseled cap transformation (Wet), and LST, respectively (Shi et al., 2021). Instead of a simple or weighted addition of the four indicators, a principal component analysis (PCA) can be utilized to compress the four indicators into one index (UESI) in order to assess the overall urban ecological status under UHI. The calculation of the UESI can be fully automated. Therefore, the UESI can be used to easily and objectively assess urban ecological status. Combined with change detection, UESI can also be used to monitor the change of the ecological status of the UHI and surrounding non-urban areas between different years. In practice, the index was successfully applied in a multitemporal ecological status assessment (Xu, 2013). Pan (2016) applied the G index spatial aggregation analysis to calculate the UHI ratio index (UHIRI) and the landscape metrics to quantify the changes of the spatial pattern of the UHI from the aspects of quantity, shape, and structure of UHI. With the help of ArcGIS software, the primitive G index of the thermal field was calculated, using the natural breakpoint method for its value clustering. Pan found that the UHI intensity had a negative linear correlation with urban vegetation coverage and a positive logarithmic correlation with urban impervious surface coverage. Bala et al. (2021) proposed the UHIRI to quantify urban heat intensity. This work analyzed the variation in LST with land cover (using land cover contribution index (LCCI)) changes in Varanasi, India, from 1989 to 2018, using Landsat images, and concluded that the replacement of vegetation with urban land cover has a severe impact on increasing UHI intensity.

9.2.3.4 Various Statistical and Machine Learning Models

Statistical and machine learning (ML) models have also been proposed to measure UHI (Rajasekar and Weng, 2009). Among these studies, a Gaussian surface model has been utilized the most because it can provide not only the intensity but also the spatial extent and the central location of the UHI (Shi et al., 2021). The kernel convolution method has also been proposed to study UHI effects because of its high efficiency in characterizing the temperature values over space in a continuous surface (Rajasekar and Weng, 2009). Chun and Guldmann (2014) explored the urban determinants of UHI using two-dimensional (2D) and three-dimensional (3D) urban information as the input for spatial statistical models. The results show that solar radiations, open spaces, vegetation, building roof-top areas, and water strongly impact LST and that spatial regressions are necessary in order to capture the neighboring effects. Khalil et al. (2021) utilized the association among LST and different influencing parameters, including elevation, road density, and EVI, to formulate a convolutional neural network (CNN) model for the prediction of LST in the past 19 years so that the UHI effect on the Lahore district could be evaluated. The results indicate that the formulated CNN model can be utilized as an essential tool for analyzing and predicting LST and thus for the evaluation of the UHI effect at any location. Soltanifard et al. (2022) and Yue et al. (2012) used principal component analysis (PCA) first to transform 16 explanatory variables (Table 9.2) and 9 variables into first five and three PCs, respectively, then applied Ordinary Least Square (OLS) and Geographically Weighted Regression (GWR) models to explore the spatially varying relationships and identify the model's efficiency at the neighborhood's scale. For the work of Soltanifard et al. (2022), it suggests that the GWR model has more ability than the OLS one to predict LST intensities and characterize spatial non-stationary (Figure 9.4). Therefore, it can be applied to be more effective strategies in planning and designing the urban neighborhoods for mitigation of the adverse heat effects. Szymanowski and Kryza (2011) addressed the issue of the potential usefulness of remotely sensed data and their derivatives for UHI modeling. Statistically significant models explained 71% to 85% of the air temperature variance. It has been stressed that remotely sensed data are important sources to model urban air temperature heat islands. However, in all of these studies, such models worked less effectively in cities frequently covered by clouds, in arid landscapes, and in urban agglomerations, so they have only been applied in a few UHI studies to date.

9.2.3.5 Spatial-Temporal Time-Series Algorithms

Advances in computing technology have fostered the development of new and powerful data fusion, gap filling, machine learning, and deep learning techniques that

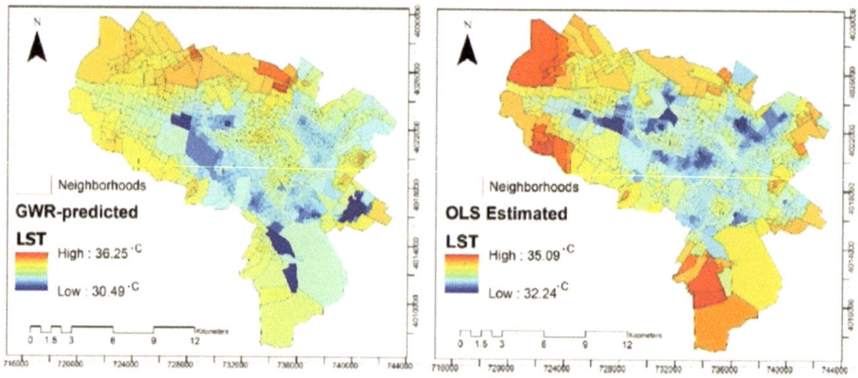

FIGURE 9.4 Predicted LST map obtained from GWR model (left) and estimated LST map generated from OLS model (right).

Source: Soltanifard et al., 2022.

have demonstrated promising results in a wide range of applications (Shi et al., 2021). Models to fuse data from multiple sensors and fill gaps can improve UHI monitoring using an ensemble of dense time series of thermal data with a high spatial resolution. For example, Zhou et al. (2020) presented a new algorithm that focuses on data gap filling using clear observations from orbit overlap regions to obtain Landsat LST data. Multiple linear regression models were established for each pixel time series to estimate the stable predictions and uncertainties. Liu et al. (2016) quantified the spatial–temporal patterns of SUHI by investigating the relationship between LST and the land cover types and the associated landscape components. Such approaches have been used to generate temporally dense and high-resolution LST over long time periods by integrating the observations of Landsat, MODIS, AVHRR, VIIRS, and ECOSTRESS (Hulley et al., 2019; Sun and Du, 2019). Such datasets facilitate subtle analyses of monthly, seasonal, and yearly trends in UHI intensity at regional levels.

More advanced spatiotemporal fusion-based methods include synergic methods (e.g., Xiao et al., 2019) by combining the kernel-driven process with the fusion-based process methods, which more emphasize on how to efficiently deal with spatial heterogeneity of landscapes (spatial variation of LST) and time change (temporal variation of LST) between any two time points. This is a key point or a research direction. Xia et al. (2019) proposed a weighted combination of kernel-driven and fusion-based methods (CKFM) to improve the spatiotemporal resolutions of time series LSTs. The CKFM is more accurate and robust compared to the kernel-driven method with an improvement of 0.1–0.6 K, and it may reconstruct more spatial details than those by a fusion-based method when it was used to downscale daily MODIS 1 km LSTs to 30 m resolution LSTs with Landsat 8 data.

Machine learning (ML) has become popular in UHI but its use has remained restricted to predicting the natural world, rather than understanding UHI itself. ML techniques may not be the solution to all the problems remotely sensed data might have. However, these techniques indeed provide a powerful set of tools that deserve serious attention to deal with some relevant UHI remotely sensed data problems. Lucas (2020) points out that ML differs from the broader field of statistics in two respects: (1) the estimation of parameters that relate to the real world is less emphasized, and (2) the driver of the predictions is expected to be the data rather than expert opinion and careful selection of plausible mechanistic models. The Google Earth engine (GEE) is a cloud-based platform for planetary-scale geospatial analysis that brings Google's massive computational capabilities to bear on a variety of high-impact societal and environmental issues (Gorelick et al., 2017). The GEE techniques for data analysis and interpretation that fully incorporate the temporal dimension remain an area of intense research and represent an important challenge for operational UHI monitoring.

9.2.4 Mitigation Strategies of the UHI Effect

Mitigation and adaptation strategies of the UHI effect (Huang and Lu, 2018) include the use of high albedo materials (Saneinejad et al., 2014), increasing evaporation from urban watersheds (Krüger and Pearlmutter, 2008) and evapotranspiration from vegetation (Kondoh and Nishiyama, 2000), roof technologies (Zinzi, 2010), and building structure and distribution adjustment (Takebayashi and Moriyama, 2012), which can be categorized into four major classes of mitigation measures: roof cooling technology, reflective cooling, vegetation cooling, and urban geometry cooling. Predictably, mitigation and adaption strategies will become a priority in future UHI research. The first strategy encourages the use of more high-albedo materials in the construction of urban areas. Such materials can dampen substantially the UHI effect through reflecting solar heat rather than absorbing such heat. To illustrate, developers can apply high-albedo materials in roofs, pavements, and road surfaces to cool the urban environment. Although albedo can be derived from satellite images (Wang et al., 2017b), their low spectral resolution is a hurdle to monitor with finer detail in urban areas. As a result, UHI studies investigating the thermal properties of materials (albedo) has widely relied on in-situ datasets rather than using remote sensing data. Because pavement covers a high percentage of urban surfaces, reflective cooling contributes greatly to UHI mitigation and thus plays an important role in all mitigation strategies. High-albedo pavements or cool pavements use reflective, permeable, and water-retentive material, which has high reflectivity in both solar radiation and thermal emittance, thus increasing the sensible heat flux to the atmosphere, yet also increases the latent heat flux from water evaporation (Santamouris, 2013).

Fostering green cities is found to be the common strategy, applicable from local to city levels, to mitigate the UHI intensity. The rationale behind this strategy is that a higher proportion of green areas can moderate the impervious surface areas. Green strategies can be implemented through planting four types of vegetation cover: urban forests (parks), street trees, private green in gardens, and green roofs or facades. Studies also show that a careful planning of vegetation cover is important based on patch size, edge, shape, and connectivity in order to maximize the benefit of this policy strategy because variation in these factors can have differential impacts on cooling effect (Chen et al., 2014c). Vegetation cooling is the most widely used and constantly increasing measurement, which comes from latent heat loss from evapotranspiration and the shading of direct solar radiation. Known as cooling islands in urban settings, the urban green spaces (UGSs) cool urban temperature via shading and evapotranspiration (Su et al., 2020). The cooling effect of UGSs on air temperature has been proven by field measurements (Hamada and Ohta, 2010). In addition, remote sensing data have confirmed the existence of the cooling effect on land surface temperature (LST) at a larger scale. Air temperature sites are often limited to their number and spatial range of measurement, whereas remote sensing techniques provide a large spatial coverage, and thus, are widely employed. Many cooling indicators have been developed to quantify the cooling effect of UGSs, including the LST difference with the surroundings, cooling intensity (Liao et al., 2021), cooling extent (Du et al., 2017), cooling efficiency (Peng et al., 2021), and cooling lapse (Du et al., 2017), etc. The cooling effect of UGSs was quantified by three metrics: Green spaces' cooling intensity (CI), cooling extent (CE), and cooling lapse (CL) (Du et al., 2016). CE (unit: m) is expressed as the maximum distance between the edge of UGS and the first turning point of the temperature curve. CI (unit: C) is the LST difference between the green space and the first turning point. CL (unit: C/km) is the LST drop with unit distance. These metrics have been widely used in previous relevant studies and have been proven as effective to quantify the cooling effect of green spaces (e.g., Du et al., 2017; Yu et al., 2018; Fan et al., 2019).

Roof cooling technology includes cool roofs using light-colored surfaces and green roofs using vegetation covering. Due to the large fraction of roofs in urban areas and the excellent operability, this type of cooling could achieve major UHI mitigation. Furthermore, urban design factors, such as building geometry, built forms, orientation, and density, could also be best designed to counteract UHIs by influencing solar radiation and air flows; and thus, there will be stable interest in this area (Yan et al., 2021).

Many studies have reported widely and successfully applied measures on mitigating the UHI effects with promising financial and environmental benefits. The possible mitigating measures could broadly be categorized as: (1) related to reducing anthropogenic heat release (e.g., switch off air conditioners); (2) related to better roof design (e.g., green roofs, roof spray cooling, reflective roofs etc.); (3) other design factors (e.g., Humidification, increased albedo, photovoltaic canopies, etc.) (Rizwan et al., 2008). Table 9.3 shows the major mitigating measures adopted or proposed by various researchers, which could also be categorized as those that could only be implemented during the design and planning stage (e.g., sky view factor and building material etc.) and those that could also be implemented after the design and planning stages (e.g., green areas and roof spray cooling) (Rizwan et al., 2008). It seems that the planting and vegetation is the most widely applied mitigation measure that could achieve huge energy savings through temperature reduction of the area. It was reported in a study conducted by Spronken-Smith et al. (2000) that parks could help control temperatures through an evaporation of more than 300% as compared to its surrounding. The mitigating measures, however, can include other design aspects with diversifying benefits. For example, in the study conducted by Kolokotroni et al. (2006), they estimated that an optimized office

TABLE 9.3

Proposed Mitigating Measures, Maximum Temperature Reduction, and Possible Energy Savings Source: Rizwan et al., 2008

Mitigation measure	Max. T. reduction (°C)	Reported Saving (%)	Reference
Vegetation, lighter color of paving and cooler roofs	3.0	–	Rosenfeld et al. (1998)
Planting and vegetation	1.3–1.6	–	Qiu et al. (2013)
Reducing anthropogenic heat and planting vegetation	1.2	40	Kikegawa et al. (2006)
Proper ventilation		10	Kolokotroni et al. (2006)
Vegetation and suitable albedo	2.0	10	Taha et al. (1999)
Off air-conditioner	1.0	6	Kikegawa et al. (2003)
Reducing anthropogenic heat and energy consumption, improvement in building design	–	–	Urano et al. (1999)
Roof spray cooling	13–17 (ceiling T)	40% ele. consumption	Jain and Rao (1974)
Flow of water over roof	–	–	Sodha et al. (1980)
Roof pond, roof spray cooling, and moving water over roof	–		Tiwari et al. (1982)
Shades, Highly reflective materials	–	–	Yamamoto et al. (2012)
Open and airy spaces, reduce heat release from buildings, etc.	–	–	
Green and highly reflective roofs	–	–	Takebayashi and Moriyama (2007)
Humidification and albedo increase	–	–	Ihara et al. (2008)
Photovoltaic canopies	–	–	Golden et al. (2007)

–: Information not available.

building in an urban area could reduce 10% cooling energy demand through proper ventilation. Urano et al. (1999) reported that anthropogenic heat release has greater potential for modifying the daytime thermal environment while the reduction of energy consumption is more effective during the nighttime. Wider buildings with varied height with associated drag effects modify the environment near the ground. Taha et al. (1999) reported that proper values of surface albedo could achieve temperature reductions and peak electric energy savings. Huang et al. (2005) also reported the effects of urban thermal environment on the comfort level of pedestrians while Rosenfeld et al. (1998) have reported ozone formation reduction of 12% after proper cooling measures are taken.

9.3 THE PROBLEMS THAT EXIST AND FUTURE DIRECTIONS IN URBAN THERMAL ENVIRONMENT STUDIES

9.3.1 Utilization of Multitemporal or Time-Series Data

There are uncertainties associated with the previous studies by the methods widely used in the currently available SUHI publications. The major issue is the analysis of limited datasets. For example, approximately 64% of the publications used single-time-point data to describe the SUHI phenomenon in a season or a year (Zhou et al., 2019). These publications focused on local-scale variability of SUHI based on Landsat and/or ASTER data. The fact is that the LST/SUHI values vary constantly and substantially over time due to changing climate/weather conditions, vegetation activities, and human activities (Quan et al., 2016). Using a single or a limited number of TIR sensor data or derived LST data may be misleading or biased especially for those studies concentrating on the temporal changes of SUHI. Given the free availability of long-term archived TIR remote sensing data, the use of a large number of images or a time-series analysis is greatly recommended and necessary in future studies. In addition, researchers should ensure that their studies use sufficient data and perform the necessary atmospheric corrections for the interpretations made.

9.3.2 Innovative Methods to Quantify SUHI Intensity

Studies have demonstrated that the methos of evaluating SUHI intensity may vary greatly due to differences in the selected urban/reference areas. With the development of new urban datasets and spatial analysis methods, more universal methods have been proposed for the calculation of SUHI intensity, with the possibility to ease the challenges to delineate the extent of urban/reference context (Zhou et al., 2019). For example, a consistent method has been proposed to determine urban and reference areas, which determines the surrounding buffer as having the same size as the urban area, by using nighttime light or land cover data (Peng et al., 2012; Zhou et al., 2016; Li et al., 2017). SUHI intensity was also assessed based on the relationship between LST and urban indicators, such as urban fraction and impervious surface area (Li et al., 2018). The relationship between SUHI and urban size can potentially be used to quantify SUHI intensity because SUHI intensity usually increases with the expansion of the urban extent. Bonafoni and Keeratikasikorn (2018) reported that they investigated the relationship between the spatial trend of LST and the distance to the city center using a ring-based buffer approach. Although the mean LST decreases away from the city core, the daytime and nighttime trends are different in both shape and values. They found that in Bangkok, Thailand the daytime LST can be modeled using an inverse S-shape function, whereas the nighttime LST can be modeled using a quadratic function. The proposed approach allows for reliable modeling and to straightforwardly infer a very accurate urban density-LST relationship. Compared with the traditional methods, these proposed methods consider more urban information and can better represent the overall characteristic of SUHI. In their study, the uncertainties due to the selection of urban/reference areas in previous methods can be mitigated to a large extent. However, further investigations would improve this study: refinements in the ring-based approach for a potential anisotropy identification, comparison studies with other megacities in different parts of the world, and analysis of the LST trend with the city topography. Therefore, the SUHI intensity calculated using these innovative new approaches can be more reliable and more comparable to one another.

9.3.3 Tools Integration

The integration of micro-scale and meso-scale models has been recently proposed (Murakami, 2006). The integrated model predicts large-scale phenomena and interactions above the surface layer using a mesoscale model. On one hand, this model simultaneously obtains the required data from surface layer using micro-scale models. Inversely, it is possible to solve a micro-scale problem and at the same time obtain the required information at top-canopy level from a meso-scale approach. On the other hand, the integrated model also provides better boundary conditions by transferring data from a larger scale to a smaller one (nesting) or vice versa (Sasaki et al., 2008). Therefore, the integration helps cover a wide range of problems. For example, the new weather research and forecasting model (WRF) is developed to integrate a meso-scale model with a canopy-scale UCM (Skamarock et al., 2005) model. This integration is not straightforward and there are a number of issues to be overcome. Unmatched temporal and spatial resolution of the scales is the main problem that may cause missing information through the data transfer between the scales (Taha, 1999). Moreover, details might be lost where data averaging is performed from finer to coarser grids or vice versa. Furthermore, as discussed previously, since difference of the domains varies in micro- and meso-scales, the continuity equation may not be satisfied at the interfaces (Mirzaei and Haghighat, 2010).

9.3.4 Advances in Modeling UHI

Our analysis indicated that determination of optimal modeling for UHI is still a central topic of UHI research. Modeling will continue to provide vital and useful results on the spatiotemporal assessment of UHI, especially when models more effectively combine TIR sensor data from multiple sensors. ML (DL) and AI are continuing to grow in evaluating UHI. For time series analyses with remote sensing data, a cloud computing platform such as GEE could bring about a substantial change in UHI analysis, as they have the capability to process big remote sensing datasets and assess the spatiotemporal dynamics of the area quickly (Shi et al., 2021). A better integration of remote sensing and station measurements into models is expected. The study also suggests that direct and indirect evaluation of UHI, especially human health issues, heat wave impacts, air pollution, and ecological security, will receive increasing scientific attention in the future. Research on mitigating and adapting to UHI impacts may warrant special attention. The intensity of UHI and its changes of LULC based on urban planning are actively being studied.

9.3.5 Height Measurement in Studying UHI

An effect of urban morphology on UHI should be further investigated (Ngie et al., 2014). More research is desirable to apply height measurements and elevation data to the UHI studies. Active remote sensing radar-generated topography data (e.g., Shuttle Radar Topography Mission (SRTM), Light Detection and Ranging (LIDAR), and Interferometric Synthetic Aperture Radar (IFSAR) data) are especially desirable to relate to urban morphology and topography for UHI studies. As the high accuracy of elevation data becomes available, the benefit with related urban morphology and topography data fusion for UHI studies will become obvious. Information on building heights is not always available. The ASTER DEM dataset contains topographic information that can be derived from the along-track, 15-m ASTER optical stereo data acquired in NIR bands 3N (nadir-viewing) and 3B (backward-viewing). These high spatial resolution digital elevation models (DEM) can be used to derive urban topography parameters: elevation, slope, and aspect over horizontal distances of more than 15 m (Welch et al., 1998). After the data quality of ASTER derived topography has been examined, the fusion of the topography data with the ASTER VNIR and SWIR datasets can be undertaken to interpret urban geometries at relevant resolution levels of 30 m so that urban structures can be related to the TIR measured LSTs.

9.4 SUMMARY

Thermal remote sensing can provide a means of observing urban thermal environmental properties spatiotemporally, such as retrieved LST and urban biophysical parameters in evaluating urban heat islands (UHIs). Therefore, in this chapter, the reasons why we need to perform studies of TIR remote sensing in urban thermal environmental properties was first introduced, then introduced mapping and analyzing UHI, including generation of UHI, factors and methods of measuring of UHI intensity, methods and techniques of evaluating UHI intensity (LST and UHI intensity calculation, comparing the difference between UHI areas and non-urban areas, UHI evaluation by using urban ecological indices, various statistical and machine learning models, and spatial-temporal time-series algorithms), and mitigation strategies of the UHI effect.

Finally, we discussed the problems that exist and future directions in urban thermal environment studies, including utilization of multitemporal or time-series data, innovative methods to quantify SUHI intensity, tools integration, advances in modeling UHI, and height measurement in studying UHI.

10 Thermal Infrared Remote Sensing Applied in Other Areas

In this final chapter, we explore three significant applications of thermal remote sensing across diverse domains. These include mapping geothermal alteration and mineralization sites (Section 10.1), monitoring active volcanoes through thermal imaging (Section 10.2), and assessing plant ecosystems using thermal remote sensing techniques (Section 10.3). Each area is comprehensively reviewed, discussing its methodologies, applications, and contributions to advancing our understanding of natural and environmental processes through thermal infrared technology.

10.1 MAPPING GEOTHERMAL ALTERATION AND MINERALIZATION SITES

10.1.1 Introduction

Geothermal systems exist in regions of anomalously high crustal heat flow that may be associated with the occurrence of young igneous bodies or hot rocks located deeper in the crust. This high geothermal heat is normally transferred to the surface by the convection of groundwater that forms hydrothermal systems. If rising hot waters reach the surface then characteristic geothermal features such as hot springs, fumaroles, and geysers may form (Figure 10.1). When the heat contained within fluid filling fractures and permeable rocks is sufficiently accessible, this energy can be exploited for power generation or direct use (geothermal energy). Studying surface geothermal activity and heat loss associated with magmatic related systems is important for monitoring of subsurface igneous activity. For example, the abundant geothermal features at Yellowstone National Park are the surface presentations of the largest active volcanic system in North America. Surface hydrothermal activity at Yellowstone has changed over time indicating variations in geothermal system activity that may be in turn related to changes in magmatic activity (Vaughan et al., 2009). In addition, the monitoring of geothermal activity has an important significance in efforts to conserve geothermal systems that are of cultural significance or are economically important tourism destinations (Haselwimmer et al., 2011) with the increasing development of geothermal resources there is potential to negatively impact these sites (Bromley et al., 2010).

Thermal infrared (TIR) remote sensing provides data with synoptic coverage for investigating the surface manifestations of geothermal systems as applied to both geothermal energy exploration and also for more fundamental research and monitoring

DOI: 10.1201/9781003276173-10

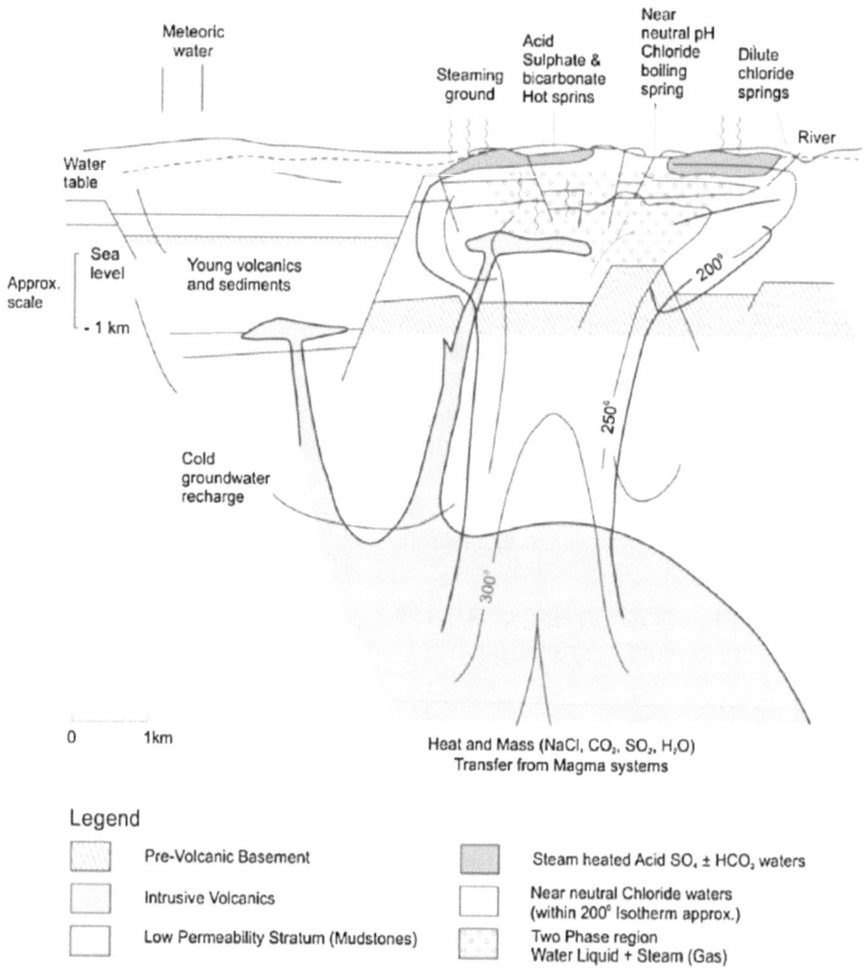

FIGURE 10.1 Conceptual geologic model of a volcanic geothermal system where recharge results from meteoric groundwater driven by heat from supplied from a buried magmatic system leading to a convective column. Steam separation results in fumaroles and steam adsorption by groundwater while hot springs occur associated with the formation of silica. Source: modified after Henley and Ellis (1983).

(Haselwimmer and Prakash, 2013). TIR remote sensing provides a method for rapid mapping and quantifying surface geothermal features in support of exploration and assessment of new resources (Allis et al., 1999; Rockwell and Hofstra, 2008; Haselwimmer et al., 2011). They are certainly rapid and are inexpensive when compared with extensive ground surveys. Well-established methods in multispectral, hyperspectral analysis, in conjunction with modern subsurface geophysics, advanced TIR imagery from space-based and airborne sensor systems, permits their direct and immediate application in geothermal energy prospecting and their evaluation using both thermal signature and spectral signatures indices (Coolbaugh et al., 2007). In addition, TIR data can be used to monitor these geothermal features at developed reservoirs that may provide indications of unsustainable resource exploration (Allis, 1980; Bromley et al., 2010) or provide estimates of surface heat loss as input to reservoir models (Bromley et al., 2011). The use of TIR data to map and quantify surface geothermal features also supports long term monitoring of magmatic-related and/or protected systems (Watson et al., 2008; Vaughan et al., 2009; Seielstad and Queen, 2009).

Remote sensing geophysical surveys are prevalent practices in exploration projects that benefit the natural resources industries such as mining and oil. The remote sensing geophysics provides diverse pieces of information concerning the geology of a particular area and measures the geophysical variations arising from the physical properties of the desired sources (Wambo et al., 2020). TIR remote sensing of mineralization mapping is one of the most important techniques at the reconnaissance stage of exploration for mineral resources related to such surveys (Pour et al., 2018). The effective management of natural resources is enhanced when detailed geological information from many sources is available, yet the efforts of exploration geologists are generally hampered by the difficulty, cost, access limits, and both natural and human risks inherent in surveying large areas. Satellite remote sensing is an essential low cost solution with large coverage at different spatial, spectral, and temporal resolutions for identifying, mapping, and collecting reliable information on prospective or target areas, and thus advanced TIR remote sensing data has been widely applied for identification and mapping of surface mineralogy in areas of suitable exposure (Van der Meer et al., 2012; Salles et al., 2016; Pour et al., 2018).

10.1.2 Thermal Image Preprocessing Techniques

The main purpose of the image preprocessing methods is to apply image processing techniques that are capable of detecting subtle hydrothermal alteration minerals and lithological units and mapping structure elements using VNIR+SWIR and TIR data. Besides image preprocessing techniques, which generally include image radiometric calibration and atmospheric correction, here we emphasize image transformation. They can be considered as image enhancement and image information extraction techniques. First, a RGB false-color-composite technique is extensively used for combining multispectral images through color coding. In fact, it is possible to create a vast number of false-color composites from the multispectral bands. For example, in the study by Sheikhrahimi et al. (2019), they used the validated ASTER band combination (7, 3, 1) in RGB for mapping hydrothermal alteration zones and lineaments associated with orogenic gold mineralization. The previous study is equivalent to the Landsat Thematic Mapper (TM) band combination (7, 4, 2) in RGB (Abdeen et al., 2001), which was selected for the global Landsat mosaic created from NASA. It is very useful for outlining lithological units, as well as structural and morphological features (Abdeen et al., 2001)—which information can be used for identifying and classifying geothermal alterations and mineralization sites—and provides striking imagery for outlining mineralization areas and drawing a base map.

Second, to reduce the effects of topography and enhance the spectral differences between bands, band ratioing techniques are used in an analysis. It is a technique where the digital number value of one band is divided by the digital number value of another band. Band ratios are very useful for highlighting hydrothermal alteration minerals. Dividing one spectral band by another produces an image that provides relative band intensities; this is able to minimize the illumination differences due to topography. Therefore, this technique is particularly applicable to highly exposed areas and rugged terrains in arid and semi-arid environments. Moreover, Red-Green-Blue (RGB) color composites technique could be easily applied to band ratios to produce image-maps of lithological units for the study area. For example, Pour and Hashim (2015) created band ratios images for constructing RGB color composite images with Landsat-8 bands based on laboratory spectra of the minerals related to hydrothermal alteration and lithological units. Different color combination images were used to enhance hydrothermally altered rocks and lithological units as well as vegetation at regional scale. Band ratios are useful for highlighting certain features or materials that cannot be seen in raw bands (Inzana et al., 2003). Band ratios derived from image spectra (4/2, 6/7, 5 and 10 in RGB) allow the identification of altered rocks, lithological units, and vegetation.

Third, image transforms are often commonly used to reduce the dimensionality of the input dataset and processing time, focus processing on the information of interest within the input file, and remove noise (Sheikhrahimi et al., 2019). Each output band (component) of a transformed image is a linear combination of every input image, thus helping identify those spectral bands that are most important for finding targets of interest or which bands contribute the most noise (Research Systems Inc., 2008). The principal component analysis (PCA) transformation was used to produce uncorrelated output bands, segregate noise components and reduce the dimensionality of datasets. This is done by determining a new set of orthogonal axes that have their origin at the data mean and that are rotated so the data variance is maximized. PC bands are linear combinations of the original spectral bands and they are also uncorrelated (Richards, 1999). The PCA is a well-known method for alteration mapping in metallogenic provinces for mineral exploration objectives (Pour and Hashim, 2011). The independent component analysis (ICA) is a statistical and computational technique for array processing and data analysis, aiming at recovering individual signals (or 'sources') from observed mixtures of the signals, exploiting only the assumption of mutual independence among the signals (Hyvärinen, 2013). It seeks signal combinations with the most non-Gaussian probability density functions. They are identified as the 'independent components' (ICs) of the observations. It is superficially related to PCA and factor analysis but in many cases is more capable of isolating underlying factors or sources in cases with realistic noise and correlation structure in the data (Hyvärinen, 2005; Comon and Jutten, 2010). In order to address that the issue of PCA-ordered PCs is not necessarily ordered by image quality, Green et al. (1988) used an approach similar to PCA, called maximum noise fraction (MNF), which was based on a different criterion, signal-to-noise ratio (SNR), to measure image quality. The MNF can produce component images in order of noise. Lee et al. (1990) re-interpreted MNF by Green et al. (1988) as a two-stage process, called noise-adjusted principal component (NAPC) transform. The two-stage process includes that whitened noise variances of each band image to unit variance are first performed and then PCA transform is carried out on the noise-whitened band images. In the study by Sheikhrahimi et al. (2019), the forward PC rotation was applied to VNIR+SWIR and TIR bands of ASTER covering the study area. A linear transform was applied to maximize the variance of the data. To separate and suppress vegetation effects, Pour et al. (2018) tested a directed principal components (DPC) analysis that is the application of optical remote sensing data for detecting mineralogy of soils and rocks in areas of the tropical/sub-tropical environment. In accomplishing this an image enhancement was based on the DPC analysis of four band ratio images, which was implemented using the following indices: vegetation index (band ratio of 4/3), clay minerals index (band ratio of 5/7), ferric iron oxide index (band ratio of 3/1), and ferrous iron oxide index (band ratio of 5/4) of ETM+ data. The input band ratio images are selected on the basis that one band ratio contains information related to the component of interest (e.g., hydrothermal alteration minerals), the spectral response of which suffers interference from the spectral response of another component (i.e., vegetation). Band ratio 4/3 serves as an excellent vegetation indicator while band ratios of 3/1, 5/4, and 5/7 are commonly used to enhance iron oxide and clay minerals, respectively. In the study of Wambo et al. (2020), ICA was applied subsequent to PCA, using images PC1 to PC8 of Landsat-8. The PCA was used to give weight to the components and remove the correlation before applying ICA for revealing hidden factors. Therefore, this fusion has a great performance to identify the pixels containing the spectral signature of the alteration minerals or mineral groups that are maximally independent of each other (Pour et al., 2019b). Components created by IC rotation were analyzed based on statistical factors, 2-D scatter plots, and visual examination of output images. Wambo et al. (2020) used the linear spectral unmixing (LSU) algorithm to map mineral prospectivity. It includes several steps, namely: (1) the Minimum Noise Fraction (MNF); (2) the Pixel Purity Index (PPI); (3) automatic end-member prediction from the n- Dimensional Visualizer; and (4) allocation of LSU. They achieved better results.

10.1.3 Identifying and Classifying Techniques/Methods of Geothermal Alteration and Mines

In the past decades, several authors have reviewed the different techniques and image processing algorithms that can be applied to geological remote sensing in general or in mineral exploration. Common image processing algorithms employed in mineral exploration are: (1) RGB combinations; (2) band ratios; (3) image transforms (e.g., PCA, ICA, and MNF); (4) matched filtering (MF); (5) linear spectral unmixing (LSU) and mixture tuned matched filtering (MTMF); (6) constrained energy minimization (CEM); (7) least square fitting method; and (8) spectral mixture analysis (SMA) and multiendmember SAM (MESAM). Recently, great attention has been given to non-parametric classifiers such as machine learning and deep learning algorithms as well as fuzzy logic modeling methods. These techniques have been applied in land-use/land-cover (LULC) problems with great success, outperforming classical algorithms (Thanh Noi and Kappas, 2017). In this subsection, machine learning (ML) and deep learning (DL) algorithms as well as fuzzy logic modeling methods are briefly introduced, reviewed, and discussed.

ML mapping geological features is a fundamental step in mineral exploration. The combined use of ML methods with remote sensing data can be considered an easy and inexpensive approach for mapping lithological units, alteration zones, structures, and indicator minerals associated with mineral deposits (Shirmard et al., 2022). ML methods are effective for remote sensing data analysis since they can automatically learn the relationship between input features such as reflectance continuum with desired outputs for prediction or classification. Moreover, ML methods are robust in spectral and ground truth measurements against noise and uncertainties. ML methods are essentially data-driven approaches that can be used in several ways, such as processing high-dimensional data into lower dimensions, predicting certain trends in the data, and identifying certain characteristics or components in the data. Hence, there is a great potential and opportunity for applying ML methods for addressing the increasingly growing size and complexity of remote sensing data. The synergy of ML and remote sensing data, including satellite, airborne, and ground-based data, could be helpful in mineral exploration. Generally speaking, ML methods are of two major types, which include supervised and unsupervised learning. Supervised ML methods (e.g., artificial neural network (ANN), random forest (RF), and support vector machine (SVM)) require labeled data that are used for regression and classification problems to model the relationship between input features and outcomes. ML-based classification methods for different datasets (problems) offer different outcomes, i.e., various classified maps. An ML method that provides the best accuracy for solving a problem may not work for another problem or dataset. Yu et al. (2012) developed an automated approach to lithological mapping in Northwestern India where different SVMs with the radial basis function kernel were trained using ASTER, ASTER-derived digital elevation model (DEM), and aeromagnetic data. The number of test input features ranged from 14 to 33, having the classification accuracy increased with the increasing number of input features. When compared with a classical classifier, SVM clearly outperformed other ML methods, achieving a much higher accuracy (Yu et al., 2012). Cracknell et al. (2014) applied RF and self-organizing maps to map lithology and volcanic-hosted massive sulfide (VHMS) alteration in Tasmania. The input data included airborne geophysical, soil geochemical, and Landsat-7 ETM+ spectral data. The approach provided new geological detail and allowed one to identify areas of interest for geological and VHMS alteration mapping. Bahrambeygi and Moeinzadeh (2017) compared SVM with the ANN method in hyperspectral mapping of ophiolite terrains in Iran using Hyperion images, the ANN having achieved a higher accuracy.

DL methods are a branch of ML methods that seek to model high-level abstractions in data by applying multiple processing layers with complex structures (Shirmard et al., 2022). DL methods such as recurrent neural networks (RNNs), convolutional neural networks (CNNs), autoencoders, deep belief networks, and restricted Boltzmann machines have been successfully applied to transform different fields (Schmidhuber, 2015; Velliangiri et al., 2019). There exist only a few studies with the application of DL in processing remote sensing data for mineral exploration. DL methods can be used for supervised and unsupervised learning. DL methods have the ability to model complex and large datasets and are loosely grouped into feedforward and recurrent neural network architectures (Shrestha and

Mahmood, 2019). Among a variety of DL methods, convolutional neural networks (CNNs) have gained more attention in remote sensing due to their success in image processing. CNNs belong to a class of deep feedforward neural networks that have been successfully applied to image processing tasks due to their ability to provide automatic feature extraction using convolutional and pooling layers. Latifovic et al. (2018) assessed the performance of convolutional neural networks (CNN) in lithological mapping in Northwestern Canada. Different data layers were used to train the network including Landsat TM/ETM+ scenes, aerial photos, and high-resolution DEM data. The CNN was evaluated for predicting surficial geological classes under two sampling scenarios. In the first scenario, a CNN used the samples obtained from the field that needed to be mapped, and in the second, the CNN trained in one field is then extended to areas where the samples available were not included in CNN model training. Finally, the fusion spatial-spectral function was primarily derived in a model by stacking local spatial characteristics obtained by a CNN-based model and spectral data. CNNs are DL methods that have the ability to model temporal data and dynamical systems. The convolutional long-short term memory (LSTM) network has been developed based on features from CNN, which has been useful for spatiotemporal datasets (Shi et al., 2015). These have been used for remote sensing applications and have potential in the scope of mineral exploration. Zhao et al. (2020) applied the CNN method on the hyperspectral data obtained in the Nevada mining region by the AVIRIS. The hierarchical spatial-spectral feature extraction with long short-term memory (HSS-LSTM) classification achieved an accuracy of 94.70%, which exceeded the most widely used methods.

Fuzzy logic modeling is based on the fuzzy set theory, which was proposed by Zadeh (1965). It is a form of many-valued logic in which the truth values of variables may be any real number between 0 and 1 both inclusive (Novák et al., 1999). Fuzzy logic modeling has been successfully applied for mineral prospectivity mapping in metallogenic provinces (Carranza, 2008). Various factors such as host rock, structures, geophysical anomalies, remote sensing studies, and documented fieldworks can be considered information layers (fuzzy members) for mineral exploration. Hence, each information layer might contain a specific weight between zero and one. It can be allocated depending on its importance and relation to ore mineralization. After determining the mineralization control factors, the data layers are integrated by the fuzzy operators. The application of fuzzy logic modeling for mineral prospectivity mapping normally includes three main stages: (1) fuzzification of evidential data; (2) logical integration of fuzzy evidential maps with the aid of an inference network and appropriate fuzzy set operations; and (3) defuzzification of fuzzy mineral prospectivity output to aid its interpretation (Carranza, 2008). A set of fuzzy membership values is expressed in a continuous series from 0 to 1. Function-member values are established for each evidence map that will be integrated. In the fuzzy logic method, a total fuzzy membership map based on the significance distance of features is weighted (for each pixel or spatial position, a particular weight between 0 and 1 is appointed) (Zimmermann and Zysno, 1980). Five operators that are useful for combining mineral exploration datasets include the fuzzy AND, fuzzy OR, fuzzy algebraic product, fuzzy algebraic sum, and fuzzy gamma (Nykänen et al., 2008). Wambo et al. (2020) reported that satellite-based mineral prospectivity maps at regional and district scales were created for the study area by implementing the fuzzy logic model to the most informative thematic layers derived from image processing results. PCA, ICA, and specialized spectral band ratios were used to extract spectral information layers related to vegetation, iron oxide/hydroxide minerals, Al–OH, Fe-Mg–OH, carbonate group minerals, and silicification using Landsat-8 data at regional scale. The satellite-based prospectivity maps are reliable for exploration of new gold prospective zones in the Ngoura-Colomines goldfield. Aali et al. (2022) used the fuzzy logic modeling technique to make the exploratory/information layers fused. The information layers fused include remote sensing data, such as Sentinel-2 and Landsat 7, aerial magnetic geophysical data, and geological data for identifying polymetallic mineralization potential zones in the Chakchak region, Yazd province, Iran. Their results indicate that some high potential zones of polymetallic mineralization were identified and verified through a detailed field campaign and drilling programs in the Chakchak region. It is concluded that the fusion of remote sensing, geophysical, and geological data using fuzzy logic modeling and the multi-class index overlap method is a robust, reliable, and low-cost approach for mining companies to explore the frontier areas with identical geologic conditions that are alleged to indicate polymetallic mineralization potential. In the by of Shirazi et al. (2022), they developed the Neuro-Fuzzy-AHP (NFAHP) technique for fusing remote sensing (i.e., ASTER alteration mineral image-maps) and geological datasets (i.e., lithological map, geochronological map, structural map, and geochemical map) to identify high potential zones of volcanic massive sulfide copper mineralization in the Sahlabad mining area, eastern Iran. They identified Argillic, phyllic, propylitic, and gossan alteration zones in the study area using band ratio and selective principal components analysis methods implemented to ASTER VNIR and SWIR bands. They concluded that the NFAHP technique shows great reliability for copper exploration in the Sahlabad mining area, and it can be extrapolated to other metallogenic provinces in Iran and other regions at the reconnaissance stage of mineral exploration.

10.1.4 Challenges and Future Directions

With the introduction of remote sensing data, supercomputers, and modeling based on ML methods, there is more potential for new resource discovery although complications and challenges exist (Shirmard et al., 2022). At present, many of the groundbreaking activities in the field of mineral discovery rely on three areas: digitization and the use of artificial intelligence, remote sensing and geophysical technologies development, and new approaches to exploration in regions where mineral deposits are under cover. The challenges of remote sensing given big data and complex ML methods face our era. Big data refers to a series of datasets so large and diverse that conventional algorithms and models for data analysis are not feasible. The challenges are in acquisition, processing, preparation, searching, sharing, transition, analysis, and interpretation of the data. The complexity of ML methods in terms of the number of model parameters and model architecture has been increasing to address different datasets. Therefore, when using remote sensing to understand geo-processes, it is critical to consider multi-source, multi-scale (resolution), high-dimensional, dynamic state, isomer, and non-linear characteristics of remote sensing data (Shirmard et al., 2022). The characteristics are fundamental assumptions and objectives when evaluating remote sensing big data and extracting information from them (Liu, 2015).

Satellite, airborne, and ground-based remote sensing data are typically very large and complicated datasets that take much time to decode and interpretate without any assurance that hidden information could be detected. Valuable information can be extracted from such dynamic signals with the help of ML methods. However, emerging ML methods have not been applied in some branches of science (Shirmard et al., 2022). Recent DL methods could be applied in mineral exploration using remote sensing data. CNNs are models ideal for automating feature extraction and capturing complex relationships in image-based data for problems in geosciences. CNN has been designed to concentrate on obvious characteristics and spatial dependence in images that can reflect valuable characteristics such as nearby regions in the case of remote sensing data. Experimental findings indicate that spectral and spatial knowledge can be exploited entirely by CNNs. In this context, Ding et al. (2017) proposed a new CNNs based hyperspectral image (HSI) classification method where the fully convolutional kernels can be automatically learned from the data through clustering without knowing the cluster number. The CNN approach achieved improved classification performance, which has been useful in identifying and discriminating target features for mineral exploration such as lithological units, alteration types, structures, and indicator minerals.

Uncertainty exists in ML methods. Although ML methods have been widely used for remote sensing data analysis, there remains an uncertainty quantification in predictions. In general, assessing the uncertainty in remote sensing-based model applications is challenging. The model validation is usually carried out via a comparison with ground truth or alternative information that presumably represents the ground truth. A number of studies reviewed remote sensing data validation methods and documented their similarities and differences (Loew et al., 2017; Wu et al., 2019c). Bayesian inference provides a principal approach to uncertainty quantification in model parameters. Recently, there has been tremendous progress in the area of Bayesian neural networks and Bayesian deep learning (Wang and Yeung, 2020); however, these methods have not been much used for remote sensing, and their application for mineral exploration is absent (Shirmard et al., 2022). These methods have the potential to provide the meaningful interpretation of remote sensing data given challenges such as noise in data, sparse datasets, and missing data. The quantification of uncertainty using Bayesian inference can be used to project the uncertainty associated with model parameters and data. Recently, Chandra and Kapoor (2020) proposed a Bayesian framework for transfer learning using neural networks that considers single and multiple sources of data. They used the Markov Chain Monte-Carlo (MCMC) method to obtain samples from the posterior distribution that consider the knowledge from remote sensing and mineral exploration given multiple sources of data and information. Bayesian framework driven by MCMC sampling has been used for geoscientific models where different sources of information have been incorporated for inference of geophysical parameters in landscape evolution models. The results show that the framework provides a robust probabilistic approach for decision making with accuracy that is similar to gradient-based learning methods. Therefore, the Bayesian approach has good potential to reduce uncertainty in modeling processes to integrate remote sensing data for exploring mineral deposits.

Transfer learning can be considered as an alternative approach of the DL method, and it can be applied in remote sensing data processing for detecting ore deposits. Transfer learning is inspired by the field of educational psychology and refers to the transfer of learning that happens when learning in one condition increases (positive transfer) or decreases (negative transfer) the performance. Transfer learning is the reuse of a pre-trained model on a new problem. It's currently very popular in DL because it can train deep neural networks with comparatively little data. This is very useful in the data science field since most real-world problems typically do not have millions of labeled data points to train such complex models (Transfer learning, 2023). There are challenges in DL methods given several sources of knowledge and data that need to be integrated for developing a robust model. Remote sensing data need to be compressed for ML methods and the use of autoencoders and variants such as deep autoencoders (Deep Generative Deconvolutional Network) is a feature where CNN can be useful for mineral exploration.

Optical multi- and hyperspectral remote sensing can aid in better understanding surface geology and mineral alteration related to geothermal activity, and thermal remote sensing allows to measure and map surface temperature distribution and heat flux, while with instruments like MODIS we gain insight into long-term thermal radiance variations. InSAR data can add information on earth surface deformation, subsidence/uplift, which can be coupled to the development of the reservoir through reservoir modeling (Van der Meer et al., 2014). However, a number of steps needs to be taken for a full uptake of remote sensing in geothermal exploration. A number of future directions for research is on geothermal systems and the present contributions of remote sensing to study these systems.

Remote sensing products are reproducible. The geologic remote sensing community has contributed to a better understanding of geothermal systems by studying surface mineralogy using both multi- and hyperspectral data. In case of multispectral data the products were mostly based on image ratio, and for hyperspectral data a wealth of spectral mapping techniques are available to map surface mineralogy. Surface temperature retrieval and the retrieval of heat flux are based on data assimilation and physical modeling, which inherently are more reproducible although subject to some boundary conditions and often empirical relationships in the models. Likewise surface deformation measurements from InSAR are based on physical principles that are 100% reproducible.

Geothermal systems can be achieved by a long-term monitoring. Acknowledging that geothermal systems are dynamic and exploration potentially accelerates certain associated processes, it is apparent that long-term monitoring is required. The present literature under exploit indicates the possibility that remote sensing offers in monitoring geothermal systems. This would be of interest as it could link geothermal systems to climate change. The hypothesis based on age dating of hydrothermal deposits suggested that elevated water table in combination with more water available from (extreme) rain events resulted in greater transfer of heat from deep heat sources to the surface. Examining long-time series of remote sensing data over selected geothermal areas in various geodynamic settings could form the basis of accepting or rejecting this hypothesis at least over decades of time spans.

10.2 THERMAL REMOTE SENSING OF ACTIVE VOLCANOES

10.2.1 Introduction

Since 1985, an increasing body of literature has demonstrated how remote measurements of heat emitted by active lava flows, lava domes, lava lakes, and erupting—or degassing—vents can provide insights into the physical processes that are both the cause and consequence of volcanic eruptions (Harris, 2013a). Observations from airborne and space-borne objects in the TIR region of the spectrum are useful when examining volcanic activity. This is because volcanic activity consists of the transfer of heat from within the Earth to the surface, and observations of this heat are frequently associated directly with the activity of a volcano at a specific time (Oppenheimer, 1998). The active volcano observed and expressed in the thermal signature might be a fresh lava flow or an active lava lake or may be more subtle, such as a degassing surface or warmed crater-lake (Blackett, 2017). Understanding the activity at a volcanic site using such TIR data is clearly useful in determining a volcano's status and in predicting future activity.

Various Earth-observing satellites have sensors capable of observing these infrared radiation/emissions and as such are routinely used for volcanic observations (Table 10.1), despite their original purposes often being far deviated from such geological applications. The temperature of volcanic features varies widely. Fresh lava may be between 1,073 and 1,273 K, although the temperature of lava lakes has been measured at up to 1,473 K, while cooler active surfaces may reach 673 K (Oppenheimer and Yirgu, 2002). At these temperatures, the main radiant emissions are in the thermal infrared (TIR), middle infrared (MIR), and shortwave infrared (SWIR) portions of the spectrum. The relationship between the spectral radiant emissions and temperature of the emitting objects, thus representing the Planck function and both the Stefan-Boltzmann and Wien displacement laws, is used (Blackett, 2017). Early satellites such as NASA's Television Infrared Observation Satellite (TIROS) series, launched throughout the 1960s, possessed TIR Earth-observation capabilities for meteorological observations. Since this period, TIR infrared observations have remained crucial for meteorological purposes and for providing an excellent means of observing terrestrial 'hot phenomena,' ranging from wildfires to active volcanoes.

More recently, Wessels et al. (2013) used a combination of satellite and airborne high-resolution visible and thermal infrared (TIR) image data to detect and measure changes at Redoubt Volcano in Alaska during the 2008–2009 unrest and eruption to isolate precursory signals of subsequent eruptions, and Carr et al. (2016) demonstrated the capability of satellite TIR observations for determining lava extrusion rates and discriminating different types of lava dome activity: low dome growth, rapid dome growth, and pyroclastic density current-producing dome collapses, at Merapi Volcano, Indonesia. Likewise, several satellite sensors have MIR

TABLE 10.1
Characteristics and Attributes of the Main Sensors Used for the Infrared Remote Sensing of Volcanoes (Blackett., 2017)

Satellite	Sensor	Full name	Characteristics
ESA/Envisat	AATSR	Advanced Along-Track Scanning Radiometer	Seven bands: VNIR-MIR-TIR, Resolution: 1,000 m
NASA/EO-1	ALI	Advanced Land Imager	Nine bands: VNIR-SWIR, Resolution: 30 m
NASA/Terra	ASTER	Advanced Spaceborne Thermal Emission and Reflection Radiometer	14 bands: VNIR-SWIR-TIR, Resolution: 15, 30, and 90 m
EAS/ERS	ATSR-1/-2	Along Track Scanning Radiometer	Four bands: SWIR-MIR-TIR, Resolution: 1,000 m
NOAA	AVHRR	Advanced Very High Resolution Radiometer	Five bands: VNIR-SWIR-MIR-TIR, Resolution: 1,100 m
NASA/Landsat-5	TM	Thematic Mapper	Seven bands: VNIR-SWIR-TIR: Resolution: 30, 120 m
NASA/Landsat-7	ETM+	Enhanced Thematic Mapper Plus	Eight bands: VNIR-SWIR-TIR, Resolution: 15, 30, and 60 m
NASA/GOES	Imager	Imager	Five bands: VNIR-MIR-TIR, Resolution: 1,000–4,000 m
NASA/EO-1	Hyperion	Hyperion	220 bands: VNIR-SWIR, Resolution: 30 m
ESA/MFGS	MVIRI	Meteosat Visible and Infrared Imager	Three bands: VIS-TIR, Resolution: 250–5,000 m
NASA/Terra/Aqua	MODIS	Moderate-Resolution Imaging Spectroradiometer	36 bands: VNIR-MIR-TIR, Resolution: 250, 500, and 1,000 m
NASA/Landsat-8	OLI	Operational Land Imager	Nine bands: VNIR-SWIR, Resolution: 15–30 m
NASA/Landsat-8	TIRS	Thermal Infrared Sensor	Two bands: TIR, Resolution: 100 m
Japan/JERS-1	OPS	Optical Sensor	Seven bands: VNIR-SWIR, Resolution: 18 m 24 m
EAS/MSGS	SEVERI	Spinning Enhanced Visible and Infrared Imager	12 bands: VNIR-TIR, Resolution: 1,000–3,000 m
NASA/S-NPP	VIIRS	Visible Infrared Imaging Radiometer Suite	22 bands: VNIR-SWIR-TIR, Resolution: 375–750 m

detection capabilities, which have been shown to be useful in monitoring both thermally anomalous volcanic surfaces and, also, those at a more ambient temperature (Blackett, 2017). Wright and Pilger (2008a, 2008b), for example, applied MIR bands from MODIS time series data to quantify the radiant emissions from a number of volcanoes and used the same data to monitor the variability in heat flux from the lava lake at Mount Erebus in Antarctica, respectively, while Coppola et al. (2016) used MODIS measured MIR data to demonstrate a volcanic activity detection system that harnesses detected MIR signals. The advantage of SWIR volcanic observations is that, in the absence of sunlight, terrestrial SWIR signals recorded from space will be purely a product of the temperature of a hot emitting surface. In contrast, MIR and TIR detections from space will be the product of emissions from both ambient and hot surface components, the former of which must therefore be removed to extract hot anomaly components. The advantage that MIR and TIR observations have over SWIR observations, however, is that they will also detect cooler volcanic surfaces, thereby without the requirement of incandescent or fresh lava to be present while facilitating the monitoring of a greater range of volcanic activity (Blackett, 2017).

The data from spaceborne sensors used to detect, monitor, and even forecast eruptions have been analyzed since the earliest days of the satellite era. Those early studies using the data available at the time focused mostly on hot spot detection and temperature measurements with TIR data (Ramsey and Flynn, 2020). The ability to extract critical TIR information from subtle phases of precursory activity in order to perform the detailed spectral mapping of the erupted products grew rapidly (Harris, 2013a). Many of these studies describe the detection of a new thermal volcanic anomaly at a quiescent status, which can make models of the sub-pixel temperature distribution. High temporal temperature data could allow more accurate modeling of lava and gas flux rates as well as chronological descriptions of each eruptive phase.

Volcanology, as is the case for many other disciplines relying on orbital image data, adapted to the fundamental technological divisions of the availability of high temporal/low spatial resolution versus that of low temporal/high spatial resolution data. Volcanological processes operating at the minute to hourly time scale (e.g., lava effusion, plume emplacement, drifting ash clouds) require data very different than those acquired on the time scale of days to weeks (Ramsey and Flynn, 2020). The former falls under a class of TIR sensors designed primarily for weather and atmospheric studies and includes sensors such as the AVHRR, the Along Track Scanning Radiometer (ATSR), the MODIS, the Visible Infrared Imaging Radiometer Suite (VIIRS), as well as instruments on the GOES satellites. These sensors are commonly designed with wide swath widths, a limited number of spectral bands, and spatial resolutions of 1.0 km/pixel or larger, which result in high temporal frequencies of minutes to hours but low spatial resolution. Modeling the data to extract information below the scale of the pixel has made these datasets invaluable for both the rapid detection of new activity as well as the analysis of time-scale dependent eruptive processes (Wright et al., 2002a). TIR data of the Earth's surface ranged from very coarse spatial resolution to the sub-100 m scale and from a single spectral channel for temperature measurements to five for the ASTER and ECOsystem Spaceborne Thermal Radiometer Experiment on Space Station (ECOSTRESS) instruments. This class of sensors includes instruments on the current Landsat series as well as older systems like ASTER and newer ones such as ECOSTRESS. These commonly have a larger number of spectral bands than the weather class of sensors and spatial resolutions of 100 m/pixel or better but a temporal frequency of days to weeks. The improved spatial and spectral resolutions do provide for studies of smaller-scale volcanic processes with detection of much smaller temperature variations. These sensors are also excellent for detecting early precursory activity despite the infrequent coverage (i.e., relatively low temporal resolution). However, the data cannot be used to describe the high-frequency changes ongoing during an eruption despite providing a detailed 'snapshot in time' of that activity.

Of the more common applications, thermal sensors' monitoring of crater lake temperatures, lava, and thermal anomalies have been conducted since the 1980s, initially using the thermal sensor on board the Landsat series of sensors (Oppenheimer, 1996; Kaneko and Wooster, 1999). The number of thermal applications has since increased with the development of more complicated techniques such as spectral unmixing analysis with multiband thermal sensors like ASTER (Pieri and Abrams, 2004; Lombardo and Buongiorno, 2006), MODIS (Wright and Flynn, 2003; Wright et al., 2005) and AVHRR (Mouginis-Mark et al., 1994; Patrick et al., 2005). The dual-band method with MIR and TIR bands is commonly applied to derive subpixel-level thermal anomalies information, though it is considered an oversimplification of reality as it assumes that only two features of different temperatures exist within a pixel, whereas realistically there may be up to seven components/features (Wright and Flynn, 2003). Alerts for thermal anomalies observed on a global scale are available freely on the internet now, utilizing both MODIS and GOES image data (Wright and Flynn, 2003; Rothery et al., 2005).

Remote sensing is commonly used to monitor ash clouds and volcanic gaseous emissions (SO_2) from both explosive and non-explosive eruptions. Large-scale eruptions can eject gases and ash into the stratosphere, which may be monitored using geostationary satellites that measure absorption in the UV, SWIR, and TIR wavelengths (e.g., TOMS, GOES). However, they often lack the spatial resolution and radiometric sensitivity to monitor smaller fluctuations in volcanic activity. The plumes may be monitored with higher resolution using ASTER data (Pieri and Abrams, 2004; Pugnaghi et al., 2006), though temporal variability monitoring is compromised with the less frequent overpasses of this sensor. SO_2 may be measured using diagnostic absorption features in either the UV or TIR region of the spectrum. Ash composition is commonly mapped using the 'split-window' technique, and it requires subtraction of the reflectance in one SWIR or TIR band from another (Prata, 1989). This technique allows distinction between volcanic ash clouds and those of meteorological origin, based on threshold values (Dean et al., 2004).

As in the listing of Kervyn et al. (2007), four main volcanological remote sensing application themes can be identified from a literature review.

Detection, analysis, and monitoring of thermal phenomena associated with active volcanic systems, including active lava flows, lava lakes, lava domes, degassing vents, fumarole fields, crater lakes, and geothermally heated ground. Detection, analysis, and monitoring of clouds of volcanic ash and gas. Mapping of volcanic deposits and morphological analysis, including spectrally based identification of mineralogical, textural, and compositional differences between different units.

Assessment of ground deformation, derivation of digital elevation models, volumetric assessment, and surface roughness mapping using radar.

In this section, volcanic hotspot detection algorithms and methods/techniques of mapping and classifying volcanos as well as challenges and future directions are briefly introduced, reviewed, and discussed.

10.2.2 Volcanic Hotspot Detection Algorithms

Steffke and Harris (2011) provide a review, classification, and appraisal of algorithms designed for volcano hot spot detection. Depending on whether a hot spot detection technique operates by assessing a pixel's radiance (or temperature) in a spectral, spatial, or temporal space, hot spot detection algorithms can be divided into three categories: (1) fixed threshold, (2) contextual, and (3) temporal as follows.

(1) Spectral character (fixed threshold algorithms). That is, does the pixel have spectral properties, i.e., a sufficiently high brightness temperature, ΔT and/or NTI, to define it as hot? This class of algorithms tends to apply single or multiple—but essentially fixed—thresholds to determine whether a pixel is hot or not. ΔT and NTI are defined, respectively, as follows:

$$\Delta T = T_{MIR} - T_{TIR} \quad (10.1)$$

$$NTI = (T_{MIR} - T_{TIR})/(T_{MIR} + T_{TIR}) \quad (10.2)$$

where NTI is the normalized thermal index; T_{MIR} and T_{TIR} are temperatures at MIR and TIR bands, respectively.

(2) Spatial context (contextual algorithms). That is, is the pixel brightness significantly different from that of its surrounding pixels? This class of algorithms assesses whether a pixel is thermally anomalous in a spatial sense (area).

(3) Temporal context (temporal algorithms). In other words, is the pixel brightness significantly different from that of its preceding history? Thus, this class of algorithms assesses whether a pixel is thermally anomalous in a temporal sense.

10.2.2.1 Fixed Threshold Algorithms

Fixed threshold algorithms use a series of tests designed to determine whether the spectral character of the pixel is consistent with the presence of a sub-pixel thermal anomaly. This is achieved by assessing whether the spectral radiance value of the pixel exceeds a fixed—or number of fixed—limits above which the spectral radiance is sufficiently high that it can only be the result of an extremely hot source residing within the pixel (Harris, 2013b). Other thresholds are often applied to assess whether the pixel is cloud-containing or contains a highly reflective source of solar light. Thresholds can be generic (i.e., set at one time and applied to multiple images) or case specific (i.e., set on a case-by-case basis using statistics taken from the image on hand).

There are six typical forms of generic tests of fixed threshold algorithms developed by the fire community between 1985 and 1995. They take a similar form, and the most complex execute up to five spectral tests to check for hot, low-reflectance, and cloud-free conditions. Thus, the following tests are carried out using a series of thresholds.

Test 1. Is the MIR temperature sufficiently high for it to contain a high temperature source? Use AVHRR's channel 3 saturation temperature to assess whether a pixel passes this test or not.

Test 2. If the background is solar heated, then the pixel will be equally hot in the TIR. Thus, this test uses the similarity between the pixel-integrated temperature in the MIR and TIR to assess solar heating effects; in effect the closer ΔT is to zero, the more likely it is that the MIR is hot simply due to solar heating.

Test 3. If the AVHRR band 4 temperature is very low, then high cold cloud may be present. The reflection from such clouds could cause T_{MIR} and ΔT to be anomalously high. Thus a cloud trap can include a T_{TIR} temperature test.

Test 4. Highly reflective conditions may also be indicative of clouds, as well as of high reflection daytime conditions that may elevate the MIR brightness temperature. Thus, a final test can be to throw away all pixels above a certain reflectance threshold, with this test being applied to the VIS and/or NIR.

The primary fixed threshold algorithm is the MODVOLC algorithm of Flynn et al. (2002) and Wright et al. (2002a). The generic form of MODVOLC was developed in a series of papers by Flynn et al. (2002), Wright et al. (2002a), and Wright et al. (2004). It was a generic fixed threshold algorithm designed to detect volcanic hot spots (Figure 10.2). The key characteristic of MODVOLC was that it was designed to provide a global detection capability using MODIS data. It became operational following the launch of MODIS on the Terra platform in February 2000. Running on the MODIS data stream at the Goddard Space Flight Center (GSFC) Distributed Active Archive Center (DAAC), MODVOLC was restricted to eight mathematical operations, a requirement imposed to minimize the drain on computer resources at the DAAC. Thus, developed by Wright et al. (2002a), the NTI concept required use of the MIR and TIR pixel radiances (R_{MIR} and R_{TIR}) in just three mathematical operations: operation (1): $R_{MIR} - R_{TIR} = X$, Operation (2): $R_{MIR} + R_{TIR} = Y$, and Operation (3): X/Y = NTI and one test: Test: IF NTI > Threshold, THEN pixel is hot. To set the threshold, with data for a variety of volcano hot spot types and sizes, from a variety of environmental settings spanning arctic, temperate, tropical, and hot desert, Wright et al. (2002a) used the histogram analysis to obtain an appropriate, global value for the NTI threshold. This value for the threshold was set to −0.8 and used to separate ambient pixels (NTI < −0.8) from hot spot pixels (NTI > −0.8). This global threshold has proved to operate with a very low rate of false detections (Wright et al., 2004), and was increased to −0.6 for daytime images. The resulting MODVOLC algorithm steps are given in Figure 10.2.

The first, robust, fixed threshold approach applied to volcano hot spot detection was that of Harris et al. (2002a) who introduced a multi-step threshold algorithm to detect hot spots in whole Earth disc images acquired by GOES. A series of tests that used the reflective, ΔT, MIR, and TIR properties of the pixel and its background were used to assign a probability of the pixel containing a hot spot. If the probability exceeded a threshold value, then the pixel was flagged as potentially hot and assigned a gray scale depending on the probability (Steffke and Harris, 2011). MODVOLC is a fixed threshold algorithm. Developed at the Hawai'i Institute of Geophysics and Planetology (HIGP), it is designed to detect volcanic thermal anomalies on a global scale using MODIS data under limited operational constraints (Wright et al., 2002a). MODVOLC is based on the ΔT principle and implements an NTI to distinguish

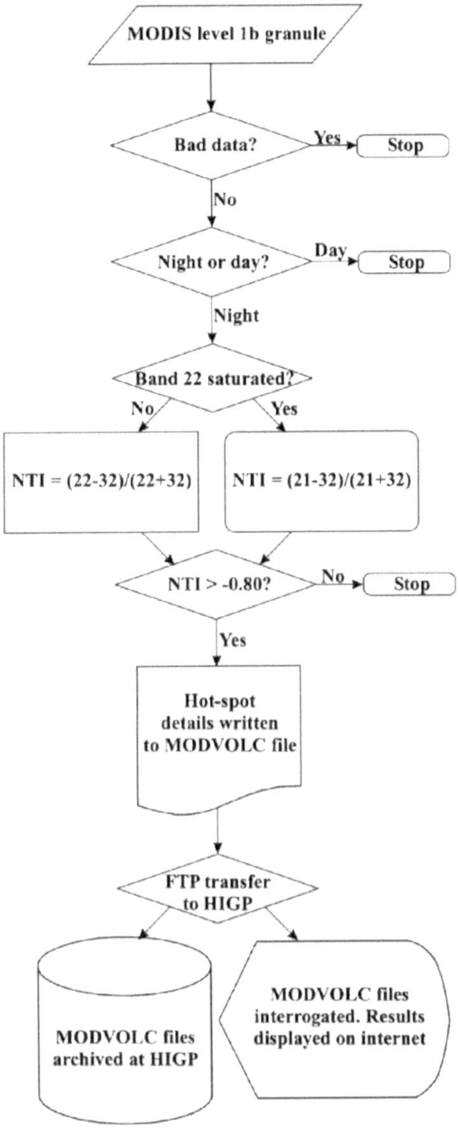

FIGURE 10.2 Flow diagram illustrating the implementation of the MODVOLC algorithm. Hotspot details written to the MODVOLC file include time detected (year/month/day/hour/minute) location (latitude/longitude), spectral radiance in Bands 21, 22, 28, 31, and 32 (in W m-2 sr-1 um-1), satellite zenith, satellite azimuth, and solar zenith (all in degrees)

Source: Wright et al., 2002a.

pixels containing sub-pixel hot spots (Flynn et al., 2002; Wright et al., 2002a). The algorithm went operational in February 2000, running at the NASA DAAC shortly after the December 1999 launch of MODIS on NASA's Terra satellite. The NTI considers the difference in radiance between MODIS bands 22 (MIR) and 32 (TIR) and then divides (normalizes) this by the sum of the radiances in the same two bands. This gives an NTI in the range of −1.0 to 1.0. Several tests were run to determine an NTI threshold that returned the minimum incidence of false positives when run on a global dataset. This was found to be −0.8 for nighttime data (daytime −0.6), so that any pixel with an NTI ≥ −0.8 was flagged as anomalous in the global hot spot searches of Flynn et al. (2002) and Wright et al. (2002a). The algorithm executes seven steps as given in Figure 10.2. Step 1 is image acquisition, followed by a data-quality check. Step 2 determines if band 22 is saturated, and step 3 computes the NTI (using band 22 or band 21 if band 22 is saturated). Step 4 checks whether the NTI is greater than −0.8. If so, step 5 writes the hot spot details to the MODVOLC data base. Step 6 then sends the pixel information (location and radiance) to HIGP via ftp, with the information being archived and posted on the MODVOLC website (http://modis.higp.hawaii.edu) as part of step 7.

Similar to MODVOLC, Middle Infrared Observations of Volcanic Activity (MIROVA) is also an automatic system for detecting thermal anomalies at high temperatures (> 500K), based on the analysis of MODIS data. The general architecture of the MIROVA system can be divided into two main parts: (1) data download and processing (developed on a local computer) and (2) data dissemination (developed on a website: www.mirovaweb.it). Unlike other systems that use MODIS data, MIROVA is not strictly a global monitoring system. In fact, although the algorithm can be applied in any environmental context (Coppola et al., 2016), the data processing chain actually operates only for a list of selected target volcanoes for which the near real time observation is requested. This architecture makes it possible to add other specific targets in case of need (i.e., unrest of a new volcano) without compromising the total system efficiency in terms of processing time, memory, storage capacity, etc. However, building a complete thermal database for all the volcanoes currently monitored by MIROVA takes time and is in the development phase.

10.2.2.2 Contextual Algorithms

Contextual algorithms apply tests that assess whether or not a pixel is anomalous in a spectral and spatial sense using thresholds set from thermal and brightness statistics calculated from the immediate background of a target pixel. Typically, the background is defined by the eight pixels surrounding a central target pixel. Next, a number of spectral parameters are defined and used to test whether the central pixel is anomalous or not. Contextual algorithms were also first developed by the wildfire community during the 1990s (e.g., Lee and Tag, 1990; Langaas, 1993). These approaches applied flexible thresholds based upon statistics calculated from pixels in the image region immediately surrounding a target pixel. Such a contextual approach was the basis of the first automated volcanic hot spot detection algorithm applied in volcanology as reported by Harris et al. (1995). Inspired by the work of Lee and Tag (1990) and Langaas (1993), the algorithm was later developed into the Volcanic Anomaly Software (VAST) algorithm of Higgins and Harris (1997). Initially, the original VAST was tested using AVHRR data for Etna, Volcano, Stromboli, and Lipari (Harris et al.,1995; Higgins and Harris, 1997), as well as wildfires in Australia (Harris, 1996) and was designed to detect thermally anomalous pixels by comparing the ΔT for each pixel with the mean from its surrounding eight pixels. The algorithm was adapted to run on GOES data for Hawai'i, whereby any pixel with ΔT that was greater than the ΔT mean plus 3.3σ for a five-pixel-wide box surrounding a 10×10-pixel target zone was flagged as a hot spot. Since VAST's development, other algorithms have been designed based on the similar principles and operations. For example, the correlation approach of Wright et al. (2002b) carried out a contextual test that correlated the spectral radiance of a pixel with that of its background. Many of these algorithms have been implemented for local-to-regional monitoring purposes including AVHRR-based monitoring of the volcanoes of Mexico, Central America, and Japan.

10.2.2.3 Temporal Algorithms

Temporal algorithms operate by comparing a pixel's radiance (or temperature) with mean values obtained for the same pixel from the time series of data. These mean values can be evaluated by assessing whether pixels in new images are thermally similar to their previous history or not. These time series allow typical pixel values for any given time of day and year to be defined and divergences from the norm to be statistically assessed, with the core method of this type in volcanology being the RAT approach of Tramutoli (1998).

Tramutoli (1998) originally proposed the technique, named Robust AVHRR Technique (RAT) to minimize the effects that different geographical and seasonal conditions can have on some fixed threshold and contextual algorithms, proposing the use of RAT in a variety of applications including detection of volcanic ash clouds, dust storms, forest fires, and earthquake-related hot spots. To achieve this, the algorithm looks through a data archive to see how the temperature for a specific pixel changes through time, thereby assessing variation caused by seasonal and atmospheric effects. The normal condition statistics were then used to assess a target pixel divergence from normal behavior, where one cause of divergence would be the development of a sub-pixel hot spot due to active lava or fire activity (Tramutoli, 1998). We here keep with the more commonly used RAT acronym. RAT relies on a data archive of MIR data to create an Absolutely Local Index of Change of Environment (ALICE (x, y, t)). This provides an estimate of how much a pixel brightness temperature diverges from its normal conditions as determined from the data time series, normalized by its natural variability in the time domain. ALICE can be mathematically defined as (Tramutoli, 1998):

$$\mathrm{ALICE}(x,y,t) = [V(x,y,t) - V_{REF}(x,y,t)]/[\sigma_V(x,y)] \tag{10.3}$$

where $V(x,y,t)$ is the brightness temperature recorded by the satellite sensor at location (x, y) and time t; $V_{REF}(x, y, t)$ is the mean brightness temperature for the same location and time; $\sigma_V(x, y)$ is the standard deviation in brightness temperature for the same location and time. A pixel can now be assessed as being anomalous depending on whether ALICE is greater than a certain value (threshold) or not. Choosing a value for the ALICE threshold corresponds to the number of standard deviations away from the mean that a value needs to be in order to be considered anomalous. Thresholds of 2.0, 2.6, and 3.0, for example, will correspond to statistical confidences of 95.4%, 99.0%, and 99.7% that the value is outside of the normal distribution of radiances for the given location and time. Pergola et al. (2004) and Di Bello et al. (2004) showed how RAT could be used with MIR (AVHRR band 3) data to detect hot spots associated with lava flow activity on Mt. Etna through application of the ALICE index. Since that time the algorithm has been used with other satellite data and has been renamed Robust Satellite Technique (RST) (Pergola et al., 2008, 2009). More recently, Yuhaniz and Vladimirova (2009) proposed a new change detection model using image tiling that enables change detection to be effective even when misregistration errors exist in the images. The method uses the tiling of a cross-referenced image time series and a 'fuzzy inference engine' to assess the degree of change in a given pixel and issue alerts if a major change occurs.

The Okmok algorithm is another temporal hot spot algorithm. It was developed at the Alaska Volcano Observatory to aid with operational hot spot detection in AVHRR data, and it is a good example of the temporal approach. By comparing the typical surface temperature for the regional background with the maximum temperature obtained from an analysis box centered over an active volcano, the Okmok algorithm provided a graphical assessment of the thermal state of the volcano. Arrival of a hot spot causes the pixel brightness temperature to diverge from the steady (annual) heating and cooling trend. Thus thermally anomalous pixels can be isolated because they exceed the typical ambient temperature range defined for any given location on any particular day.

10.2.2.4 Algorithms' Reliability and Advantages and Disadvantages

Steffke and Harris (2011) carried out a comparison of the main contextual, temporal, and fixed threshold algorithms used in volcanology: VAST, RAT, and MODVOLC. The test involved visually examining all AVHRR and MODIS images acquired for five volcano hot spot test cases of varying intensity to assess based on following criteria: (1) how many visually identified hot spots were found by each algorithm, (2) how many pixels were missed, and (3) how many false detections occurred. The conclusion was that no algorithm was perfect and all algorithms detected and missed anomalous pixels. All algorithms also experienced false detections and performed differently for different hot spot cases. They suggested that an automated algorithm could be used to flag hot spots, but if the user requires identification of all anomalous pixels, with the certainty that there are no false positives, manual cross-checking always needs to be carried out.

Each algorithm type also has a number of advantages and disadvantages that need to be considered. (1) Fixed threshold algorithms can be applied to single images, require little computational or processing power, and can thus be applied easily and quickly to any single image. It may be quite easy, for example, to use the band math tool in ENVI to create a ΔT or NTI image and then apply a fixed threshold using the masking procedures. The same simplicity can be used for a way to deal with global datasets. (2) Contextual algorithms can operate on a single image but can be computationally needed and may be difficult to apply to global datasets. (3) Temporal algorithms require availability of a large image archive (time series data) and exacting geolocation; if a pixel is slightly mislocated the statistics will be affected. This may necessitate access to such an archive, plus considerable storage and computer processing power while the statistics are being generated. The user also needs to assess the importance of the number of detections versus the number of missed and false cases in considering commission error and omission error. While MODVOLC, for example, will yield a low-to-zero rate of false detection, VAST and RAT will detect many more hot spot images and flag many more pixels within those images, but with a higher rate of false alarm.

10.2.3 Mapping and Classification of Volcanos

Characterization of the morphology and physical characteristics (e.g., grain size, density) of volcanic deposits such as lava flows, tephra, and pyroclastic density currents (PDCs) is fundamental to the ability to understand the eruption and emplacement processes that produced the deposits (Carr et al., 2021). To address the ability, hot spot algorithms discussed in the last subsection are necessary. In addition, mapping and classification of active volcanos also is needed to understand volcano physical type and property. Among these mapping techniques/methods, machine learning (ML) algorithms including deep learning (DL) algorithms have been widely used to automatically process remote sensing data in volcanic applications (Li et al., 2014; Zhang et al., 2021). It has been evidenced that automatic detection of hot lava flows in NRT (near-real-time) can be achieved by using an unsupervised ML classifier exploiting NHI (normalized hotspot indices) from any available satellite sensor between the S2-MSI and L8-OLI acquired at a time interval close to the start of the eruption. For example, a K-means algorithm has been used as an unsupervised classifier to easily detect hot/incandescent pixels with NHI as the input features. Unfortunately, the weakness of such an approach is that subtle anomalies may not be detected because they do not produce high enough changes in spectral radiances in SWIR and NIR bands compared with other phenomena affecting the same bands, such as clouds or snow. Thus, a supervised data-driven approach random forest (RF), one of the ML methods, would be more suitable to enhance the accuracy level. Since traditional fixed-threshold approaches rely on a set of decision rules, a supervised technique replicating the if-else logic based on the input features and the target during the training phase may represent the best candidate in this sense (James et al., 2013; Bonaccorso, 2017). According to this perspective, the decision tree algorithm (RF) may represent a useful tool. In fact, the decision tree replicates the if-else logic based on the input features and the target during the training phase (James et al., 2013; Bonaccorso, 2017). RF models are an ensemble learning method for classification, regression, and other tasks that operate by constructing a multitude of decision trees at training time. For classification tasks, the output of the RF is the class selected by most trees (Schonlau and Zou, 2020). It has been widely used as a tool to detect anomalies because the separation between anomalies and normal behavior is made easier due to the shortest path in the trees than normal instances. Corradino et al. (2019) used the artificial neural network (ANN), another of the ML methods, to map recent lava flows at Mount Etna using multispectral Sentinel-2 images. Compared with ground-based measurements and actual lava flows of Mount Etna placed in 2017 and 2018, our automatic ML classified result provides excellent results in terms of accuracy, precision, and sensitivity.

Contemporary image recognition techniques based on convolutional neural network (CNN), as a widely used DL algorithm to classify images exploring spatially discriminative shape/texture features, are especially efficient at automatically classifying heat sources semantically on a global scale (Ma et al., 2019). The CNN-based classification could potentially be an alternative to visual inspection by human experts. For example, Kate and Shukla (2021) applied the CNN-based and the decision tree-based classifications independently and then combined their results to increase classification reliability. Both classification methods need training samples. They selected a variety of different hotspots from global regions. The different hotspots were classified into eight categories—volcano, factory, oil platform, fire (forest), fire (grass), fire (urban), false detection by specular reflection, and non-typable—by visually inspecting satellite images from before and after the detection of the hotspot and maps and high-resolution aerial photos on Google Maps. Validation work using 10,959 classification results corresponding to hotspots acquired from May 2017 to July 2019 indicates that the two classification results were in 79.7% agreement.

Manual mapping through visual interpretation is a practical approach to obtain the extent of a lava flow on remotely sensed data (Smets et al., 2010). The resulting lava flow boundaries are usually quite accurate when combining visual interpretation of remotely sensed data with readily available geological maps and field work. Usually being in vector format, the result of the mapping can be directly imported into a geographical information system (GIS) for spatial analysis and for a geological map update. In addition to the manual mapping approach, pixel-based classification is a frequently used technique for volcanic mapping. Pixel-based classification of lava flows is based on spectral differences between lava flows and their surroundings—such as vegetation and urban infrastructure—and between lava surfaces of various age (Li et al., 2017). In many cases, low spatial resolution imagery leads to a large number of mixed pixels, which may result in inaccurate classification. Pixels constituting a mixture of lava flow and vegetation or a mixture of other lava flows of different ages are indeed more likely to be misclassified. In addition, spectral heterogeneity exists among pixels that are part of single lava flows, in which the local environment will control the weathering and vegetation evolution of a lava surface, which may differ along the single lava surface type. Thanks to the homogenous mixing, these challenges might be solved by using the geographical object-based image analysis (GEOBIA) approach. Object-based classification has been applied for mapping volcanic deposits from SPOT imagery in recent studies focusing on lahars, yet the potential of GEOBIA for automated mapping of lava flow fields has so far not been specifically investigated. In the study of Li et al. (2017), the RF classifier was tested to map lava flows based on pixels and objects. Image classification was conducted for the 20 individual flows and for eight groups of flows of similar age using a Landsat 8 image and a DEM of the volcano, both at 30-meter spatial resolution. Results demonstrate that object-based classification produces maps with continuous and homogeneous lava surfaces, in agreement with the physical characteristics of lava flows, while lava flows mapped through the pixel-based classification are heterogeneous and fragmented including much 'salt and pepper noise.'

10.3 THERMAL REMOTE SENSING OF PLANT ECOSYSTEMS

10.3.1 Introduction

Remote sensing of plant ecosystem studies have been widely carried out in the VNIR (0.3–1.0 μm) and SWIR (1.0–2.5 μm) regions with a focus on extracting biochemical and biophysical vegetation properties (Matson et al., 1994; Schlerf and Atzberger, 2012; Pu and Cheng, 2015; Pu et al., 2018). However, spectral data from the VNIR and SWIR domains do not adequately describe all structural, chemical, and functional characteristics of vegetation. Specific vegetation components such as polysaccharides (e.g., cellulose) and leaf surface properties (e.g., waxes and hairs) have their primary absorption features located in the TIR range (Ribeiro da Luz and Crowley, 2007). Technological advance in the development of TIR spectrometers and sensors for laboratory, field, airborne, and spaceborne platforms has further enabled the acquisition of high spatial and spectral resolution data in the MWIR and LWIR domains for plant ecosystem studies. TIR remote sensing fundamentally influences a pretty wide range of biological processes, from enzymatic reactions to ecosystem biogeochemistry to species and biome distributions (Still et al., 2019). The majority of research into the effects of temperature on ecological processes is based on air temperature measurements, which are commonly available and typically collected several meters above ground level/canopy top. However, since the surface temperature (T_{surf}) of an object like a leaf can diverge substantially from air temperature (T_{air}), measuring only T_{air} is insufficient for studying plant ecosystem characteristics.

There is a certain history of using thermal infrared (TIR, ~8–14 μm wavelength range) measurements to evaluate relative temperature differences in human-dominated systems (e.g., agricultural environment). Many applications in precision agriculture include detection of plant stress using measurements of crop canopy temperatures as indicators of plant water or nutrient stresses or for assessing changes in stomatal conductance (Jones, 2004; Jones et al., 2009; Maes and Steppe, 2012). For example, Siebert et al. (2014) demonstrated the value of TIR measurements for assessing crop stress and yields. They reported that crop canopy T_{leaf} deviated strongly from T_{air} measured just above the canopy; this difference was driven by soil moisture; and T_{leaf} was frequently colder than above canopy T_{air} due to large evapotranspiration cooling. In addition, Leuzinger et al. (2010) analyzed how different tree species display a range of leaf temperatures related to leaf size and shape and showed the effect of the surrounding environment (park versus urban) on temperatures of broadleaf tree species. Despite the increasing use and success of thermal imaging in human-dominated settings, applications in natural ecosystems are relatively rare. Previous studies in forests showed consistent daytime leaf temperature (T_{leaf}) higher than air temperatures (T_{air}) among a range of broadleaf and needleleaf species, with a general trend of larger-leaved deciduous canopies having higher T_{leaf} than smaller-leaved species. The study illustrated that mean leaf temperatures in forest trees were not adequately explained by either stomatal conductance or leaf size but strongly depend on canopy architecture (leaf area density, branching habits) in combination with leaf traits (Leuzinger and K€orner, 2007). At the landscape scale, thermal imaging measurements of T_{surf} indicated the importance of topographic position, spatial heterogeneity, and the separating between surface and air temperatures in an alpine system (Scherrer and K€orner, 2010).

In general, the TIR spectral range has hardly been explored in plant ecosystem studies. The spectral behavior of plants has been misinterpreted to some extent in the TIR region due to some challenges, such as the complexity of vegetation spectral features and the lack of suitable instruments with relatively high spatial and spectral resolutions and a low signal to noise performance (Ribeiro da Luz and Crowley, 2010; Ullah et al., 2012a, 2012b). In the past few decades, these aforementioned limitations have primarily been overcome, which has resulted in an increasing number of studies, mainly at the leaf level, on the characterization, identification, and discrimination of plant species (Ribeiro da Luz and Crowley, 2007; Ullah et al., 2012a, 2012b; Rock et al., 2016) and a certain number of studies at canopy level (Ribeiro da Luz and Crowley, 2010; Meerdink et al., 2019) on the retrieval of vegetation biochemical and biophysical parameters (Ullah et al., 2014; Neinavaz et al., 2016a, 2017), as well as on water stress detection (Sepulcre-Cantó et al., 2006; Jones et al., 2009). In this section, we present a comprehensive review of the research progress in TIR remote sensing of plant ecosystem structures and functions. It should be noted that the focus of this section is on assessing the use of TIR remote sensing data for vegetation and ecosystem studies rather than the methodology used in this field.

10.3.2 Terrestrial Ecosystem Structure

The biological factors affecting leaf/canopy energy balance and ultimately leaf/canopy temperature, including transpiration, leaf angle, pubescence, leaf size, leaf shape and canopy position, number of leaves per stem, and canopy structure: canopy density and height, vary within and across plant species and ambient conditions. The ecosystem structure variables may include plant species and composition, leaf area index (LAI), canopy density and height, etc. The first attempt to better identify plant species differences at leaf level using TIR data has been made by (Gates and Tantraporn, 1952), and they suggested that the upper leaf surfaces and the old leaves tend to have significantly higher reflectance values than the lower leaf surfaces and young leaves, respectively. Yet, Wong and Blevin (1967) found that the reflectance spectra of leaves of the same species but of different age (i.e., juvenile, mature, and old yellowed leaves) are very similar in the TIR domain. In spite of these results, Wong and Blevin's (1967) findings can be considered initial research, broadening the belief that plant species are featureless in the TIR domain and considering them being close to blackbody emitters in the TIR domain. However, in the 1980s, Salisbury (1986) and Salisbury and Milton (1987, 1988) took TIR reflectance measurements of fresh leaves of 13 deciduous tree species and found that the different species have distinctive spectra. They concluded that due to the unique composition and structure of epidermal plant cells of different plant species (e.g., cuticular waxes), the reflectance peaks locate different positions between the wavelengths 8–14 μm (Salisbury and Milton, 1987, 1988). Arp and Phinney (1980) discovered some physiological adaptation in plant species related to their radiational environments (desert, tropics, and temperate regions) in the TIR region.

Additionally, Salisbury and Milton (1987) found that the TIR spectral signature varies seasonally and are different between spring and summer leaves, whereas the difference is slight between summer and fall. They also revealed higher reflectance values in the leaves of trees that grew in metal-enriched soils than in the leaves from trees in the control area, while the reflectance spectral features remained similar. It was also revealed that plant leaves exhibit significant differences in their emittance (Idso and Jackson, 1969). Furthermore, Elvidge (1988) demonstrated that plant spectral changes could also be related to plant maturity. Significant differences in emissivity were also revealed for plants from different ecological groups and species, as well as for diverse plant materials (e.g., bark, woody materials, etc.) from individual species (Elvidge, 1988). For instance, Raven et al. (2002) found a difference in the directional reflectance and emissivity of the leaves of two different plant species over between 2 and 15 μm.

Ullah et al. (2012a, 2012b) utilized thermal hyperspectral data measured at the laboratory level and proposed important wavebands for discrimination of plant species in the MWIR (i.e., 3.34, 4.19, and 4.60 μm) and the LWIR (i.e., 9.44, 12.71, and 13.70 μm) domains, based on their study of 13 plant species. However, a different set of species under investigation may have resulted in different wavebands being proposed. This issue was later transferred to a field experiment by Rock et al. (2016) using an imaging system (i.e., emissive thermal infrared imaging spectroscopy) that allows, in comparison to the time required when undertaking directional hemispherical reflectance (DHR) laboratory measurements, fast acquisition of high spectral resolution TIR images. Based on field collected TIIR image data, Rock et al. (2016) demonstrated that classification accuracies similar to the laboratory spectra, which outperform VNIR/SWIR reflectance spectra in terms of classification accuracy, could be achieved with this field images. However, Rock et al. (2016) also revealed that signal to noise ratio is critical for species discrimination, with TIR data outperforming VNIR-SWIR only at a high signal to noise ratio. Ribeiro da Luz and Crowley (2010) used real high spatial and spectral resolution thermal infrared imagery (8.0–13.5 μm) from the SEBASS airborne sensor to analyze and map tree species canopy spectral features at the State Arboretum of Virginia, near Boyce, Virginia. Fifty tree species were analyzed and mapped and about half were directly identified with varying degrees of success based on spectral matched filtering that utilized laboratory-measured species leaf spectra as the target signatures. They found that best results were obtained from species having relatively strong spectral contrast, wide and flat leaves, closed planophile canopies, and/or large canopy areas. Tree species with small leaves or unfavorable leaf orientations showed spectral attenuation likely resulting from cavity blackbody effects. The results suggested that TIR remote sensing data provide an important new opportunity for mapping plant species and for investigating the ecosystems that they support.

In general, the emissivity value of plant species at the canopy level is higher than for any separate part of the plant due to plant structure and the cavity effect (Rubio et al., 1997). Emissivity responds to geometry and patterns between plant canopies. Therefore, the effect of the vegetation geometric structure and canopy components on the TIR data should always be considered in the upscaling of the emissivity from leaf to canopy level. Consequently, the emissivity of dense canopies is higher than the emissivity of an individual leaf (Rubio et al., 1997). The emissivity value of green and senescent vegetation also is higher at canopy level compared to leaf level (Palluconi et al., 1990).

Prediction of vegetation biophysical variables, LAI, one of the most important vegetation biophysical parameters, has been widely investigated using data from VNIR/SWIR regions. It has also been successfully retrieved using hyperspectral TIR data under controlled laboratory conditions applying various vegetation indices and a non-parametric statistical approach (Neinavaz et al., 2016a). The results of Neinavaz et al. (2016b) showed that LAI might be successfully retrieved at the high value of approximately 5.5 when TIR hyperspectral data were used. The LAI of wheat was accurately mapped using TIR data under different moisture stress conditions using thermal imaging and a support vector machine classifier (Banerjee et al. 2018). The results clearly showed that thermal image analysis can be applied as a non-destructive, rapid technique to characterize the temperature of crop canopy and then to estimate the LAI of wheat grown under different moisture stress conditions. In addition, a combination of LSE and spectral reflectance from VNIR and SWIR could boost the retrieval accuracy of LAI (R^2_{CV} = 0.81, $RMSE_{CV}$ = 0.63 m^2m^{-2}) in a mixed temperate forest (Neinavaz et al., 2019). Stobbelaar et al. (2022) also found that a combining VNIR data with LSE significantly improved the LAI retrieval accuracy (R^2 = 0.7907, $RMSE_{CV}$ = 0.8351). These findings all indicating that combining LSE with VNIR data can increase the prediction accuracy of LAI.

10.3.3 Terrestrial Ecosystem Function

TIR data have been widely used for deriving canopy temperature as an indicator of water stress, canopy conductance, and transpiration for irrigation scheduling (Idso, 1982; Sepulcre-Cantó et al., 2006). The terrestrial ecosystem functions that may be characterized with TIR data may be related to water stress, net primary product, ecosystem thermodynamics, etc.

Several temperature-based indices have been developed to normalize leaf temperatures to current environmental conditions for a more quantitative estimation of water stress. Initially, differences between canopy and air temperatures were considered the first indicators of crop water stress, namely SDD (Idso et al., 1977). A further improvement was the now well-known Crop Water Stress Index (CWSI) (Jackson et al., 1981). The Water Deficit Index (WDI) developed by Moran et al. (1994) and the thermal index of relative stomatal conductance (Jones, 1999) is alternative temperature-based indices that take into account partially vegetated sites and show a direct linear relationship between canopy temperature and stomatal conductance. Besides, the thermal variations within the canopy are associated with water stress and thus could be used for water stress detection at the tree level, based on TIR imagery data (Sepulcre-Cantó et al., 2006). The Standard Deviation of Canopy Temperature (CTSD) was successfully applied to evaluate water stress in crops. The results showed that this index has a strong potential for monitoring water stress in crops since it only depends on the canopy temperature (Han et al., 2016). The CTSD outperformed the CWSI for monitoring moderately stressed crops within the field (González-Dugo et al., 2006). Nonetheless, the impact of the canopy structure and view angles on canopy temperature should have been taken into account (Jones et al., 2009; Hu et al., 2019). In addition, the normalized relative canopy temperature is associated closely and significantly with canopy water content and relative water content under arid and semi-arid conditions using hyperspectral VNIR/SWIR data and thermal imaging data.

Two types of parameterization, determined empirically from vapor pressure deficit (VPD) (CWSI_e) and theoretically from the Penman-Monteith equation (CWSI_t), are evaluated on the potential of indicating plant water stress based on T_c and concurrent air temperature (T_a) measurements on three sites under a Mediterranean climate. Banerjee et al. (2018) demonstrated that the mean canopy temperature is raised with increasing moisture stress levels during various crop growth stages. They showed the efficiency of using thermal images compared to digital images for discriminating between leaves and soil under different moisture stress conditions. In summary, temperature-based indices have an enormous potential for pre-visual detection of plant water stress and have, therefore, been applied in agricultural research and practice (Khanal et al., 2017). Additionally, Liu et al. (2018d) discovered a difference in canopy temperature between lodging and non-lodging agricultural areas using UAV data, with lodged areas having a higher canopy temperature. It was also revealed by Sagan et al. (2019) that thermal infrared imagery obtained from UAV is efficient for plant phenotyping.

The possibility of retrieval of the leaf water content has also been investigated using data from the MWIR to LWIR regions (Ullah et al., 2012a). However, Ullah et al. (2014) demonstrated that the MWIR and SWIR regions seem to be more sensitive spectral regions than the LWIR domain concerning the retrieval of leaf water content as the fundamental water absorption features lie in the MWIR region and not in the LWIR region. The fuel moisture content and equivalent water thickness as a mass-based and area-based vegetation water content indicator could also be predicted at canopy level using hyperspectral TIR data. In addition, plant mass may play a more significant role in determining spectral emissivity than plant area (Neinavaz et al., 2017). Moreover, the integration of TIR data and VNIR/SWIR data could improve the prediction accuracy of foliar traits, including cellulose, lignin, leaf mass per area, nitrogen, as well as water content (Meerdink et al., 2016).

Thermal remote sensing can provide environmental measuring means with capabilities for measuring ecosystem development and integrity. Recent advances in applying principles of nonequilibrium thermodynamics to ecology provide fundamental insights into energy partitioning in ecosystems (Quattrochi and Luvall, 2009). Ecosystems are nonequilibrium systems, open to material and energy flows in an ecosystem, which grow and develop structures and processes to increase energy degradation. More developed terrestrial ecosystems will be more effective at dissipating the solar energy.

TIR energy theory has been developed as a result of research to reveal principles of ecological development. The research in ecological thermodynamics has focused on linking physics and systems sciences with biology, especially linking the science of ecology with the laws of thermodynamics. Furthermore, recent advances of development in nonequilibrium thermodynamics related to the investigation of self-organizing phenomena in quite different types of systems (from simple convection cell systems to forested ecosystems) have revealed that all self-organizing ecosystem phenomena (including ecosystem development) involve similar processes, processes that are mandated by the second law of thermodynamics.

Applying recent advances in thermodynamics, specifically exergy or second law analysis, the second law of thermodynamics can be extended so as to apply to nonequilibrium regions and processes in ecosystems. In this circumstance, systems can be described in terms of the exergy fluxes setting up gradients (e.g., temperature and pressure differences in classical thermodynamic systems). With the development of these exergy flux gradients, the system is no longer in equilibrium. The system responds to these imposed gradients by self-organizing in a way that resists the ability of the exergy fluxes to develop gradients, and hence move the system to nonequilibrium. More formally, the thermodynamic principle (the restated second law) that governs the behavior of systems states that, as systems are moved away from equilibrium, they will utilize all avenues available to counter the applied gradients. As a system self-organizes it will become more effective at exergy utilization.

Terrestrial ecosystems' surface temperatures have been measured using TIMS. Luvall and his collaborators (Luvall and Holbo, 1989; Luvall et al., 1990) have calculated many ecosystem energy budgets, including tropical forests, mid-latitude varied ecosystems, and semi-arid ecosystems. These data showed that within a given biome type and under similar environmental conditions (air temperature, relative humidity, winds, and solar irradiance), the more developed the ecosystem, the cooler its surface temperature and the more degraded the quality of its reradiated energy. In the case of nonequilibrium thermodynamic theory, three hypotheses about ecosystem development can be created that associate with the use of thermal remote sensing measures (Kay and Schneider, 1994; Schneider and Kay, 1994a; Schneider and Sagan, 2005):

The ratio (Rn/K*) of net all-wave radiation (Rn) to net short-wave radiation (K*), received at the surface will be larger for more developed ecosystems and these systems will have a lower surface temperature.

Spatial variation of surface temperature (T) will be less for more developed ecosystems. The spatial variation can be indexed using the beta index (Holbo and Luvall, 1989). The more developed ecosystems will have a larger beta index.

More developed ecosystems will exhibit a smaller temperature change in response to a given amount of energy input (net radiation). This can be measured using TRN (Luvall and Holbo, 1989). The more developed ecosystems should have a larger TRN.

The investigations discussed earlier suggest that ecosystems develop structure and function that degrades the quality of the incoming energy more effectively, that is they degrade more exergy. Furthermore, those investigations demonstrate that analysis of airborne collected radiated energy fluxes is a valuable tool for measuring the energy budget and energy transformations in terrestrial ecosystems. Given the previous hypothesis, formulated from first principles of nonequilibrium thermodynamics of self-organizing systems, that a more developed ecosystem degrades more exergy, the ecosystem temperature, Rn/K*, beta index, and TRN are excellent candidates for indicators of ecological integrity. The potential applications for these methods for remote sensed ecosystem classification and ecosystem health/integrity evaluation is apparent.

10.4 SUMMARY

In this chapter, there are three areas of TIR remote sensing applied that are briefly introduced and reviewed. In the first area, it is thermal remote sensing applied to mapping geothermal alteration and mineralization sites. It is composed of thermal image preprocessing techniques: a RGB false-color-composite technique, band ratioing technique, and image transforms techniques including principal component analysis (PCA), independent component analysis (ICA), and maximum noise fraction (MNF); identifying and classifying techniques/methods of geothermal alteration and mines: machine learning (ML) and deep learning (DL) algorithms as well as fuzzy logic modeling methods; and challenges and future directions, which have been reviewed and discussed. In the second part thermal remote sensing of active volcanoes has been reviewed and discussed. In this area, volcanic hotspot detection algorithms including three categories: (1) fixed threshold, (2) contextual, and (3) temporal are first introduced. Then advanced mapping and classification of volcano methods are introduced and reviewed. In the third area, thermal remote sensing of plant ecosystems was introduced and discussed, including terrestrial ecosystem structure variables, plant species and leaf area index (LAI) mapping, and the terrestrial ecosystem function factors water stress and ecosystem thermodynamics.

References

Aali, A. A., A. Shirazy, A. Shirazi, A. B. Pour, A. Hezarkhani, A. Maghsoudi, M. Hashim, and S. Khakmardan, 2022. Fusion of remote sensing, magnetometric, and geological data to identify polymetallic mineral potential zones in chakchak region, Yazd, Iran. *Remote Sens.*, 14: 6018. https://doi.org/10.3390/rs14236018.

AATSR, 2022 *About AATSR*. https://earth.esa.int/eogateway/instruments/aatsr, accessed on December , 2022.

Abdeen, M. M., A. K. Thurmond, M. G. Abdelsalam, and R. J. Stern, 2001. Application of ASTER band-ratio images for geological mapping in arid regions: The Neoproterozoic Allaqi Suture, Egypt. In *Proceedings of GSA 2001 Annual Meeting*. Boston.

Abdikan, S., C. Bayik, A. Sekertekin, F. Bektas Balcik, S. Karimzadeh, M. Matsuoka, and F. B. Sanli, 2022. Burned area detection using multi-sensor SAR, optical, and thermal data in mediterranean pine forest. *Forests*, 13: 347. https://doi.org/10.3390/f13020347.

Agam, N., W. P. Kustas, M. C. Anderson, F. Li, and C. M. U. Neale, 2007. A vegetation index-based technique for spatial sharpening of thermal imagery. *Remote Sens. Environ.*, 107: 545–558.

Agathangelidis, I., and C. Cartalis, 2019. Improving the disaggregation of MODIS land surface temperatures in an urban environment: A statistical downscaling approach using high-resolution emissivity. *Int. J. Remote Sens.*, 40(13): 5261–5286. https://doi.org/10.1080/01431161.2019.1579386.

Ahn, K.-H., and R. Palmer, 2016. Regional flood frequency analysis using spatial proximity and basin characteristics: Quantile regression vs. parameter regression technique. *J. Hydrol.*, 540: 515–526.

Aires, F., A. Chédin, N. A. Scott, and W. B. Rossow, 2002. A regularized neural net approach for retrieval of atmospheric and surface temperatures with the IASI instrument. *J. Appl. Meteorol.*, 41: 144–159. https://doi.org/10.1175/1520-0450(2002) 041 <0144:ARNNAF>2.0.CO;2.

Akagi, S. K., R. J. Yokelson, C. Wiedinmyer, M. J. Alvarado, J. S. Reid, T. Karl, J. D. Crounse, and P. O. Wennberg, 2011. Emission factors for open and domestic biomass burning for use in atmospheric models. *Atmos. Chem. Phys.*, 11(9): 4039–4072.

Ali, I., F. Greifeneder, J. Stamenkovic, M. Neumann, and C. Notarnicola, 2015. Review of machine learning approaches for biomass and soil moisture retrievals from remote sensing data. *Remote Sens.*, 7(12): 16398–16421.

Allen, R. G., L. S. Pereira, D. Raes, and M. Smith, 1998. Crop evapotranspiration-guideline for computing crop water requirement. *FAO Irrigation and Drainage Paper 56*. Rome: FAO, 3006541.

Allen, R. G., M. Tasumi, A. Morse, and R. Trezza, 2005. A Landsat-based energy balance and evapotranspiration model in Western US water rights regulation and planning. *Irri. Drain. Sys.*, 19: 251–268.

Allen, R. G., M. Tasumi, and R. Trezza, 2007. Satellite-based energy balance for mapping evapotranspiration with internalized calibration (METRIC)-model. *J. Irrig. Drain. Eng.*, 133: 380–394.

Allis, R. G., 1980. Changes in heat flow associated with exploitation of Wairakei geothermal field, New Zealand. *N. Z. J. Geol. Geophys.*, 24: 1–19.

Allis, R. G., G. D. Nash, and S. D. Johnson, 1999. Conversion of thermal infrared surveys to heat flow: Comparisons from Dixie Valley, Nevada, and Wairakei, New Zealand. *Geotherm. Resour. Counc. Trans.*, 23: 499–504.

Amani, M., S. Parsian, S. M. MirMazloumi, and O. Aieneh, 2016. Two new soil moisture indices based on the NIR-red triangle space of Landsat-8 data. *Int. J. Appl. Earth Obs. Geoinf.*, 50: 176–186.

Amazirh, A., O. Merlinb, and S. Er-Raki, 2019. Including Sentinel-1 radar data to improve the disaggregation of MODIS land surface temperature data. *ISPRS J. Photogramm. Remote Sens.*, 150: 11–26. https://doi.org/10.1016/j.isprsjprs.2019.02.004.

Anderson, M. C., R. G. Allen, A. Morse, and W. P. Kustas, 2012. Use of Landsat thermal imagery in monitoring evapotranspiration and managing water resources. *Remote Sens. Environ.*, 122: 50–65.

Anderson, M. C., J. M. Norman, G. R. Diak, W. P. Kustas, and J. R. Mecikalski, 1997. A two-source time-integrated model for estimating surface fluxes using thermal infrared remote sensing. *Remote Sens. Environ.*, 60(2): 195–216.

Anderson, M. C., J. M. Norman, W. P. Kustas, R. Houborg, P. J. Starks, and N. Agam, 2008. A thermal-based remote sensing technique for routine mapping of land-surface carbon, water and energy fluxes from field to regional scales. *Remote Sens. Environ.*, 112(12): 4227–4241.

Andris, C., D. Cowen, and J. Wittenbach, 2013. Support vector machine for spatia variation. *Trans. GIS*, 17: 41–61.

Arino, O., I. Piccolini, F. Siegert, H. Eva, E. Chuvieco, P. Martin, R. H. Fraser, D. Roy, and J. Pereira, 1999. Burn scars mapping methods. In *Forest Fire Monitoring and Mapping: A Component of Global Observation of Forest Cover. Report of a Workshop* (pp. 3–5).

Armijo, L., 1966. Minimization of functions having Lipschitz continuous first partial derivatives. *Pac. J. Math.*, 16: 1–3.

Arnfield, A. J., 2003. Two decades of urban climate research: A review of turbulence, exchanges of energy and water, and the urban heat island. *Int. J. Climatol.*, 23: 1–26.

Arp, G. K., and D. E. Phinney, 1980. Ecological variations in thermal infrared emissivity of vegetation. *Environ. Exp. Bot.*, 20(2): 135–148. https://doi.org/10.1016/0098-8472(80)90006-4.

Avdan, U., and U. G. Jovanovska, 2016. Algorithm for automated mapping of land surface temperature using LANDSAT 8 satellite data. *J. Sens.*, 2016: 1–8.

Bachour, R., W. R. Walker, A. M. Ticlavilca, M. McKee, and I. Maslova, 2014. Estimation of spatially distributed evapotranspiration using remote sensing and a relevance vector machine. *J. Irrig. Drain. Eng.*, 140: 04014029.

Bahrambeygi, B., and H. Moeinzadeh, 2017. Comparison of support vector machine and neutral network classification method in hyperspectral mapping of ophiolite mélanges–A case study of east of Iran. *Egypt. J. Remote Sens. Space Sci.*, 20: 1–10.

Bai, L., D. Long, and L. Yan, 2019. Estimation of surface soil moisture with downscaled land surface temperatures using a data fusion approach for heterogeneous agricultural land. *Water Resour. Res.*, 55: 1105–1128. https://doi.org/10.1029/2018WR024162.

Bala, R., R. Prasad, and V. P. Yadav, 2020. Thermal sharpening of MODIS land surface temperature using statistical downscaling technique in urban areas. *Theor. Appl. Climatol.*, 141: 935–946.

Bala, R., R. Prasad, and V. P. Yadav, 2021. Quantification of urban heat intensity with land use/land cover changes using Landsat satellite data over urban landscapes. *Theor. Appl. Climatol.*, 145: 1–12.

Balew, A., and F. Semaw, 2022. Impacts of land-use and land-cover changes on surface urban heat islands in Addis Ababa city and its surrounding. *Environ., Dev. Sustain.*, 24: 832–866. https://doi.org/10.1007/s10668-021-01472-3.

Banerjee, K., P. Krishnan, and N. Mridha, 2018. Application of thermal imaging of wheat crop canopy to estimate leaf area index under different moisture stress conditions. *Biosyst. Eng.*, 166: 13–27.

Bartholie, J. F., L. N. Namken, and C. L. Wiegand, 1972. Aerial thermal scanner to determine temperatures of soils and of crop canopies differing in water stress. *Agron. J.*, 64: 603–608.

Bartkowiak, P., M. Castelli, and C. Notarnicola, 2019. Downscaling land surface temperature from MODIS dataset with random forest approach over alpine vegetated areas. *Remote Sens.*, 11(11): 1319.

Bastiaanssen, W. G. M., 1995. *Regionalization of Surface Flux Densities and Moisture Indicators in Composite Terrain.* Ph.D. Thesis, Wageningen Agricultural University, Wageningen, The Netherlands.

Bastiaanssen, W. G. M., M. Menenti, R. A. Feddes, and A. A. M. Holtslag, 1998a. A remote sensing surface energy balance algorithm for land (SEBAL): 1. Formulation. *J. Hydrol.*, 212–213: 198–212.

Bastiaanssen, W. G. M., H. Pelgrum, J. Wang, Y, Ma, J. F. Moreno, G. J. Roerink, and T. Van der Wal, 1998b. A remote sensing surface energy balance algorithm for land (SEBAL): 2. Validation. *J. Hydrol.*, 212–213: 213–229.

Bechtel, B., K. Zakšek, and G. Hoshyaripour, 2012. Downscaling land surface temperature in an urban area: A case study for Hamburg, Germany. *Remote Sens.* 4: 3184–3200. https://doi.org/10.3390/rs4103184.

Becker, F., and Z.-L. Li, 1990. Towards a local split window method over land surfaces. *Int. J. Remote Sens.*, 11: 369–393. https://doi.org/10.1080/01431169008955028.

Becker, F., and Z.-L. Li, 1995. Surface temperature and emissivity at various scales: Definition, measurement and related problems. *Remote Sens. Rev.*, 12: 225–253. https://doi.org/10.1080/02757259509532286.

Berk, A., G. P. Anderson, P. K. Acharya, L. S. Bernstein, L. Muratov, J. Lee, M. J. Fox, S. M. Adler-Golden, J. H. Chetwynd, M. L. Hoke, and R. B. Lockwood, 2005. MODTRAN5: A reformulated atmospheric band model with auxiliary species and practical multiple scattering options. In *Multispectral and Hyperspectral Remote Sensing Instruments and Applications II* (Vol. 5655, pp. 88–95). SPIE.

Berk, A., P. Conforti, R. Kennett, T. Perkins, F. Hawes, and J. Van Den Bosch, 2014. MODTRAN® 6: A major upgrade of the MODTRAN® radiative transfer code. In *2014 6th Workshop on Hyperspectral Image and Signal Processing: Evolution in Remote Sensing (WHISPERS)* (pp. 1–4). IEEE.

Blackett, M., 2017. An overview of infrared remote sensing of volcanic activity. *J. Imaging*, 3(2): 13.

Bonaccorso, G., 2017. *Machine Learning Algorithms.* Birmingham: Packt Publishing Ltd.

Bonafoni, S., 2016. Downscaling of Landsat and MODIS land surface temperature over the heterogeneous urban area of Milan. *IEEE J. Sel. Top. Appl. Earth Obs. Remote Sens.*, 9(5): 2019–2027.

Bonafoni, S., and C. Keeratikasikorn, 2018. Land surface temperature and urban density: Multiyear modeling and relationship analysis using MODIS and Landsat data. *Remote Sens.*, 10: 1471.

Bonafoni, S., and G. Tosi, 2017. Downscaling of land surface temperature using airborne high-resolution data: A case study on Aprilia, Italy. *IEEE Geosci. Remote Sens. Lett.*, 14(1): 107–111.

Bond, W. J., and B. W. Van Wilgen, 1996. *Fire and Plants.* New York: Chapman & Hall.

Brenning, A., M. A. Pena, S. Long, and A. Soliman, 2012. Thermal remote sensing of ice-debris landforms using ASTER: An example from the Chilean Ande. *The Cryosphere*, 6: 367–382. https://doi.org/10.5194/tc-6-367-2012.

Bromley, C. J., S. M. van Manen, and D. Graham, 2010. Monitoring surface geothermal features using time series of aerial and ground-based photographs. *American Geophysical Union*, Fall Meeting 2010, abstract#IN33B-1308. San Francisco.

Bromley, C. J., S. M. van Mane, and W. Mannington, 2011. Heat flux from steaming ground: Reducing uncertainties. *Thirty-Sixth Workshop on Geothermal Reservoir Engineering*, Stanford University, Stanford, CA.

Bucher, T. and F. Lehmann, 2000. Fusion of HyMap hyperspectral with HRSC-A multispectral and DEM data for geoscientific and environmental applications. In *IGARSS 2000. IEEE 2000 International Geoscience and Remote Sensing Symposium. Taking the Pulse of the Planet: The Role of Remote Sensing in Managing the Environment. Proceedings (Cat. No. 00CH37120)* (Vol. 7, pp. 3234–3236). IEEE.

Budzier, H., and G. Gerlach, 2011. *Thermal Infrared Sensors: Theory, Optimization and Practice* (pp. 149–171). John Wiley & Sons Ltd. ISBN: 978-0-470-87192-8.

Cai, G., Y. Xue, Y. Hu, Y. Wang, J. Guo, Y. Luo, C. Wu, S. Zhong, and S. Qi, 2007. Soil moisture retrieval from MODIS data in Northern China Plain using thermal inertia model. *Int. J. Remote Sens.*, 28: 3567–3581.

Cao, C., X. Lee, S. Liu, N. Schultz, W. Xiao, M. Zhang, and L. Zhao, 2016. Urban heat islands in China enhanced by haze pollution. *Nat. Commun.*, 7(1): 1–7.

Cardille, J. A., and M. G. Turner, 2017. Understanding landscape metrics. In *Learning Landscape Ecology: A Practical Guide to Concepts and Techniques* (pp.45–63).

Carlson, T., 2007. An overview of the "triangle method" for estimating surface evapotranspiration and soil moisture from satellite imagery. *Sensors*, 7: 1612–1629. http://doi.org/10.3390/s7081612.

Carlson, T. N., and D. A. Ripley, 1997. On the relation between NDVI, fractional vegetation cover, and leaf area index. *Remote Sens. Environ.*, 62: 241–252.

Carnahan, W. H., and R. C. Larson, 1990. An analysis of an urban heat sink. *Remote Sens. Environ.*, 33(1): 65–71.

Carr, B. B., A. B. Clarke, and L. Vanderkluysen, 2016. The 2006 lava dome eruption of Merapi Volcano (Indonesia): Detailed analysis using MODIS TIR. *J. Volcanol. Geotherm. Res.*, 311: 60–71.

Carr B. B., E. Lev, T. Sawi, K. A. Bennett, C. S. Edwards, S. A. Soule, S. V. Varga, and G. I. Marliyan, 2021. Mapping and classification of volcanic deposits using multi-sensor unoccupied aerial systems. *Remote Sens. Environ.*, 264: 112581.

Carranza, E. J. M., 2008. *Geochemical Anomaly and Mineral Prospectivity Mapping in GIS* (Vol. 11). Elsevier.

Casells, V., J. A. Sobrino, and C. Coll, 1992a. On the use of satellite thermal data for determining evapotranspiration in partially vegetated areas. *Int. J. Remote Sens.*, 13(14): 2669–2682.

Casells, V., J. A. Sobrino, and C. Coll, 1992b. A physical model for interpreting the land surface temperature obtained by remote sensors over incomplete canopies. *Remote Sens. Environ.*, 39(3): 203–211.

CBERS, 2022. *Satellite Programme: China-Brazil Earth Resources Satellite*. https://space.oscar.wmo.int/satelliteprogrammes/view/cbers, accessed on December 7, 2022.

Chandra, R., and A. Kapoor, 2020. Bayesian neural multi-source transfer learning. *Neurocomputing*, 378: 54–64. https://doi.org/10.1016/j.neucom.2019.10.042.

Chen, A., X. A. Yao, R. Sun, and I. Chen, 2014c. Effect of urban green patterns on surface urban cool islands and its seasonal variations. *Urban For. Urban Green.*, 13: 646–654.

Chen, L., R. Sun, and Y. Lu, 2019. A conceptual model for a process-oriented landscape pattern analysis. *Sci. China Earth Sci.*, 62: 2050–2057.

Chen, X., W. Li, J. Chen, Y. Rao, and Y. Yamaguchi, 2014a. A combination of TsHARP and thin plate spline interpolation for spatial sharpening of thermal imagery. *Remote Sens.*, 6: 2845–2863. https://doi.org/10.3390/rs6042845.

Chen, X., W. Li, J. Chen, W. Zhan, and Y. Rao, 2014b. A simple error estimation method for linear-regression-based thermal sharpening techniques with the consideration of scale difference. *Geo-Spat. Inf. Sci.*, 17(1): 54–59. https://doi.org/10.1080/10095020.2014.889546.

Chen, X. H., Y. Yamaguchi, J. Chen, and Y. S. Shi, 2012. Scale effect of vegetation-index-based spatial sharpening for thermal imagery: A simulation study by ASTER data. *IEEE Geosci. Remote Sens. Lett.*, 9(4): 549–553.

Chen, X. L., H. M. Zhao, P. X. Li, and Z. Y. Yin, 2006. Remote sensing image-based analysis of the relationship between urban heat island and land use/cover changes. *Remote Sens. Environ.*, 104: 133–146.

Cheruy, F., J. L. Dufresne, S. A. Mesbah, J. Y. Grandpeix, and F. Wang, 2017. Role of soil thermal inertia in surface temperature and soil moisture-temperature feedback. *J. Adv. Modeling Earth Syst.*, 9: 2906–2919.

Choudhury, B. J., and M. Menenti, 1993. Parameterization of land surface evaporation by means of location dependent potential evaporation and surface temperature range. *Depart. Environ. Food Rural Aff. (Defra): London, UK*, 212: 561–568.

Christensen, N. L., 1993. Fire regimes and ecosystem dynamics. In *Fire in the Environment* (pp. 233–244). Wiley.

Chun, B., and J. M. Guldmann, 2014. Spatial statistical analysis and simulation of the urban heat island in high-density central cities. *Landsc. Urban Plan.*, 125: 76–88.

Chybicki, A., and Z. Łubniewski, 2017. Optimized AVHRR land surface temperature downscaling method for local scale observations: Case study for the coastal area of the Gulf of Gdańsk. *Open Geosci.*, 9: 419–435. https://doi.org/10.1515/geo-2017-0032.

Ciappa, A., and G. Budillon, 2012. The Terra Nova Bay (Antarctica) polynya observed by MODIS ice surface temperature imagery from May to June 2009. *Int. J. Remote Sens.*, 33(14): 4567–4582.

Ciraolo, G., C. Cammalleri, F. Capodici, G. D'Urso, and A. Maltese, 2012. Mapping evapotranspiration on vineyards: A comparison between Penman-Monteith and Energy Balance approaches for operational purposes. In *Proceedings of SPIE 8531, Remote Sensing for Agriculture, Ecosystems, and Hydrology XIV*, 85310Q. https://doi.org/10.1117/12.974967.

Coheur, P. F., L. Clarisse, S. Turquety, D. Hurtmans, and C. Clerbaux, 2009. IASI measurements of reactive trace species in biomass burning plumes. *Atmos. Chem. Phys.*, 9(15): 5655–5667.

Coll, C. and V. Caselles, 1997. A split-window algorithm for land surface temperature from advanced very high resolution radiometer data: Validation and algorithm comparison. *J. Geophys. Res.: Atmos.*, 102(D14): 16697–16713.

Coll, C., V. Caselles, J. A. Sobrino, and E. Valor, 1994. On the atmospheric dependence of the split-window equation for land surface temperature. *Remote Sens.*, 15(1): 105–122.

Colwell, R. N., F. T. Ulaby, D. S. Simonett, J. E. Estes, and G. A. Thorley, 1983. *Manual of Remote Sensing* (Theory, instrument and techniques, Vol. 1). Falls Church: American Society of Photogrammetry.

Comon, P., and C. Jutten, 2010. *Handbook of Blind Source Separation*. New York: Academic Press.

Cook, B. D., L. A. Corp, R. F. Nelson, E. M. Middleton, D. C. Morton, J. T. McCorkel, J. G. Masek, K. J. Ranson, V. Ly, and P. M. Montesano, 2013. NASA Goddard's LiDAR, hyperspectral and thermal (G-LiHT) airborne imager. *Remote Sens.*, 5(8): 4045–4066.

Coolbaugh, M. F., C. Kratt, A. Fallacaro, W. M. Calvin, and J. V. Taranik, 2007. Detection of geothermal anomalies using advanced spaceborne thermal emission and reflection radiometer (ASTER) thermal infrared images at Bradys Hot Springs, Nevada, USA. *Remote Sens. Environ.*, 106(3): 350–359.

Coppola, D., M. Laiolo, C. Cigolini, D. D. Donne, and M. Ripepe, 2016. Enhanced volcanic hot-spot detection using MODIS IR data. In *Detecting, Modelling and Responding to Effusive Eruptions* (pp. 181–205). London: Geological Society of London.

Corradino, C., G. Ganci, A. Cappello, G. Bilotta, A. Hérault, and C. D. Negro, 2019. Mapping recent lava flows at mount etna using multispectral Sentinel-2 images and maching learning techniques. *Remote Sens.*, 11: 1916. https://doi.org/10.3390/rs11161916.

Courault, D., B. Seguin, and A. Olioso, 2005. Review on estimation of evapotranspiration from remote sensing data: From empirical to numerical modeling approaches. *Irrig. Drain. Syst.*, 19: 223–249.

Cracknell, M. J., A. M. Reading, and A. W. McNeill, 2014. Mapping geology and volcanic-hosted massive sulfide alteration in the Hellyer–Mt Charter region, Tasmania, using Random Forests—and Self-Organising Maps. *Aust. J. Earth Sci.*, 61: 287–304.

Dash, P., F.-M. Gottsche, F.-S. Olesen, and H. Fischer, 2002. Land surface temperature and emissivity estimation from passive sensor data: Theory and practice current trends. *Int. J. Remote Sens.*, 23(13): 2563–2594.

De Noronha Vaz, E., P. Cabral, M. Caetano, P. Nijkamp, and M. Painho, 2012. Urban heritage endangerment at the interface of future cities and past heritage: A spatial vulnerability assessment. *Habitat Int.*, 36: 287–294.

Dean, K., J. Dehn, K. P. Papp, S. Smith, P. Izbekov, R. Peterson, C. Keamey, and A. Steffke, 2004. Integrated satellite observations of the 2001 eruption of Mt Cleveland, Alaska. *J. Volcanol. Geotherm. Res.*, 135: 51–73.

Deilami, K., M. Kamruzzaman, and Y. Liu, 2018. Urban heat island effect: A systematic review of spatio-temporal factors, data, methods, and mitigation measures. *Int. J. Appl. Earth Obs. Geoinf.*, 67: 30–42.

Del Frate, F., and D. Solimini, 2004. On neural network algorithms for retrieving forest biomass from SAR data. *IEEE Trans. Geosci. Remote Sens.*, 42(1): 24–34.

Deng, C., and C. Wu, 2013a. Examining the impacts of urban biophysical compositions on surface urban heat island: A spectral unmixing and thermal mixing approach. *Remote Sens. Environ.*, 131: 262–274.

Deng, C., and C. Wu, 2013b. Estimating very high resolution urban surface temperature using a spectral unmixing and thermal mixing approach. *Int. J. Appl. Earth Obs. Geoinf.*, 23: 155–164.

Dennison, P. E., K. Charoensiri, D. A. Roberts, S. H, Peterson, and R. O. Green, 2006. Wildfire temperature and land cover modeling using hyperspectral data. *Remote Sens. Environ.*, 100(2): 212–222.

Di Bello, G., C. Filizzolam, T. Lacava, F. Marchese, N. Pergola, C. Pietrapertosa, S. Piscitelli, I. Scaffidi, and V. Tramutoli, 2004. Robust satellite technique for volanic and seismic hazard monitoring. *Ann. Geophys.*, 47(1): 49–64.

Ding, C., Y. Li, Y. Xia, W. Wei, L. Zhang, and Y. Zhang, 2017. Convolutional neural networks based hyperspectral image classification method with adaptive kernels. *Remote Sens.*, 9(6): 618.

Dominguez, A., J. Kleissl, J. C. Luvall, and D. L. Rickman, 2011. High-resolution urban thermal sharpener (HUTS). *Remote Sens. Environ.*, 115: 1772–1780. https://doi.org/10.1016/j.rse.2011.03.008.

Dozier, J., 1981. A method for satellite identification of surface temperature fields of subpixel resolution. *Remote Sens. Environ.*, 11: 221–229.

Du, H., W. Cai, Y. Xu, Z. Wang, Y. Wang, and Y. Cai, 2017. Quantifying the cool island effects of urban green spaces using remote sensing data. *Urban For. Urban Green.*, 27: 24–31.

Du, H., X. Song, H. Jiang, Z. Kan, Z. Wang, and Y. Cai, 2016. Research on the cooling island effects of water body: A case study of Shanghai, China. *Ecol. Indic.*, 67: 31–38.

Ebrahimy, H., and M. Azadbakht, 2019. Downscaling MODIS land surface temperature over a heterogeneous area: An investigation of machine learning techniques, feature selection, and impacts of mixed pixels. *Comput. Geosci.*, 124: 93–102. https://doi.org/10.1016/j.cageo.2019.01.004.

Elvidge, C. D., 1988. Thermal infrared reflectance of dry plant materials: 2.5–20.0 μm. *Remote Sens. Environ.*, 26: 265–285.

Essa, W., B. Verbeiren, J. van der Kwast, T. Van de Voorde, and O. Batelaan, 2012. Evaluation of the DisTrad thermal sharpening methodology for urban areas. *Int. J. Appl. Earth Obs. Geoinf.*, 19: 163–172.

Eswar, R., M. Sekhar, and B. K. Bhattacharya, 2016. Disaggregation of LST over India: Comparative analysis of different vegetation indices. *Int. J. Remote Sens.*, 37(5): 1035–1054. http://doi.org/10.1080/01431161.2016.1145363.

Fan, H., Z. Yu, G. Yang, T. Y. Liu, T. Y. Liu, C. H. Hung, and H. Vejre, 2019. How to cool hot-humid (asian) cities with urban trees? A optimal landscape size perspective. *Agric. For. Meteorol.*, 265: 338–348.

Fassnacht, K. S., S. T. Gower, M. D. MacKenzie, E. V. Nordheim, and T. M. Lillesand, 1997. Estimating the leaf area index of north central Wisconsin forests using the Landsat Thematic Mapper. *Remote Sens., Environ.*, 61(2): 229–245.

Feng, H., C. Chen, H. Dong, J. Wang, and Q. Meng, 2013. Modified shortwave infrared perpendicular water stress index: A farmland water stress monitoring method. *J. Appl. Meteorol. Climatol.*, 52(9): 2024–2032.

Fensholt, R., and I. Sandholt, 2003. Derivation of a shortwave infrared water stress index from MODIS near-and shortwave infrared data in a semiarid environment. *Remote Sens. Environ.*, 87(1): 111–121.

Fisher, J. B., K. P. Tu, and D. D. Baldocchi, 2008. Global estimates of the land–atmosphere water flux based on monthly AVHRR and ISLSCP-II data, validated at 16 FLUXNET sites. *Remote Sens. Environ.*, 112(3), 901–919.

Flannigan, M. D., and T. H. V. Haar, 1986. Forest fire monitoring using NOAA satellite AVHRR. *Can. J. For. Res.*, 16(5): 975–982.

Flasse, S. P., and P. Ceccato, 1996. A contextual algorithm for AVHRR fire detection. *Int. J. Remote Sens.*, 17: 419–424.

Flynn L. P., R. Wright, H. Garbeil, H. J. L. Harris, and E. Pilger, 2002. A global thermal alert using MODIS: Initial results from 2000–2001. *Adv. Environ. Monit. Mod.*, 1: 37–69.

Franc, G. B., and A. P. Cracknell, 1994. Retrieval of land and sea surface temperature using NOAA-11 AVHRR-data in north-eastern Brazil. *Int. J. Remote Sens.*, 15(8): 1695–1712.

Francois, C., and C. Ottlé, 2002. Atmospheric corrections in the thermal infrared: Global and water vapor dependent split-window algorithms-applications to ATSR and AVHRR data. *IEEE Trans. Geosci. Remote Sens.*, 34(2): 457–470.

Fraser, R. H., Z. Li, and J. Cihlar, 2000. Hotspot and NDVI differencing synergy (HANDS): A new technique for burnt area mapping, *Remote Sens. Environ.*, 74: 362–376.

Freeborn, P. H., M. J. Wooster, W. M. Hao, C. A. Ryan, B. L. Nordgren, and S. P. Baker, 2008. Relationships between energy release, fuel mass loss, and trace gas and aerosol emissions during laboratory biomass fires. *J. Geophys. Res.*, 113(D1): D01102.

Friedl, M. A., 2002. Forward and inverse modeling of land surface energy balance using surface temperature measurements. *Remote Sens. Environ.*, 79(2–3): 344–354.

Friedman, J. H., 2001. Greedy function approximation: A gradient boosting machine. *Ann. Stat.*, 29: 1189–1232.

Fromm, M., J. Alfred, K. Hoppel, J. Hornstein, R. Bevilacqua, E. Shettle, R. Servranckx, Z. Li, and B. Stocks, 2000. Observations of boreal forest fire smoke in the stratosphere by POAM III, SAGE II, and lidar in 1998. *Geophys. Res. Lett.*, 27: 1407–1410.

Fu, P., Y. Xie, Q. Weng, S. Myint, K. Meacham-Hensold, and C. Bernacchi, 2019. A physical model-based method for retrieving urban land surface temperatures under cloudy conditions. *Remote Sens. Environ.*, 230(26). https://doi.org/10.1016/j.rse.2019.05.010.

Gallo, K. P., and T. W. Owen, 1998. Assessment of urban heat islands: A multi-sensor perspective for the Dallas-Ft. worth, USA region. *Geocarto Int.*, 13: 35–41.

Gallo, K. P., J. D. Tarpley, A. L. McNab, and T. R. Karl, 1995. Assessment of urban heat islands: A satellite perspective. *Atmos. Res.*, 37: 37–43.

Galve, J. M., C. Coll, V. Caselles, and E. Valor, 2008. An atmospheric radiosounding database for generating land surface temperature algorithms. *IEEE Trans. Geosci. Remote Sens.*, 46(5): 1547–1557.

Gao, B. C., 1996. NDWI—A normalized difference water index for remote sensing of vegetation liquidwater from space. *Remote Sens. Environ.*, 58: 257–266.

Gao, B. C., C. O. Davis, and A. F. H. Goetz, 2006a. A review of atmospheric correction techniques for hyperspectral remote sensing of land surfaces and ocean color. In *IEEE International Geoscience and Remote Sensing Symposium, 2006 (IGARSS 2006)*, Denver, CO, pp. 1979–1981.

Gao, B. C., M. J. Montes, C. O. Davos, and A. F. H. Goetz. 2009. Atmospheric correction algorithms for hyperspectral remote sensing data of land and ocean. *Remote Sens. Environ.*, 113(supplement 1): S17–S24.

Gao, F., W. P. Kustas, and M. C. Anderson, 2012. A data mining approach for sharpening thermal satellite imagery over land. *Remote Sens.*, 4: 3287–3319. https://doi.org/10.3390/rs4113287.

Gao, F., J. Masek, M. Schwaller, and F. Hall, 2006b. On the blending of the Landsat and MODIS surface reflectance: Predicting daily Landsat surface reflectance. *IEEE Trans. Geosci. Remote Sens.*, 44(8): 2207–2218.

Gao, L., W. Zhan, F. Huang, J. Quan, X. Lu, F. Wang, W. Ju, and J. Zhou, 2017. Localization or globalization? Determination of the optimal regression window for disaggregation of land surface temperature. *IEEE Trans. Geosci. Remote Sens.*, 55(1): 477–490.

Garg, R. D., 2022. *Thermal Remote Sensing. Indian Institute of Remote Sensing (NRSA)*, Dehradun. https://www.lkouniv.ac.in/site/writereaddata/siteContent/202004021910157352ajay_misra_geo_Thermal_RS.pdf, accessed on April 30, 2022.

Garrigues, S., D. Allard, and F. Baret, 2006b. Influence of the spatial heterogeneity on the non-linear estimation of Leaf Area Index from moderate resolution remote sensing data. *Remote Sens. Environ.*, 106: 286–298.

Garrigues, S., D. Allard, F. Baret, and J. Morisette, 2008. Multivariate quantification of landscape spatial heterogeneity using variogram models. *Remote Sens. Environ.*, 112: 216–230.

Garrigues, S., D. Allard, F. Baret, and M. Weiss, 2006a. Quantifying spatial heterogeneity at the landscape scale using variogram models. *Remote Sens. Environ.*, 103: 81–96.

Gates, D. M., and W. Tantraporn, 1952. The reflectivity of deciduous trees and herbaceous plants in the infrared to 25 microns. *Science*, 115: 613–616.

Gerhards, M., M. Schlerf, K. Mallick, and T. Udelhoven, 2019. Challenges and future perspectives of multi-/hyperspectral thermal infrared remote sensing for crop water-stress detection: A review. *Remote Sens.*, 11: 1240. https://doi.org/10.3390/rs11101240.

Ghosh, A., and P. K. Joshi, 2014. Hyperspectral imagery for disaggregation of land surface temperature with selected regression algorithms over different land use land cover scenes. *ISPRS J. Photogramm. Remote Sens.*, 96: 76–93.

Ghulam, A., Qin, Q., Wang, L., Zhan, Z. and Wang, D., 2004, September. Development of broadband Albedo based ecological safety monitoring index. In IGARSS 2004. *2004 IEEE International Geoscience and Remote Sensing Symposium* (Vol. 6, pp. 4115–4118). IEEE.

Ghulam, A., Q. Qin, and Z. Zhan, 2007. Designing of the perpendicular drought index. *Environ. Geol.*, 52(6): 1045–1052.

Giglio, L., J. Descloitres, C. O. Justice, and Y. J. Kaufman, 2003. An enhanced contextual fire detection algorithm for MODIS. *Remote Sens. Environ.*, 87(2–3): 273–282.

Giglio, L., and C. O. Justice, 2003. Effect of wavelength selection on wildfire characterization using the Dozier retrieval. *Int. J. Remote Sens.*, 24: 3515–3520.

Giglio, L., and J. D. Kendall, 2001. Application of the Dozier retrieval towildfire characterization—A sensitivity analysis. *Remote Sens. Environ.*, 77: 34–49.

Giglio, L., J. T. Randerson, G. R. van der Werf, P. S. Kasibhatla, G. J. Collatz, D. C. Morton, and R. S. DeFries, 2010. Assessing variability and long-term trends in burned area by merging multiple satellite fire products. *Biogeosciences*, 7: 1171–1186.

Giglio, L., T. Loboda, D. P. Roy, B. Quayle, and C. O. Justice, 2009. An active-fire based burned area mapping algorithm for the MODIS sensor. *Remote Sens., Environ.*, 113(2): 408–420.

Giglio, L., W. Schroeder, and C. O. Justice, 2016. The collection 6 MODIS active fire detection algorithm and fire products. *Remote Sens. Environ.*, 178: 31–41.

Giglio, L., G. R. Van der Werf, J. T. Randerson, G. J. Collatz, and P. Kasibhatla, 2006. Global estimation of burned area using MODIS active fire observations. *Atmos. Chem. Phys.*, 6: 957–974. www.atmos-chem-phys.net/6/957/2006/.

Gill, M. K., M. W. Kemblowski, and M. McKee, 2007. Soil moisture data assimilation using support vector machines and ensemble kalman filter. *J. Am. Water Resour. Assoc.*, 43: 1004–1015.

Gillespie, A., S. Rokugawa, T. Matsunaga, J. S. Cothern, S. Hook, and A. B. Kahle, 1998. A temperature and emissivity separation algorithm for Advanced Spaceborne Thermal Emission and Reflection Radiometer (ASTER) images. *IEEE Trans. Geosci. Remote Sens.*, 36: 1113–1126. https://doi.org/10.1109/36.700995.

Gillespie, A. R., S. Rokugawa, S. J. Hook, T. Matsunaga, and A. B. Kahle, 1996. *Temperature/Emissivity Separation Algorithm Theoretical Basis Document*, Version 2.4. (pp. 1–64). Maryland: NASA/GSFC.

Golden, J. S., J. Carlson, K. E. Kaloush, and P. Phelan, 2007. A comparative study of the thermal and radiative impacts of photovoltaic canopies on pavement surface temperatures. *Solar Energy*, 81(7): 872–883.

Gong, P., R. Pu, Z. Li, J. Scarborough, N. Clinton, and L. M. Levien, 2006. An integrated approach to wildland fire mapping of California, USA using NOAA/AVHRR data. *Photogramm. Eng. Remote Sens.*, 72(2): 139–150.

González-Dugo, M., M. Moran, L. Mateos, and R. Bryant, 2006. Canopy temperature variability as an indicator of crop water stress severity. *Irrig. Sci.*, 24: 233–240.

Gorelick, N., M. Hancher, M. Dixon, S. Ilyushchenko, D. Thau, and R. Moore, 2017. Google Earth Engine: Planetary-scale geospatial analysis for everyone. *Remote Sens. Environ.*, 202: 18–27.

Goward, S. N., Y. Xue, and K. P. Czajkowski, 2002. Evaluating land surface moisture conditions from the remotely sensed temperature/vegetation index measurements: An exploration with the simplified biosphere model. *Remote Sens. Environ.*, 79(2–3): 225–242.

Gowda, P. H., J. L. Chávez, T. A. Howell, T. H. Marek, and L. L. New, 2008. Surface energy balance based evapotranspiration mapping in the Texas High Plains. *Sensors*, 8: 5186–5201.

Green, A. A., M. Berman, P. Switzer, and M. D. Craig, 1988. A transformation for ordering multispectral data in terms of image quality with implications for noise removal. *IEEE Trans. Geosci. Remote Sens.*, 26: 65–74.

Groenevelt, P. H., and C. D. Grant, 2004. Analysis of soil shrinkage data. *Soil Till. Res.*, 79: 71–77.

Guo, G., Z. Wu, and Y. Chen, 2014. Estimation of subpixel land surface temperature using Landsat TM imagery: A case examination over a heterogeneous urban area. In *2014 Second International Workshop on Earth Observation and Remote Sensing Applications*. 2014 IEEE Xplore.

Gustafson, W. T., A. R. Gillespie, and G. J. Yamada, 2006. Revisions to the ASTER temperature/emissivity separation algorithm. In *2nd International Symposium on Recent Advances in Quantitative Remote Sensing*, Torrent (Valencia), Spain.

Gustafson, W. T., R. N. Handcock, A. R. Gillespie, and H. Tonooka, 2003. Image sharpening method to recover stream temperatures from ASTER images. In *Remote Sensing for Environmental Monitoring, GIS Applications, and Geology II* (Vol. 4886, pp. 72–83). SPIE.

Hamada, S., and T. Ohta, 2010. Seasonal variations in the cooling effect of urban green areas on surrounding urban areas. *Urban For. Urban Green.*, 9: 15–24.

Han, M., H. Zhang, K. C. DeJonge, L. H. Comas, and T. J. Trout, 2016. Estimating maize water stress by standard deviation of canopy temperature in thermal imagery. *Agric. Water Manag.*, 177: 400–409.

Hang, H. T., and A. Rahman, 2018. Characterization of thermal environment over heterogeneous surface of National Capital Region (NCR), India using LANDSAT-8 sensor for regional planning studies. *Urban Clim.*, 24: 1–18.

Hansen, P. C., and D. P. O'Leary, 1993. The use of the L-curve in the regularization of discrete ill-posed problems. *SIAM J. Sci. Comput.*, 14(6): 1487–1503.

Harris, A., 2013a. *Thermal Remote Sensing of Active Volcanoes: A User's Manual.* Cambridge: Cambridge University Press.

Harris, A. (Ed.), 2013b. Ch5: Hot spot detection. In *Thermal Remote Sensing of Active Volcanoes: A User's Manual.* Cambridge University Press.

Harris, A. J. L., 1996. Toward automated fire monitoring from space: Semi-automated mapping of the January 1994 New South Wales wildfires using AVHRR Data. *Int. J Wildland Fire*, 6(3): 107–116.

Harris, A. J. L., E. Pilger, and L. P. Flynn, 2002a. Web-based hot spot monitoring using GOES: What it is and how it works. *Adv. Environ. Monit. Model.*, 1(3): 5–36.

Harris, A. J. L., D. A. Rothery, R. W. Carlton, S. Langaas, and H. Mannstein, 1995. Non-zero saturation of AVHRR thermal channels over high temperature targets: Evidence from volcano data and a possible explanation. *Int. J. Remote Sens.*, 16(1): 189–196.

Hart, M. A., and D. J. Sailor, 2009. Quantifying the influence of land-use and surface characteristics on spatial variability in the urban heat island. *Theor. Appl. Climatol.*, 95: 397–406.

Haselwimmer, C. E., A. Prakash, and G. Holdmann, 2011. Geothermal exploration at pilgrim hot springs, alaska using airborne thermal infrared remote sensing. In *Geothermal Resource Council Annual Meeting, 2011.* San Diego. https://energy-alaska.wdfiles.com/local--files/pilgrim-hot-springs-project-phase-1/6_GRC_2011_article.pdf.

Haselwimmer, C. and Prakash, A., 2013. Thermal infrared remote sensing of geothermal systems. In *Thermal infrared remote sensing: sensors, methods, applications* (pp. 453–473). Dordrecht: Springer Netherlands.

Hassan-Esfahani, L., A. Torres-Rua, and M. McKee, 2015. Assessment of optimal irrigation water allocation for pressurized irrigation system using water balance approach, learning machines, and remotely sensed data. *Agric. Water Manag.*, 153: 42–50.

Henley, R. W., and A. J. Ellis, 1983. Geothermal systems ancient and modern: A geochemical review. *Earth Sci. Rev.*, 19(1): 1–50.

Herrero-Huerta, M., S. Lagüela, S. M. Alfieri, and M. Menenti, 2019. Generating high-temporal and spatial resolution TIR image data. *Int. J. Appl. Earth Obs. Geoinf.*, 78: 149–162. https://doi.org/10.1016/j.jag.2019.01.016.

Higgins, J., and A. J. L. Harris, 1997. VAST: A program to locate and analyse volcanic thermal anomalies automatically from remotely sensed data. *Comput Geosci.*, 23(6): 627–645.

Holbo, H., and J. Luvall, 1989. Modeling surface temperature distributions in forest landscapes. *Remote Sens. Environ.*, 27: 11–24.

Hook, S. J., A. R. Gabell, A. A. Green, and P. S. Kealy, 1992. A comparison of techniques for extracting emissivity information from thermal infrared data for geologic studies. *Remote Sens. Environ.*, 42(2): 123–135.

Hook, S. J., R. G. Vaughan, H. Tonooka, and S. G. Schladow, 2007. Absolute radiometric in-flight validation of mid infrared and thermal infrared data from ASTER and MODIS on the Terra spacecraft using the lake Tahoe, CA/NV, USA, Automated Validation Site. *IEEE Trans. Geosci. Remote Sens.*, 45: 1798–1807.

Hu, L. Q., and N. A. Brunsell, 2015. A new perspective to assess the urban heat island through remotely sensed atmospheric profiles. *Remote Sens. Environ.*, 158: 393–406.

Hu, L., and M. Delichatsios, 2019. *Thermal radiation. Encyclopedia of Wildfires and Wildland-Urban Interface (WUI) Fires*, S. L. Manzello (Ed.), Springer Nature Switzerland AG. https://doi.org/10.1007/978-3-319-51727-8_78-1.

Hu, T., H. Li, B. Cao, A. I. van Dijk, L. J. Renzullo, Z. Xu, J. Zhou, Y. Du, and Q. Liu, 2019. Influence of emissivity angular variation on land surface temperature retrieved using the generalized split-window algorithm. *Int. J. Appl. Earth Obs. Geoinf.*, 82: 101917.

Huang, B., J. Wang, H. Song, D. Fu, and K. Wong, 2013. Generating high spatiotemporal resolution land surface temperature for urban heat island monitoring. *IEEE Geosci. Remote Sens. Lett.*, 10(5): 1011–1015. https://doi.org/10.1109/LGRS.2012.2227930.

Huang, H., R. Ooka, and S. Kato, 2005. Urban thermal environment measurements and numerical simulation for an actual complex urban area covering a large district heating and cooling system in summer. *Atmos. Environ.*, 39: 6362–6375.

Huang, L., J. Li, D. Zhao, and J. Zhu, 2008. A fieldwork study on the diurnal changes of urban microclimate in four types of ground cover and urban heat island of Nanjing, China. *Build. Environ.*, 43: 7–17.

Huang, Q., and Y. Lu, 2018. Urban heat island research from 1991 to 2015: A bibliometric analysis. *Theor. Appl. Climatol.*, 131: 1055–1067.

Huang, S.-X., and J. Li, 2000. Theoretical and numerical analysis of inversion of satellite remote sensing. In H. G. W. Begehr, R. P. Gilbert, and J. Kajiwara, (Eds.), *Proceedings of the 2nd ISAAC Congress* (Vol. 1, pp. 129–136). Boston, MA: Springer.

Hulley, G., 2013. A water vapor scaling (WVS) method for improving atmospheric correction of thermal infrared (TIR) data. In C. Kuenzer, and S. Dech (Eds.), *Thermal Infrared Remote Sensing, Sensors, Methods, Applications* (pp. 125–165). Springer.

Hulley, G., S. Shivers, E. Wetherley, and R. Cudd, 2019. New ECOSTRESS and MODIS land surface temperature data reveal fine-scale heat vulnerability in cities: A case study for Los Angeles County, California. *Remote Sens.*, 11: 2136.

Hulley, G. C, and S. J. Hook, 2011. Generating consistent land surface temperature and emissivity products between ASTER and MODIS data for earth science research. *IEEE Trans. Geosci. Remote Sens.*, 49: 1304–1315.

Hulley, G. C., S. T. Hook, and A. M. Baldridge, 2009. Validation of the North American ASTER Land Surface Emissivity Database (NAALSED) version 2.0 using pseudo-invariant sand dune sites. *Remote Sens. Environ.*, 113: 2224–2233.

Hutengs, C., and M. Vohland, 2016. Downscaling land surface temperatures at regional scales with random forest regression. *Remote Sens. Environ.*, 178: 127–141.

Hyvärinen, A., 2005. A unifying model for blind separation of independent sources. *Signal Process.*, 85: 1419–1427.

Hyvärinen, A., 2013. Independent component analysis: Recent advances. *Phil Trans. R. Soc. A.*, 371: 20110534. https://doi.org/10.1098/rsta.2011.0534.

Ichoku, C., L. Giglio, M. J. Wooster, and L. A. Remer, 2008. Global characterization of biomass-burning patterns using satellite measurements of fire radiative energy. *Remote Sens. Environ.*, 112(6): 2950–2962.

Ichoku, C., and Y. J. Kaufman, 2005. A method to derive smoke emission rates from MODIS fire radiative energy measurements. *IEEE Trans. Geosci. Remote Sens.*, 43(11): 2636–2649.

Idso, S. B., 1982. Non-water-stressed baselines: A key to measuring and interpreting plant water stress. *Agric. Meteorol.*, 27: 59–70.

Idso, S. B., and R. D. Jackson, 1969. Thermal radiation from the atmosphere. *J. Geophys. Res.*, 74(23): 5397–5403. https://doi.org/10.1029/JC074i023p05397.

Idso, S. B., R. D. Jackson, and R. J. Reginato, 1977. Remote-sensing of crop yields. *Science*, 196: 19–25.

Ihara, T., Y. Kikegawa, K. Asahi, Y. Genchi, and H. Kondo, 2008. Changes in year-round air temperature and annual energy consumption in office building areas by urban heat-island countermeasures and energy-saving measures. *Appl. Energy*, 85(1): 12–25.

Imhoff, M. L., P. Zhang, R. E. Wolfe, and L. Bounoua, 2010. Remote sensing of the urban heat island effect across biomes in the continental USA. *Remote Sens. Environ.*, 114(3): 504–513. https://doi.org/10.1016/j.rse.2009.10.008.

Inamdar, A. K., and A. French, 2009. Disaggregation of GOES land surface temperatures using surface emissivity. *Geophys. Res. Lett.*, 36: L02408. https://doi.org/10.1029/2008GL036544.

Inzana, J., T. Kusky, G. Higgs, and R. Tucker, 2003. Supervised classification of Landsat TM band ratio images and Landsat TM band ratio image with radar for geological interpretations of central Madagascar. *J. Afr. Earth Sci.*, 37: 59–72.

Ismanto, H., H. Hartono, and M. A. Marfai, 2020. Classification tree analysis (CTA) of smoke detection using Himawari_8 satellite data over Sumatera–Borneo Island, Indonesia. *SN Appl. Sci.*, 2: 1520. https://doi.org/10.1007/s42452-020-03310-z.

Jackson, R. D., S. B. Idso, R. J. Reginato, and P. J. Pinter, 1981. Canopy temperature as a crop water stress indicator. *Water Resour. Res.*, 17: 1133–1138.

Jackson, R. D., R. J. Reginato, and S. B. Idso, 1977. Wheat canopy temperature: A practical tool for evaluating water requirements. *Water Resour. Res.*, 13(3): 651–656.

Jain, S. P., and K. R. Rao, 1974. Experimental study on the effect of roof spray cooling on unconditioned and conditioned buildings. *Build. Sci.*, 9(1): 9–16.

James, G., D. Witten, T. Hastie, and R. Tibshirani, 2013. *An Introduction to Statistical Learning* (Vol. 112, p. 18). New York: Springer.

Jawak, S. D., S. F. Wankhede, and A. J. Luis, 2018. Exploration of glacier surface facies mapping techniques using very high-resolution Worldview-2 satellite data. *Multidiscip. Digit. Publ. Inst. Proc.*, 2: 339.

Jawak, S. D., S. F. Wankhede, and A. J. Luis, 2019. Explorative study on mapping surface facies of selected glaciers from Chandra Basin, Himalaya using WorldView-2 data. *Remote Sens.*, 11(10): 1207.

Jeganathan, C., N. A. S. Hamm, S. Mukherjee, P. M. Atkinson, P. L. N. Raju, and V. K. Dadhwal, 2011. Evaluating a thermal image sharpening model over a mixed agricultural landscape in India. *Int. J. Appl. Earth Obs. Geoinf.*, 13(2): 178–191.

Jensen, J. R. 2007. *Remote Sensing of the Environment: An Earth Resource Perspective* (2nd Edition, 592 p). Upper Saddle River: Prentice Hall, ISBN: 0-13-188950-8.

Jensen, J. R., 2015. *Introductory Digital Image Processing: A Remote Sensing Perspective* (4th Edition, 632 p). Upper Saddle River: Prentice Hall.

Jia, H., D. Yang, W. Deng, Q. Wei, and W. Jiang, 2021. Predicting land surface temperature with geographically weighed regression and deep learning. *Wires. Data Min. Knowl.*: 1–19. https://doi.org/10.1002/WIDM.1396.

Jia, L., Z. Su, B. van den Hurk, M. Menenti, A. Moene, H. A. R. De Bruin, J. J. B. Yrisarry, M. Ibanez, and A. Cuesta, 2003. Estimation of sensible heat flux using the Surface Energy Balance System (SEBS) and ATSR measurements. *Phys. Chem. Earth*, 28: 75–88.

Jiang, Y., P. Fu, and Q. Weng, 2015. Downscaling GOES land surface temperature for assessing heat wave health risks. *IEEE Geosci. Remote Sens. Lett.*, 12(8): 1605–1609. https://doi.org/10.1109/LGRS.2015.2414897.

Jiang, Y., and Q. Weng, 2013. Estimating LST using a vegetation-cover-based thermal sharpening technique. *IEEE Geosci. Remote Sens. Lett.*, 10(5): 1249–1252. https://doi.org/10.1109/LGRS.2013.2257667.

Jiao, W., L. Zhang, Q. Chang, D. Fu, Y. Cen, and Q. Tong, 2016. Evaluating an enhanced vegetation condition index (VCI) based on VIUPD for drought monitoring in the continental United States. *Remote Sen.*, 8(3): 224.

Jiménez-Muñoz, J. C., and J. A. Sobrino, 2003. A generalized single-channel algorithm for retrieving land surface temperature from remote sensing data. *J. Geophys. Res.*, 108(D22): 4688. https://doi.org/10.1029/2003JD003480.

Jiménez-Muñoz, J. C., and J. A. Sobrino, 2006. Error sources on the land surface temperature retrieved from thermal infrared single channel remote sensing data. *Int. J. Remote Sens.*, 27(5): 999–1014.

Jiménez-Muñoz, J. C., J. A. Sobrino, D. Skoković, C. Mattar, and J. Cristobal, 2014. Land surface temperature retrieval methods from Landsat-8 thermal infrared sensor data. *IEEE Geosci. Remote Sens. Lett.*, 11(10): 1840–1843.

Jones, H. G., 1999. Use of infrared thermometry for estimation of stomatal conductance as a possible aid to irrigation scheduling. *Agric. For. Meteorol.*, 95: 139–149.

Jones, H. G., 2004. Application of thermal imaging and infrared sensing in plant physiology and ecophysiology. *Adv. Bot. Res.*, 41: 107–163.

Jones, H. G., R. Serraj, B. R. Loveys, L. Xiong, A. Wheaton, and A. H. Price, 2009. Thermal infrared imaging of crop canopies for the remote diagnosis and quantification of plant responses to water stress in the field. *Funct. Plant Biol.*, 36: 978–989.

Kaneko, T., and M. J. Wooster, 1999. Landsat infrared analysis of fumarole activity at Unzen Volcano: Time-series comparison with gas and magma fluxes. *J. Volcanol. Geotherm. Res.*, 89: 57–64.

Kang, Y.-X., G.-H. Lu, Z.-Y. Wu, and H. He, 2009. Comparison and analysis of bare soil evaporation models combined with ASTER data in Heihe River Basin. *Water Sci. Eng.*, 2(4): 16–27. https://doi.org/10.3882/J.Issn.1674-2370.2009.04.002.

Karimi, N., A. Farokhnia., L. Karimi, M. Eftekhari, and H. Ghalkhani, 2012. Combining optical and thermal remote sensing data for mapping debris-covered glaciers (Alamkouh Glaciers, Iran). *Cold Reg. Sci. Technol.*, 71: 73–83.

Kate, V., and P. Shukla, 2021. A 3 tier CNN model with deep discriminative feature extraction for discovering malignant growth in multi-scale histopathology images. *Inf. Med. Unlocked*, 24: 100616.

Kaufman, Y. J., C. J. Tucker, and I. Fung, 1990. Remote sensing of biomass burning in the tropics. *J. Geophys. Res.*, 95: 9927–9939.

Kaufman, Y. J., C. O. Justice, L. P. Flynn, J. D. Kendall, E. M. Prins, L. Giglio, D. E. Ward, W. P. Menzel, and A. W. Setzer, 1998. Potential global fire monitoring from EOS-MODIS. *J. Geophys. Res.: Atmos.*, 103(D24): 32215–32238.

Kay, J. J., and E. D. Schneider, 1994. Embracing complexity, the challenge of the ecosystem approach. *Alternatives*, 20(3): 32–38.

Kaya, G. T., 2014. Introduction to optical & thermal remote sensing. *ITU Institute of Earthquake Engineering and Disater Management.* https://bte.bilgem.tubitak.gov.tr/sites/images/g1-6.pdf, accessed on April 25, 2022.

Kennedy, P. J., A. S. Belward, and J. M. Gregoire, 1994. An improved approach to fire monitoring in West Africa using AVHRR data. *Int. J. Remote Sens.*, 15(11): 2235–2255.

Keramitsoglou, I., C. T. Kiranoudis, and Q. Weng, 2013. Downscaling geostationary land surface temperature imagery for urban analysis. *Geosci. Remote Sens. Lett.*, 10(5): 1253–1257. https://doi.org/10.1109/LGRS.2013.2257668.

Kerr, Y. H., J. P. Lagouarde, and J. Imbernon, 1992. Accurate land surface temperature retrieval from AVHRR data with use of an improved split window algorithm. *Remote Sens. Environ.*, 41(2–3): 197–209.

Kervyn, M., F. Kervyn, R. Goossens, S. K. Rowland, and G. G. J. Ernst, 2007. Mapping volcanic terrain using high-resolution and 3D satellite remote sensing. *Geol. Soc. Spec. Publ.*, 283: 5–30.

Key, C. H., and N. C. Benson, 2003. The normalized burn ratio (NBR): A land sat TM radiometric measure of burn severity. *US Geological Survey Northern Rocky Mountain Science Center.* https://www.frames.gov/catalog/5860.

Khalil, U., B. Aslam, U. Azam, and H. M. D. Khalid, 2021. Time series analysis of land surface temperature and drivers of urban heat island effect based on remotely sensed data to develop a prediction model. *Appl. Artif. Intell.*, 35(15): 1803–1828. https://doi.org/10.1080/08839514.2021.1993633.

Khanal, S., J. Fulton, and S. Shearer, 2017. An overview of current and potential applications of thermal remote sensing in precision agriculture. *Comput. Electron. Agric.*, 139: 22–32.

Khoob, A. R., 2008. Artificial neural network estimation of reference evapotranspiration from pan evaporation in a semi-arid environment. *Irrig. Sci.*, 27(1): 35–39. https://doi.org/10.1007/S00271-008-0119-Y.

Kikegawa, Y., Y. Genchi, H. Kondo, and K. Hanaki, 2006. Impacts of city-block-scale countermeasures against urban heat-island phenomena upon a building's energy-consumption for air-conditioning. *Appl. Energy*, 83(6): 649–668.

Kikegawa, Y., Y. Genchi, H. Yoshikado, and H. Kondo, 2003. Development of a numerical simulation system toward comprehensive assessments of urban warming countermeasures including their impacts upon the urban buildings' energy-demands. *Appl. Energy*, 76(4): 449–466.

Kim, Y., and J. Baik, 2002. Maximum urban heat island intensity in Seoul. *J. Appl. Meteorol.*, 41: 651–659.

Kim, Y., and J. Baik, 2005. Spatial and temporal structure of the urban heat island in Seoul. *J. Appl. Meteorol.*, 44: 591–605.

Klysik, K., and K. Fortuniak, 1999. Temporal and spatial characteristics of the urban heat island of Lodz, Poland. *Atmos. Environ.*, 33: 3885–3895.

Kneizys, F, X., L. W. Abreu, G. P. Anderson, J. H. Chetwynd, E. P. Shettle, A. Berk, L. S, Bernstein, D. C. Robertson, P. K. Acharya, L. A. Rothman, J. E. A. Selby, W. O. Gallery, and S. A. Clough, 1996. *The MODTRAN 2/3 Report & LOWTRAN 7 Model*, F19628-91-C-0132. Phillips Laboratory Hanscom AFB, Bedford.

Koetz, B., W. Bastiaanssen, M. Berger, P. Defourney, U. Del Bello, M. Drusch, M. Drinkwater, R. Duca, V. Fernandez, D. Ghent, and R. Guzinski, 2018. High spatio-temporal resolution land surface temperature mission-a copernicus candidate mission in support of agricultural monitoring. In *Igarss 2018-2018 IEEE international geoscience and remote sensing symposium* (pp. 8160–8162). IEEE.

Kogan, F. N., 1990. Remote sensing of weather impacts on vegetation in non-homogeneous areas. *Int. J. Remote Sens.*,11: 1405–1419.

Kogan, F. N., 1995. Application of vegetation index and brightness temperature for drought detection. *Adv. Space Res.*, 15: 91–100.

Kohavi, R., 1995. A study of cross-validation and bootstrap for accuracy estimation and model selection. *Int. Jt. Conf. Artif. Intell.*, 14: 1137–1143.

Kolokotroni, M., I. Giannitsaris, and R. Watkins, 2006. The effect of the London Urban Heat Island on building summer cooling demand and night ventilation strategies. *Sol. Energy*, 80(4): 383–392.

Kondoh, A., and J. Nishiyama, 1999. Changes in hydrological cycle due to urbanization in the suburb of Tokyo Metropolitan Area, Japan. *Adv. Space Res.*, 26: 1173–1176.

Kondoh, A., and J. Nishiyama, 2000. Changes in hydrological cycle due to urbanization in the suburb of Tokyo Metropolitan area, Japan. *Adv. Space Res.*, 26: 1173–1176.

Koutsias, N., M. Karteris, and E. Chuvico, 2000. The use of intensity-hue-saturation transformation of Landsat-5 Thematic Mapper data for burned land mapping. *Photogramm. Eng. Remote Sens.*, 66(7): 829–840.

Kramer, H. J., 2002. *Observation of the Earth and Its Environment, Survey of Missions and Sensors* (4th Edition, 1955 p). Berlin: Springer-Verlag.

Krüger, E. L., and D. Pearlmutter, 2008. The effect of urban evaporation on building energy demand in an arid environment. *Energ Build.*, 40: 2090–2098.

Kubelka, P., and F. Munk, 1931. Ein Beitrag zur Optik der Farbanstriche. *Zeitschrift für Technische Physik*, 12: 593–601.

Kustas, W., and M. Anderson, 2009. Advances in thermal infrared remote sensing for land surface modeling. *Agric. For. Meteorol.*, 149: 2071–2081.

Kustas, W. P., and C. S. Daughtry, 1990. Estimation of the soil heat flux/net radiation ratio from spectral data. *Agric. For. Meteorol.*, 49(3): 205–223.

Kustas, W. P., and J. M. Norman, 1996. Use of remote sensing for evapotranspiration monitoring over land surfaces. *Hydrol. Sci. J.*, 41(4): 495–516.

Kustas, W. P., and J. M. Norman, 1997. A two-source approach for estimating turbulent fluxes using multiple angle thermal infrared observations. *Water Resour. Res.*, 33(6): 1495–1508.

Kustas, W. P., and J. M. Norman, 1999. Evaluation of soil and vegetation heat flux predictions using a simple two-source model with radiometric temperatures for partial canopy cover. *Agric. For. Meteorol.*, 94(1): 13–29.

Kustas, W. P., J. M. Norman, M. C. Anderson, and A. N. French, 2003. Estimating subpixel surface temperatures and energy fluxes from the vegetation index–radiometric temperature relationship. *Remote Sens. Environ.*, 85(4): 429–440.

Lan, X., E. Zhao, P. Leng, Z.-L. Li, J. Labed, F. Nerry, X. Zhang, and G. Shang, 2022. Alternative physical method for retrieving land surface temperatures from hyperspectral thermal infrared data: Application to IASI observations. *Trans. Geosci. Remote Sens.*, 60: 4500812.

Langaas, S., 1993. A parametrised model for savanna fire detection using AVHRR night images. *Int. J. Remote Sens.*, 14(12): 2245–2262.

Larson, R. C., and W. H. Carnahan, 1997. The influence of surface characteristics on urban radiant temperatures. *Geocarto Int.*, 12(3): 5–16.

Latifovic, R., D. Pouliot, and J. Campbell, 2018. Assessment of convolution neural networks for surficial geology mapping in the South Rae geological region, Northwest Territories, Canada. *Remote Sens.*, 10: 307. https://doi.org/10.3390/rs10020307.

Le, M. S., and Y. A. Liou, 2021. Temperature-soil moisture dryness index for remote sensing of surface soil moisture assessment. *IEEE Geosci. Remote Sens. Lett.*, 19: 1–5.

Lee, J. B., A. S. Woodyatt, and M. Berman, 1990. Enhancement of high spectral resolution remote sensing data by a noise-adjusted principal components transform. *IEEE Trans. Geosci. Rem. Sens.*, 28(3): 295–304.

Lee, S., and J. Choi, 2022. A snow cover mapping algorithm based on a multitemporal dataset for GK-2A imagery. *GIScі. Remote Sens.*, 59(1): 1078–1102. https://doi.org/10.1080/15481603.2022.2097395.

Lee, T. F., and P. M. Tag, 1990. Improved detection of hotspots using the AVHRR 3.7-μm channel. *Bull. Am. Meteorol. Soc.*, 71(12): 1722–1730.

Leuzinger, S., and C. K orner, 2007. Tree species diversity affects canopy leaf temperatures in a mature temperate forest. *Agric. For. Meteorol.*, 146: 29–37.

Leuzinger, S., R. Vogt, and C. K orner, 2010. Tree surface temperature in an urban environment. *Agri. For. Meteorol.*, 150: 56–62.

Levine, J. (Ed.), 1991. *Global Biomass Burning: Atmospheric, Climatic, and Biospheric Implications*. MIT Press.

Li, C., J. Wang, L. Wang, L. Hu, and P. Gong, 2014. Comparison of classification algorithms and training sample sizes in urban land classification with Landsat thematic mapper imagery. *Remote Sens.*, 6: 964–983.

Li, F., W. P. Kustas, J. H. Prueger, C. M. U. Neale, and T. J. Jackson, 2005. Utility of remote sensing based two-source energy balance model under low and high vegetation cover conditions. *J. Hydrometeorol.*, 6: 878–891.

Li, F., and T. J. Lyons, 1999. Estimation of regional evapotranspiration through remote sensing. *J. Appl. Meteorol.*, 38(11): 1644–1654.

Li, H., G. Wu, F. Xu, and S. Li, 2021a. Landsat-8 and Gaofen-1 image-based inversion method for the downscaled land surface temperature of rare earth mining areas. *Infrared Phys. Technol.*, 113: 103658. https://doi.org/10.1016/j.infrared.2021.103658.

Li, H., Y. Zhou, X. Li, L. Meng, X. Wang, S. Wu, and S. Sodoudi, 2018a. A new method to quantify surface urban heat island intensity. *Sci. Total Environ.*, 624: 262–272.

Li, J., C. Song, L. Cao, F. Zhu, X. Meng, and J. Wu, 2011. Impacts of landscape structure on surface urban heat islands: A case study of Shanghai, China. *Remote Sens Environ.*, 115(12): 3249–3263. https://doi.org/10.1016/j.rse.2011.07.008.

Li, L., C. Solana, F. Canters, and M. Kervyn, 2017a. Testing random forest classification for identifying lava flows and mapping age groups on a single Landsat 8 image. *J. Volcanol. Geotherm. Res.*, 345: 109–124.

Li, R. R., Y. J. Kaufman, W. M. Hao, J. M. Salmon, and B. C. Gao, 2004. A technique for detecting burn scars using MODIS data. *IEEE Trans. Geosci. Remote Sens.*, 42: 1300–1308.

Li, T., and Q. Meng, 2018. A mixture emissivity analysis method for urban land surface temperature retrieval from Landsat 8 data. *Landsc. Urban Plan.*, 179: 63–71.

Li, T., Y. Xu, and L. Yao, 2021b. Detecting urban landscape factors controlling seasonal land surface temperature: From the perspective of urban function zones. *Environ. Sci. Pollut. Res.*, 28: 41191–41206. https://doi.org/10.1007/s11356-021-13695-y.

Li, W., L. Nic, Z.-L. Li, and H. Wu, 2018b. Downscaling land surface temperature by using random forest regression algorithm. *IGARSS 2018, 2527–2530*, IEEE Xplore.

Li, W., H. Wu, S.-B. Duan, Z.-L. Li, and Q. Liu, 2019. Selection of predictor variables in downscaling land surface temperature using random forest algorithm. *IGARSS 2019 1817–1820*, IEEE Xplore.

Li, X., W. Song, L. Lian, and X. Wei, 2015. Forest fire smoke detection using back-propagation neural network based on MODIS data. *Remote Sens.*, 7: 4473–4498. https://doi.org/10.3390/rs70404473.

Li, X., X. Xin, J. Jiao, Z. Peng, H. Zhang, S. Shao, and Q. Liu, 2017b. Estimating subpixel surface heat fluxes through applying temperature-sharpening methods to MODIS data. *Remote Sens.*, 9: 836. https://doi.org/10.3390/rs9080836.

Li, X., Y. Zhou, G. R. Asrar, M. Imhoff, and L. Li, 2017c. The surface urban heat island response to urban expansion: A panel analysis for the conterminous United States. *Sci. Total Environ.*, 605: 426–435.

Li, Z., A. Khananian, R. H. Fraser, and J. Cihlar, 2001. Automatic detection of fire smoke using artificial neural networks and threshold approaches applied to AVHRR imagery. *IEEE Trans. Geosci. Remote Sens.*, 39(9): 1859–1870.

Li, Z., S. Nadon, and J. Cihlar, 2000. Satellite detection of Canadian boreal forest fires: Development and application of an algorithm. *Int. J. Remote Sens.*, 21: 3057–3069.

Li, Z.-L., and F. Becker, 1990. Properties and comparison of temperature-independent thermal infrared spectral indices with NDVI for HAPEX data. *Remote Sens. Environ.*, 33: 165–182.

Li, Z.-L., and F. Becker, 1993. Feasibility of land surface temperature and emissivity determination from AVHRR data. *Remote Sens. Environ.*, 43: 67–85.

Li, Z.-L., B. H. Tang, H. Wu, H. Ren, G. Yan, Z, Wan, I. F. Trigo, and J. A. Sobrino, 2013. Satellite-derived land surface temperature: Current status and perspectives. *Remote Sens. Environ.*, 131: 14–37.

Li, Z.-L., R. Tang, Z. Wan, Y. Bi, C. Zhou, B. Tang, G. Yan, and X. Zhang, 2009. A review of current methodologies for regional evapotranspiration estimation from remotely sensed data. *Sensors*, 9: 3801–3853. https://doi.org/10.3390/s90503801.

Liang, S., 2001. Narrowband to broadband conversions of land surface albedo I Algorithms. *Remote Sens. Environ.*, 76: 213–238.

Liao, W., Z. Cai, Y. Feng, D. Gan, and X. Li, 2021. A simple and easy method to quantify the cool island intensity of urban greenspace. *Urban For. Urban Green.*, 62: 127173.

Lillesand, T. M., R. W. Kiefer, and J. W. Chipman, 2008. *Remote Sensing and Image Interpretation* (6th Edition, p. 359). New York: John Wiley and Sons Inc.

Lillo-Saavedra, M., A. García-Pedrero, G. Merino, and C. Gonzalo-Martín, 2018. TS2uRF: A new method for sharpening thermal infrared satellite imagery. *Remote Sens.*,10: 249. https://doi.org/10.3390/rs10020249.

Lin, T.-H., G.-L. Liu, and Y.-C. Chenc, 2010. Remote sensing of smoke plumes with moderate resolution imaging spectroradiometer reflectance measurements. *J. Appl. Remote Sens.*, 4: 041876.

Liou, Y.-A., and S. K. Kar, 2014. Evapotranspiration estimation with remote sensing and various surface energy balance algorithms—a review. *Energies*, 7: 2821–2849. https://doi.org/10.3390/en7052821.

Liou, Y.-A., M. S. Le, and H. Chien, 2019. Normalized difference latent heat index for remote sensing of land surface energy fluxes. *IEEE Trans. Geosci. Remote Sens.*, 57(3): 1423–1433.

Liu, C., R. Xu, F. Xie, J. Jin, L. Yuan, G. Lv, Y. Wang, C. Li, and J. Wang, 2020. New airborne thermal-infrared hyperspectral imager system: Initial validation. *IEEE J. Sel. Top. Appl. Earth Obs. Remote Sens.*, 13: 4149–4165. https://doi.org/10.1109/JSTARS.2020.3010092.

Liu, H., and Q. Weng, 2018. Scaling effect of fused ASTER-MODIS land surface temperature in an urban environment. *Sensors*, 18: 4058. https://doi.org/10.3390/s18114058.

Liu, K., and H. Su, 2016. Enhancing the spatial resolution of satellite-derived land surface temperature (LST) using an operational trapezoid interpolation disaggregation model (OTI_DisTrad). In *2016 Fourth International Workshop on Earth Observation and Remote Sensing Applications 978-1-5090-1479-8/16/$31.00 ©2016 IEEE*, IEEE Xplore, 5 p.

Liu, K., H. Su, and X. Li, 2016a. Estimating high-resolution urban surface temperature using a hyperspectral thermal mixing (HTM) approach. *IEEE J. Sel. Top. Appl. Earth Obs. Remote Sens.*, 9(2): 804–815. https://doi.org/10.1109/JSTARS.2015.2459375.

Liu, K., H. Su, X. Li, S. Chen, R. Zhang, W. Wang, L. Yang, H. Liang, and Y. Yang, 2018a. A thermal disaggregation model based on trapezoid interpolation. *IEEE J. Sel. Top. Appl. Earth Obs. Remote Sens.*, 11(3): 808–820. https://doi.org/10.1109/JSTARS.2018.2790002.

Liu, K., H. Su, J. Tian, X. Li, W. Wang, L. Yang, and H. Liang, 2018b. Assessing a scheme of spatial-temporal thermal remote-sensing sharpening for estimating regional evapotranspiration. *Int. J. Remote Sens.*, 39(10): 3111–3137. https://doi.org/10.1080/01431161.2018.1434326.

Liu, K., H. B. Su, X. K. Li, W. M. Wang, L. J. Yang, and H. Liang, 2016b. Quantifying spatial-temporal pattern of urban heat island in Beijing: An improved assessment using land surface temperature (LST) time series observations from LANDSAT, MODIS, and Chinese new satellite GaoFen-1. *IEEE J. Sel. Top. Appl. Earth Obs. Remote Sens.*, 9: 2028–2042.

Liu, K., S. Wang, X. Li, and T. Wu, 2019. Spatially disaggregating satellite land surface temperature with a nonlinear model across agricultural areas. *J. Geophys. Res. Biogeosciences*, 124(11): 3232–3251.

Liu, P., 2015. A survey of remote-sensing big data. *Front. Environ. Sci.*, 3. https://doi.org/10.3389/fenvs.2015.00045.

Liu, T., R. Li, X. Zhong, M. Jiang, X. Jin, P. Zhou, S. Liu, C. Sun, and W. Guo, 2018d. Estimates of rice lodging using indices derived from UAV visible and thermal infrared images. *Agric. For. Meteorol.*, 252: 144–154.

Liu, W., Y. Xie, Z. Wan, and H. Zheng, 2012. Analysis on urban heat island effect of the different dynamics of urban surface biophysical descriptors in Shanghai city. *Remote Sens. Technol. Appl.*, 27(5): 797–803.

Liu, X., Y. Wang, and Q. Liu, 2018c. PSGAN: A generative adversarial network for remote sensing image pan-sharpening. In *Proceedings of 25th IEEE International Conference on Image Processing (ICIP)*, Athens, Greece, October 7–10, 873–877. https://doi.org/10.1109/ICIP.2018.8451049.

Liu, Y., J. Qian, and H. Yue, 2021. Combined Sentinel-1awith Sentinel-2a to estimate soil moisture in farmland. *IEEE J. Sel. Top. Appl. Earth Obs. Remote Sens.*, 14: 1292–1310.

Loboda, T., K. J. O'Neal, and I. Csiszar, 2007. Regionally adaptable dNBR-Based algorithm for burned area mapping from MODIS data. *Remote Sens. Environ.*, 109: 429–442.

Loew, A., W. Bell, L. Brocca, C. E. Bulgin, J. Burdanowitz, X. Calbet, R. V. Donner, D. Ghent, A. Gruber, T. Kaminski, J. Kinzel, C. Klepp, J. C. Lambert, G. Schaepman-Strub, M. Schröder, and T. Verhoelst, 2017. Validation practices for satellite-based earth observation data across communities. *Rev. Geophys.*, 55: 779–817. https://doi.org/10.1002/2017RG000562.

Lombardo, V., and M. F. Buongiorno, 2006. Lava flow thermal analysis using three infrared bands of remote-sensing imagery: A study case from Mount Etna 2001 eruption. *Remote Sens. Environ.*, 101: 141–149.

Lovejoy, T. E., 1994. The quantification of biodiversity: An esoteric quest or a vital component of sustainable development? *Philos. Trans. Royal Soc. Lond. Ser. B: Biol. Sci.*, 345(1311): 81–87.

Lu, D., and Q. Weng, 2006. Spectral mixture analysis of ASTER images for examining the relationship between urban thermal features and biophysical descriptors in Indianapolis, Indiana, USA. *Remote Sens. Environ.*, 104(2): 157–167.

Lu, S., Z. Ju, T. Ren, and R. Horton, 2009. A general approach to estimate soil water content from thermal inertia. *Agric. For. Meteorol.*, 149: 1693–1698.

Lucas, T. C., 2020. A translucent box: Interpretable machine learning in ecology. *Ecol. Monogr.*, 90(4): e01422.

Luo, D., X. Wen, S. Li, and J. Cao, 2022. The ATI-ET triangle model: A novel approach to estimate soil moisture applied to modis data. *Sensors*, 22: 7926. https://doi.org/10.3390/s22207926

Luo, X., Y. Chen, Z. Wang, H. Li, and Y. Peng, 2021. Spatial downscaling of MODIS land surface temperature based on a geographically and temporally weighted autoregressive model. *IEEE J. Sel. Top. Appl. Earth Obs. Remote Sens.*, 14: 7637–7653. https://doi.org/10.1109/JSTARS.2021.3094184.

Luvall, J. C., and H. R. Holbo, 1989. Measurements of short-term thermal responses of coniferous forest canopies using thermal scanner data. *Remote Sens. Environ.*, 27: 1–10.

Luvall, J. C., D. Lieberman, M. Lierberman, G. S. Hartshor, and R. Peralta, 1990. Estimation of tropical forest canopy temperatures, thermal response numbers, and evapotranspiration using an aircraft-based thermal sensor. *Photogramm. Eng. Remote Sens.*, 56(10): 1393–1401.

Ma, J., W. Zhang, A. Marinoni, L. Gao, and B. Zhang, 2018. An improved spatial and temporal reflectance unmixing model to synthesize time series of Landsat-like images. *Remote Sens.*, 10(9): 1388.

Ma, L., Y. Liu, X. Zhang, Y. Ye, G. Yin, and B. A. Johnson, 2019. Deep learning in remote sensing applications: A meta-analysis and review. *ISPRS J. Photogramm. Remote Sens.*, 152: 166–177.

Ma, Y., O. Tsukamoto, and H. Ishikawa, 2002. Remote sensing parameterization of the processes of energy and water cycle over desertification areas. *Sci. China Ser. D: Earth Sci.*, 45: 47–53.

Ma, X. L., Z. Wan, C. C. Moeller, W. P. Menzel, L. E. Gumley, and Y. Zhang, 2000. Retrieval of geophysical parameters from moderate resolution imaging spectroradiometer thermal infrared data: Evaluation of a two-step physical algorithm. *Appl. Opt.*, 39(20): 3537.

Maeda, E. E., 2014. Downscaling MODIS LST in the East African mountains using elevation gradient and land-cover information. *Int. J. Remote Sens.*, 35(9): 3094–3108. http://doi.org/10.1080/01431161.2014.903442.

Maes, W. H., and K. Steppe, 2012. Estimating evapotranspiration and drought stress with ground-based thermal remote sensing in agriculture: A review. *J. Exp. Bot.*, 63: 4671–4712.

Malakar, N. K., and G. C. Hulley, 2016. A water vapor scaling model for improved land surface temperature and emissivity separation of MODIS thermal infrared data. *Remote Sens. Environ.*, 182: 252–264.

Manoli, G., S. Fatichi, M. Schläpfer, K. Yu, T. W. Crowther, N. Meili, P. Burlando, G. G. Katul, and E. Bou-Zeid, 2019. Magnitude of urban heat islands largely explained by climate and population. *Nature*, 573(7772): 55–60.

Mao, K., J. Shi, J. Tang, Z. Li, X. Wang, and K. Chen, 2008. A neural network technique for separating land surface emissivity and temperature from ASTER imagery. *IEEE Trans. Geosci. Remote Sens.*, 46(1): 200–208. https://doi.org/10.1109/TGRS.2007.907333.

Mao, Q., J. Peng, and Y. Wang, 2021. Resolution enhancement of remotely sensed land surface temperature: Current status and perspectives. *Remote Sens.*, 13: 1306. https://doi.org/10.3390/rs13071306.

Martin-Vide, J., P. Sarricolea, and M. C. Moreno-García, 2015. On the definition of urban heat island intensity: The "rural" reference. *Front. Earth Sci.*, 3: 1–3. https://doi.org/10.3389/feart.2015.00024.

Masiello, G., and C. Serio, 2013. Simultaneous physical retrieval of surface emissivity spectrum and atmospheric parameters from infrared atmospheric sounder interferometer spectral radiances. *Appl. Opt.*, 52(11): 2428–2446.

Masiello, G., C. Serio, S. Venafra, G. Liuzzi, L. Poutier, and F.-M. Göttsche, 2018. Physical retrieval of land surface emissivity spectra from hyper-spectral infrared observations and validation with in situ measurements. *Remote Sens.*, 10(6): 976.

Matson, M., and J. Dozier, 1981. Identification of subresolution high temperature sources using a thermal IR sensor. *Photogramm. Eng. Remote Sens.*, 47(9): 1311–1318.

Matson, M., and B. Holben, 1986. Satellite detection of tropical burning in Brazil. *Int. J. Remote Sens.*, 8(3): 509–516. https://doi.org/10.1080/01431168708948657.

Matson, P. A., L. F. Johnson, C. B. Billow, J. R. Miller, and R. Pu, 1994, Seasonal patterns and remote spectral estimation of canopy chemistry across the Oregon Transect. *Ecol. Appl.*, 4(2): 280–298.

Mauser, W., and S. Schädlich, 1998. Modelling the spatial distribution of evapotranspiration on different scales using remote sensing data. *J. Hydrol.*, 212: 250–267.

McIntire, J., X. Xiong, J. J. Butler, A. Angal, D. Moyer, Q. Ji, T. Schwarting, D. Link, and C. Sun, 2022. An overall assessment of JPSS-3 VIIRS radiometric performance based on pre-launch testing. *Remote Sens.*, 14(9): 1999.

McMillin, L. M., 1975. Estimation of sea surface temperature from two infrared window measurements with different absorptions. *J. Geophys. Res.*, 80: 5113–5117.

Mecikalski, J. R., G. R. Diak, M. C. Anderson, and J. M. Norman, 1999. Estimating fluxes on continental scales using remotely sensed data in an atmospheric-land exchange model. *J. Appl. Meteorol.*, 38(9): 1352–1369.

Meerdink, S., D. Roberts, G. Hulley, P. Gader, J. Pisek, K. Adamson, J. King, and S. J. Hook, 2019. Plant species' spectral emissivity and temperature using the hyperspectral thermal emission spectrometer (HyTES) sensor. *Remote Sens. Environ.*, 224: 421–435.

Meerdink, S. K., D. A. Roberts, J. Y. King, K. L. Roth, P. E. Dennison, C. H. Amaral, and S. J. Hook, 2016. Linking seasonal foliar traits to VSWIR-TIR spectroscopy across California ecosystems. *Remote Sens. Environ.*, 186: 322–338.

Menaut, J. C., L. Abbadie, and P. M. Vitousek, 1993. Nutrient and organic matter dynamics in tropical ecosystems. In *Fire in the Environment* (pp. 215–230). Wiley.

Menenti, M., and B. Choudhury, 1993. Parameterization of land surface evaporation by means of location dependent potential evaporation and surface temperature range. *Proc. IAHS Conf. Land Surf. Process.* IAHS Publ., 212: 561–568.

Mirzaei, P. A., and F. Haghighat, 2010. Approaches to study Urban Heat Island—Abilities and limitations. *Build. Environ.*, 45: 2192–2201.

Mitkari, K. V., M. K. Arora, and R. K. Tiwari, 2017. Extraction of glacial lakes in gangotri glacier using object-based image analysis. *IEEE J. Sel. Top. Appl. Earth Obs. Remote Sens.*, 10: 5275–5283.

Mitraka, Z., N. Chrysoulakis, G. Doxani, F. D. Frate, and M. Berger, 2015. Urban surface temperature time series estimation at the local scale by spatial-spectral unmixing of satellite observations. *Remote Sens.*, 7: 4139–4156. https://doi.org/10.3390/rs70404139.

Mohamadi, B., S. Chen, T. Balz, K. Gulshad, and S. C. McClure, 2019. Normalized method for land surface temperature monitoring on coastal reclaimed areas. *Sensors*, 19(22): 4836.

Monteith, J. L., 1965. Evaporation and environment. *Symp. Soc. Explor. Biol.*, 19: 205–234.

Moosavi, V., A. Talebi, M. H. Mokhtari, and M. R. Hadian, 2016. Estimation of spatially enhanced soil moisture combining remote sensing and artificial intelligence approaches. *Int. J. Remote Sens.*, 37: 5605–5631.

Moosavi, V., A. Talebi, M. H. Mokhtari, S. R. F. Shamsi, and Y. Niazi, 2015. A wavelet-artificial intelligence fusion approach (WAIFA) for blending Landsat and MODIS surface temperature. *Remote Sens. Environ.*, 169: 243–254. http://doi.org/10.1016/j.rse.2015.08.015.

Moran, M. S., 1998. NASA's airborne terrestrial applications sensor (ATLAS): A customer's perspective. In *NASA Commercial Remote Sensing Verification and Validation Symposium, August 4–6, 1998*, John C. Stennis Space Center, MS, 14 p.

Moran, M. S., T. R. Clarke, Y, Inoue, and A. Vidal, 1994. Estimating crop water deficit using the relation between surface-air temperature and spectral vegetation index. *Remote Sens. Environ.*, 49(3): 246–263.

Morris, C. J. G., I. Simmonds, and N. Plummer, 2001. Quantification of influences of wind and cloud on the nocturnal urban heat island of a large city. *J. Appl. Meteorol.*, 40: 169–182.

Mouginis-Mark, P., H. Garbeil, and P. Flament, 1994. Effects of viewing geometry on AVHRR observations of volcanic thermal anomalies. *Remote Sens. Environ.*, 48: 51–60.

Mouillot, F., M. G. Schultz, and C. Yue, 2014. Ten years of global burned area products from spaceborne remote sensing—a review: Analysis of user needs and recommendations for future developments. *Int. J. Appl. Earth Obs. Geoinf.*, 26: 64–79.

Mukherjee, S., P. Joshi, and R. Garg, 2014. A comparison of different regression models for downscaling Landsat and MODIS land surface temperature images over heterogeneous landscape. *Adv. Space Res.*, 54: 655–669. http://doi.org/10.1016/j.asr.2014.04.013.

Murakami, S., 2006. Environmental design of outdoor climate based on CFD. *Fluid Dyn. Res.*, 38: 108e26.

Murray, T., and A. Verhoef, 2007. Moving a more mechanistic approach in the determination of soil heat flux from remote measurements. I. A universal approach to calculate thermal inertia. *Agric. For. Meteorol.*, 147: 80–87.

NASA-ASTER, 2022. *ASTER, Advanced Spaceborne Thermal Emission and Reflection Radiometer.* https://asterweb.jpl.nasa.gov/mission.asp, accessed on June 14, 2022.

NASA-MODIS, 2022. *MODIS, The Moderate Resolution Imaging Spectrometer.* https://modis.gsfc.nasa.gov/data/, accessed on June 14, 2022.

Neinavaz, E., R. Darvishzadeh, A. K. Skidmore, and H. Abdullah, 2019. Integration of Landsat-8 thermal and visible-short wave infrared data for improving prediction accuracy of forest leaf area index. *Remote Sens.*, 11: 390.

Neinavaz, E., M. Schlerf, R. Darvishzadeh, M. Gerhards, and A. K. Skidmore, 2021. Thermal infrared remote sensing of vegetation: Current status and perspectives. *Int. J. Appl. Earth Obs. Geoinf.*, 102: 102415.

Neinavaz, E., A. K. Skidmore, R. Darvishzadeh, and T. A. Groen, 2016a. Leaf area index retrieved from thermal hyperspectral data. *Int. Arch. Photogramm. Remote Sens. Spatial Inf. Sci.*, XLI(B7): 99–105.

Neinavaz, E., A. K. Skidmore, R. Darvishzadeh, and T. A. Groen, 2016b. Retrieval of leaf area index in different plant species using thermal hyperspectral data. *ISPRS J. Photogramm. Remote Sens.*, 119: 390–401.

Neinavaz, E., A. K. Skidmore, R. Darvishzadeh, and T. A. Groen, 2017. Retrieving vegetation canopy water content from hyperspectral thermal measurements. *Agric. For. Meteorol.*, 247: 365–375.

Ngie, A., K. Abutaleb, F. Ahmed, A. Darwish, and A. Mahmoud, 2014. Assessment of urban heat island using satellite remotely sensed imagery: A review. *S. Afr. Geogr. J.*, 96(2): 198–214. https://doi.org/10.1080/03736245.2014.924864.

Nichol, J. E., 1996. High-resolution surface temperature patterns related to urban morphology in a tropical city: A satellite-based study. *J. Appl. Meteorol.*, 35(1): 135–146. https://doi.org/10.1175/1520-0450(1996)035\0135:HRSTPR[2.0.CO;2.

Nishii, R., S. Kusanobu, and S. Tanaka, 1996. Enhancement of low spatial resolution image based on high resolution-bands. *IEEE Trans. Geosci. Remote Sens.*, 34(5): 1151–1158.

NOAA-AVHRR, 2022. *NOAA Comprehensive Large Array-data Stewardsship System (CLASS).* https://www.avl.class.noaa.gov/release/data_available/avhrr/index.htm, accessed on June 12, 2022.

Norman, J. M., and F. Becker, 1995. Terminology in thermal infrared remote sensing of natural surfaces. *Agric. For. Meteol.*, 77: 153–166.

Norman, J. M., W. P. Kustas, and K. S. Humes, 1995. A two-source approach for estimating soil and vegetation energy fluxes from observations of directional radiometric surface temperature. *Agric. For. Meteorol.*, 77: 263–293.

Norman, J. M., W. P. Kustas, J. H. Prueger, and G. R. Diak, 2000. Surface flux estimation using radiometric temperature: A dual-temperature-difference method to minimize measurement errors. *Water Resour. Res.*, 36(8): 2263–2274.

Novák, V., I. Perfilieva, and J. Močkoř, 1999. *Mathematical Principles of Fuzzy Logic.* Dordrecht: Kluwer Academic. ISBN 978-0-7923-8595-0.

NRC, National Research Council. 2007. *Earth Science and Applications from Space: National Imperatives for the Next Decade and Beyond.* Washington, DC: The National Academies Press. https://doi.org/10.17226/11820.

Nykänen, V., D. L. Groves, V. J. Ojala, P. Eilu, and S. J. Gardoll, 2008. Reconnaissance-scale conceptual fuzzy-logic prospectivity modelling for iron oxide copper—gold deposits in the northern Fennoscandian Shield. Finland. *Aust. J. Earth Sci.*, 55(1): 25–38.

Oke, T. R., 1982. The energetic basis of the urban heat island. *Q. J. R. Meteorol. Soc.*, 108: 1–24.

Oki, T., and S. Kanae, 2006. Global hydrological cycles and world water resources. *Science*, 313(5790): 1068–1072.

Olivera-Guerra, L., C. Mattar, O. Merlin, C. Durán-Alarcón, A. Santamaría-Artigas, and R. Fuster, 2017. An operational method for the disaggregation of land surface temperature to estimate actual evapotranspiration in the arid region of Chile. *ISPRS J. Photogramm. Remote Sens.*, 128: 170–181. http://doi.org/10.1016/j.isprsjprs.2017.03.014.

Oppenheimer, C., 1996. Crater lake heat losses estimated by remote sensing. *Geophys. Res. Lett.*, 23: 1793–1796.

Oppenheimer, C., 1998. Volcanological applications of meteorological satellites. *Int. J. Remote Sens.*, 19(15): 2829–2864.

Oppenheimer, C., and G. Yirgu, 2002. Thermal imaging of an active lava lake: Erta 'ale volcano, Ethiopia. *Int. J. Remote Sens.*, 23: 4777–4782.

Otón, G., R. Ramo, J. Lizundia-Loiola, and E. Chuvieco, 2019. Global detection of long-term (1982–2017) burned area with AVHRR-LTDR data. *Remote Sens.*, 11(18): 2079.

Ottlé, C., and M. Stoll, 1993. Effect of atmospheric absorption and surface emissivity on the determination of land surface temperature from infrared satellite data. *Int. J. Remote Sens.*, 14: 2025–2037.

Ottlé, C., and D. Vidal-Madjar, 1992. Estimation of land surface temperature with NOAA9 data. *Remote Sens. Environ.*, 40(1): 27–41.

Palluconi, F., A. B. Kahle, G. Hoover, and J. F. Conel, 1990. The spectral emissivity of prairie and pasture grasses at Konza Prairie, Kansas. In: *Processing of the American Meteorological Society Symposium on the First ISLSCP Field Experiment (FIFE)*, American Meteorological Society, pp. 77–78.

Pan, J. H., 2016. Area delineation and spatial-temporal dynamics of urban heat island in Lanzhou City, China using remote sensing imagery. *J. Indian Soc. Remote Sens.*, 44: 111–127.

Pan, T., C. Zhang, W. Kuang, G. Luo, G. Du, and Z. Yin. 2020. Large-scale rain-fed to paddy farmland conversion modified land-surface thermal properties in Cold China. *Sci. Total Environ.*, 722: 137917.

Patrick, M. R., J. Dehn, and K. Dean, 2005. Numerical modelling of lava flow cooling applied to the 1997 Okmok eruption: Comparison with advanced very high resolution radiometer thermal imagery. *J. Geophys. Res. Solid Earth*, 110: B02210. https://doi.org/02210.01029/02003JB002538.

Paul, S., S. Willmes, O. Gutjahr, A. Preußer, and G. Heinemann, 2015a. Spatial feature reconstruction of cloud-covered areas in daily MODIS composites. *Remote Sens. (Basel)*, 7(5): 5042–5056.

Paul, S., S. Willmes, and G, Heinemann, 2015b. Long-term coastal-polynya dynamics in the southern Weddell Sea from MODIS thermal-infrared imagery. *Cryosphere*, 9(6): 2027–2041.

Pedregosa, F., G. Varoquaux, A. Gramfort, V. Michel, B. Thirion, O. Grisel, M. Blondel, P. Prettenhofer, R. Weiss, V. Dubourg, et al., 2011. Scikit-learn: Machine learning in python. *J. Mach. Learn. Res.*, 12: 2825–2830.

Peng, J., Y. Dan, R. Qiao, Y. Liu, J. Dong, and J. Wu, 2021. How to quantify the cooling effect of urban parks? Linking maximum and accumulation perspectives. *Remote Sens. Environ.*, 252: 112135.

Peng, J., R. Qiao, Y. Liu, T. Blaschke, S. Li, J. Wu, Z. Xu, and Q. Liu, 2020. A wavelet coherence approach to prioritizing influencing factors of land surface temperature and associated research scales. *Remote Sens. Environ.*, 246: 111866.

Peng, S., S. Piao, P. Ciais, P. Friedlingstein, C. Ottle, F.-M. Breon, H. Nan, L. Zhou, and R. B. Myneni, 2012. Surface urban heat island across 419 global big cities. *Environ. Sci. Technol.*, 46: 696–703.

Peng, Y., W. Li, X. Luo, and H. Li, 2019. A geographically and temporally weighted regression model for spatial downscaling of MODIS land surface temperatures over urban heterogeneous regions. *IEEE Trans. Geosci. Remote Sens.*, 57(7): 5012–5027.

Pereira, J., E. Chuvieco, A. Beudoin, and N. Desbois, 1997. *Remote Sensing of Burned Areas: A Review. A Review of Remote Sensing Methods for the Study of Large Wildland Fires.* Departamento de Geografa, Universidad de Alcal, pp. 127–184.

Pereira, J. M. C., A. M. O. Sousa, and A. C. L. Sá, 1999. *Regional-Scale Burnt Area Mapping in Southern Europe Using NOAA-AVHRR 1 km Data.* Remote Sensing of Large Wildfires, pp. 139–155.

Pereira, O. J. R., A. J. Melfi, C. R. Montes, and Y. Lucas, 2018. Downscaling of ASTER thermal images based on geographically weighted regression kriging. *Remote Sens.*, 10: 633. https://doi.org/10.3390/rs10040633.

Pergola, N., D. A. Giuseppe, M. Lisi, F. Marchese, G. Mazzeo, and V. Tramutoli, 2009. Time domain analysis of robust satellite technique (RST) for near real-time monitoring of active volcanoes and thermal precursor identification. *Phys. Chem. Earth*, 34: 84–94.

Pergola, N., F. Marchese, and V. Tramutoli, 2004. Automated detection of thermal features of active volcanoes by means of infrared AVHRR records. *Remote Sens. Environ.*, 93: 311–327.

Pergola, N., F. Marchese, V. Tramutoli, C. Filizzola, and M. Ciampa, 2008. Advanced satellite technique for volcanic activity monitoring and early warning. *Ann. Geophys.*, 51: 287–301.

Pieri, D., and M. Abrams, 2004. ASTER watches the world's volcanoes: A new paradigm for volcanological observations from orbit. *J. Volcanol. Geotherm. Res.*, 135: 13–28.

Platt, N. H., and V. P. Wright, 1991. *Lacustrine Carbonates: Facies Models, Facies Distributions and Hydrocarbon Aspects* (pp. 57–74). Lacustrine Facies Analysis.

Pour, A. B., M. Hashim, Y. Park, and J. K. Hong, 2018. Mapping alteration mineral zones and lithological units in Antarctic regions using spectral bands of ASTER remote sensing data. *Geocarto. Int.*, 32(12): 1281–1306.

Pour, A. B., and M. Hashim, 2011. Identification of hydrothermal alteration minerals for exploring of porphyry copper deposit using ASTER data. *SE Iran. J. Asian Earth Sci.*, 42: 1309–1323.

Pour, A. B., and M. Hashim, 2015. Regional hydrothermal alteration mapping using Landsat-8 data. In *2015 International Conference on Space Science and Communication (IconSpace)* (pp. 199–202). IEEE.

Pour, A. B., Y. Park, L. Crispini, A. Läufer, J. K. Hong, T.-Y. S. Park, B. Zoheir, B. Pradhan, A. M. Muslim, M. S. Hossain, and O. Rahmani, 2019b. Mapping listvenite occurrences in the damage zones of Northern Victoria Land, antarctica using ASTER Satellite Remote Sensing Data. *Remote Sens.*, 11: 1408. https://doi.org/10.3390/rs11121408.

Prata, A. J., 1989. Infrared radiative transfer calculations for volcanic ash clouds. *Geophys. Res. Lett.*, 16: 1293–1296.

Prata, A. J., 1993. Land surface temperatures derived from the advanced very high resolution radiometer and the along-track scanning radiometer 1: Theory. *J. Geophys. Res.*, 98(D9): 16689–16702.

Prata, A. J., V. Caselles, C. Coll, J. A. Sobrino, and C. Ottlé, 1995. Thermal remote sensing of land surface temperature from satellites: Current status and future prospects. *Remote Sens. Rev.*, 12: 175–224.

Price, J. C., 1977. Thermal inertia mapping: A new view of the Earth. *J. Geophys. Res.*, 82: 2582–2590.

Price, J. C., 1980. The potential of remotely sensed thermal infrared data to infer surface soil moisture and evaporation. *Water Resour. Res.*, 16(4): 787–795.

Price, J. C., 1982. On the use of satellite data to infer surface fluxes at meteorological scales. *J. Appl. Meteorol.*, 21: 1111–1122.

Price, J. C., 1983. Estimating surface temperatures from satellite thermal infrared data. A simple formulation for the atmospheric effect. *Remote Sens. Environ.*, 13: 353–361.

Price, J. C., 1984. Land surface temperature measurements from the split window channels of the NOAA 7 advanced very high resolution radiometer. *J. Geophys. Res. Atmos.*, 89: 7231–7237. https://doi.org/10.1029/JD089iD05p07231.

Priestley, C. H. B., and R. J. Taylor, 1972. On the assessment of surface heat flux and evaporation using large-scale parameters. *Monthly Weather Rev.*, 100(2): 81–92.

Pu, R., 2017. *Hyperspectral Remote Sensing: Fundamentals and Practices* (p. 466). CRC Press, Taylor & Francis Group.

Pu, R., 2021. Assessing scaling effect in downscaling land surface temperature in a heterogenous urban environment. *Int. J. Appl. Earth Obs. Geoinf.*, 96: 102256.

Pu, R., and S. Bonafoni, 2021. Reducing scaling effect on downscaled land surface temperature maps in heterogenous urban environments. *Remote Sens.*, 13(24): 5044.

Pu, R., and S. Bonafoni, 2023. Thermal infrared remote sensing data downscaling investigations: An overview on current status and perspectives. *Remote Sens., Appl.: Soc. Environ.*, 29: 100921.

Pu, R., and J. Cheng, 2015. Mapping forest leaf area index using reflectance and textural information derived from worldview-2 imagery in a mixed natural forest area in Florida, USA. *Int. J. Appl. Earth Obs. Geoinf.*, 42: 11–23.

Pu, R., P. Gong, and G. S. Biging, 2003. Simple calibration of AVIRIS data and LAI mapping of forest plantation in southern Argentina. *Int. J. Remote Sens.*, 24(23): 4699–4714.

Pu, R., P. Gong, R. Michishita, and T. Sasagawa, 2006. Assessment of multi-resolution and multi-sensor data for urban surface temperature retrieval. *Remote Sens. Environ.*, 104: 211–225.

Pu, R., S. Landry, and Q. Yu, 2018. Assessing the potential of multi-seasonal high resolution pléiades satellite imagery for mapping urban tree species. *Int. J. Appl. Earth Obs. Geoinf.*, 71: 144–158.

Pu, R., Z. Li, P. Gong, I. Csiszar, R. Fraser, W.-M. Hao, S. Kondragunta, and F. Weng, 2007. Development and analysis of a 12-year daily 1-km forest fire data across the North America from NOAA/AVHRR data. *Remote Sens. Environ.*, 108: 198–208.

Pugnaghi, S., G. Gangale, S. Corradini, and M. F. Buongiorno, 2006. Mt Etna sulfur dioxide flux monitoring using ASTER-TIR data and atmospheric observations. *J. Volcanol. Geotherm. Res.*, 152(1–2): 74–90.

Qin, Q., C. Jin, N. Zhang, and X. Yang, 2010. A two-dimensional spectral space-based model for drought monitoring and its re-examination. In *Geoscience and Remote Sensing Symposium (IGARSS), 2010 IEEE International*. IEEE, pp. 3869–3872.

Qin, Z., A. Karnieli, and P. Berliner, 2001b. A mono-window algorithm for retrieving land surface temperature from Landsat TM data and its application to the Israel-Egypt border region. *Int. J. Remote Sens.*, 22(18): 3719–3746.

Qin, Z., G. D. Olmo, A. Karnieli, and P. Berliner, 2001a. Derivation of split window algorithm and its sensitivity analysis for retrieving land surface temperature from NOAA-advanced very high-resolution radiometer data. *J. Geophys. Res.*, 106(D19): 22655–22670.

Qiu, G. Y., H. Y. Li, Q. T. Zhang, C. H. E. N. Wan, X. J. Liang, and X. Z. Li, 2013. Effects of evapotranspiration on mitigation of urban temperature by vegetation and urban agriculture. *J. Integr. Agric.*, 12(8): 1307–1315.

Quan, J., W. Zhan, Y. Chen, M. Wang, and J. Wang, 2016. Time series decomposition of remotely sensed land surface temperature and investigation of trends and seasonal variations in surface urban heat islands. *J. Geophys. Res. Atmos.*, 121: 2638–2657.

Quattrochi, D. A., and J. C. Luvall, 1999. Thermal infrared remote sensing for analysis of landscape ecological processes: Methods and applications. *Landscape Ecol.*, 14: 577–598.

Quattrochi, D. A., and J. C. Luvall, 2009. Thermal remote sensing in earth science research. In *The SAGE Handbook of Remote Sensing* (pp. 64–78). SAGE Publications, Inc.

Quattrochi, D. A., and N. S. Goel, 1995. Spatial and temporal scaling of thermal remote sensing data. *Remote Sens. Rev.*, 12(3–4): 255–286.

Quintano, C., and A. Fernández-Manso, 2011. Estimation of area burned by forest fires in Mediterranean countries: A remote sensing data mining perspective. *For. Ecol. Manag.*, 262(8): 1597–1607.

Rajasekar, U., and Q. H. Weng, 2009. Urban heat island monitoring and analysis using a non-parametric model: A case study of Indianapolis. *ISPRS J. Photogramm. Remote Sens.*, 64: 86–96.

Ramsey, M. S., and I. T. W. Flynn, 2020. The Spatial and Spectral resolution of ASTER infrared image data: A paradigm shift in volcanological remote sensing. *Remote Sens.*, 12: 738. https://doi.org/10.3390/rs12040738.

Randriambelo, T., S. Baldy, M. Bessafi, M. Petit, and M. Despinoy, 1998. An improved detection and characterization of active fires and smoke plumes in south-eastern Africa and Madagascar. *Int. J. Remote Sens.*, 19(14): 2623–2638.

Rao, P. K., 1972. Remote sensing of urban "heat islands" from an environmental satellite. *Bull. Am. Meteorol. Soc.*, 53: 647–648.

Ravail, N., and San-Miguel-Ayanz, 2002. Evaluation of the performance of the world fire web contextual fire detection algorithm over Spain for the years 1997–98, *Eur Report*, Office for Official Publications of the European Communities, L-2985, Luxemburg.

Raven, P. N., D. L. Jordan, and C. E. Smith, 2002. Polarized directional reflectance from laurel and mullein leaves. *Opt. Eng.*, 41: 1002–1012.

Razafimpanilo, H., R. Frouin, S. F. Iacobellis, and R. C. Somerville, 1995. Methodology for estimating burned area from AVHRR reflectance data. *Remote Sens. Environ.*, 54(3): 273–289.

Research Systems Inc, 2008. *ENVI Tutorials*. Research Systems Inc, Boulder CO.

Reutter, H., F. S. Olesen, and H. Fischer, 1994. Distribution of the brightness temperature of land surfaces determined from AVHRR data. *Int. J. Remote Sens.*, 15: 95–104.

Ribeiro da Luz, B., and J. K. Crowley, 2007. Spectral reflectance and emissivity features of broad leaf plants: Prospects for remote sensing in the thermal infrared (8.0–14.0 μm). *Remote Sens. Environ.*, 109: 393–405.

Ribeiro da Luz, B., and J. K. Crowley, 2010. Identification of plant species by using high spatial and spectral resolution thermal infrared (8.0–13.5 μm) imagery. *Remote Sens. Environ.*, 114: 404–413.

Richards, J. A., 1999. *Remote Sensing Digital Image Analysis: An Introduction* (p. 240). Berlin, Germany: Springer-Verlag.

Rizwan, A. M., Y. C. L. Dennis, and C. Liu, 2008. A review on the generation, determination and mitigation of Urban Heat Island. *J. Environ. Sci.*, 20: 120–128.

Robinson, J. M., 1991. Fire from space: Global evaluation using infrared remote sensing. *Int. J. Remote Sens.*, 12: 3–24.

Rock, G., M. Gerhards, M. Schlerf, C. Hecker, and T. Udelhoven, 2016. Plant species discrimination using emissive thermal infrared imaging spectroscopy. *Int. J. Appl. Earth Obs. Geoinf.*, 53: 16–26.

Rockwell, B. W., and A. H. Hofstra, 2008. Identification of quartz and carbonate minerals across northern Nevada using ASTER thermal infrared emissivity data—Implications for geologic mapping and mineral resource investigations in well-studied and frontier areas. *Geosphere*, 4(1): 218–246.

Rosenfeld, A. H., H. Akbari, and J. J. Romm, 1998. Cool communities: Strategies for heat island mitigation and smog reduction. *Energy Build.*, 28: 51–62.

Rothery, D. A., D. Coppola, and C. Saunders, 2005. Analysis of volcanic activity patterns using MODIS thermal alerts. *Bull. Volcanol.*, 67: 539–556.

Roy, D. P., P. E. Lewis, and C. O. Justice, 2002. Burned area mapping using multi-temporal moderate spatial resolution data—a bi-directional reflectance model-based expectation approach. *Remote Sens. Environ.*, 83: 263–286.

Roy, D. P., Y. Jin, P. E. Lewis, and C. O. Justice, 2005. Prototyping a global algorithm for systematic fire-affected area mapping using MODIS time series data. *Remote Sens. Environ.*, 97(2): 137–162.

Rubio, E., V. Caselles, and C. Badenas, 1997. Emissivity measurements of several soils and vegetation types in the 8–14, μm Wave band: Analysis of two field methods. *Remote Sens. Environ.*, 59: 490–521.

Rudjord, Q., R. Solberg, G. Spreen, and S. Gerland, 2022. Estimating thin ice thickness around Svalbard using MODIS satellite imagery. *Geogr. Ann. A Phys. Geogr.*, 104(2): 127–149. https://doi.org/10.1080/04353676.2022.2070158.

Sabins, F. F., 1986. *Remote Sensing: Principle and Interpretation*. New York: W. H. Freeman.

Sabins, F. F., Jr., 1997. *Remote Sensing Principles and Interpretation* (p. 494). New York: Freeman.

Sadeghi, M., E. Babaeian, M. Tuller, and S. B. Jones, 2017. The optical trapezoid model: A novel approach to remote sensing of soil moisture applied to Sentinel-2 and Landsat-8 observations. *Remote Sens. Environ.*, 198: 52–68.

Sadeghi, M., S. B. Jones, and W. D. Philpot, 2015. A linear physically-based model for remote sensing of soil moisture using short wave infrared bands. *Remote Sens. Environ.*, 164: 66–76.

Sagan, V., M. Maimaitijiang, P. Sidike, K. Eblimit, K. T. Peterson, S. Hartling, F. Esposito, K. Khanal, M. Newcomb, and D. Pauli, 2019. UAV-based high resolution thermal imaging for vegetation monitoring, and plant phenotyping using ICI 8640 P, FLIR Vue Pro R 640, and thermomap cameras. *Remote Sens.*, 11: 330.

Salisbury, J. W., 1986. Preliminary measurements of leaf spectral reflectance in the 8–14 μm region. *Int. J. Remote Sens.*, 7: 1879–1886.

Salisbury, J. W., and N. M. Milton, 1987. Preliminary measurements of spectral signatures of tropical and temperate plants in the thermal infrared. *NASA Technical Report Server*, No 88A32909. https://ntrs.nasa.gov/citations/19880045682.

Salisbury, J. W., and N. M. Milton, 1988. Thermal infrared (2.5-to 13.5-μm) directional hemispherical reflectance of leaves. *Photogramm. Eng. Remote Sens.*, 54(9): 1301–1304.

Salisbury, W., and D. M. D'Aria, 1992. Emissivity of terrestrial materials in the 8–14 μm atmospheric window. *Remote Sens. Environ.*, 42: 83–106.

Salles, R. R., C. R. S. Filho, T. Cudahy, L. E. Vicente, and L. V. S. Monteiro, 2016. Hyperspectral remote sensing applied to uranium exploration: A case study at the Mary Kathleen metamorphic-hydrothermal U-REE deposit, NW, Queensland, Australia. *J. Geochem. Explor.*, 179: 36–50.

Sandholt, I., K. Rasmussen, and J. Andersen, 2002. A simple interpretation of the surface temperature/vegetation index space for assessment of surface moisture status. *Remote Sens. Environ.*, 79(2–3): 213–224.

Saneinejad, S., P. Moonen, and J. Carmeliet, 2014. Comparative assessment of various heat island mitigation measures. *Build Environ.*, 73: 162–170.

Santamouris, M., 2013. Using cool pavements as a mitigation strategy to fight urban heat island—a review of the actual developments. *Renew. Sustain. Energ. Rev.*, 26: 224–240.

Santos, C., I. J. Lorite, M. Tasumi, R. G. Allen, and E. Fereres, 2008. Integrating satellite-based evapotranspiration with simulation models for irrigation management at the scheme level. *Irrig. Sci.*, 26: 277–288.

Sasaki, K., A. Mochida, H. Yoshino, H. Watanabe, and T. Yoshida, 2008. A new method to select appropriate countermeasures against heat-island effects according to the regional characteristics of heat balance mechanism. *J. Wind Eng. Ind. Aerodyn.*, 96: 1629e39.

Sattari, F., M. Hashim, and A. B. Pour, 2018. Thermal sharpening of land surface temperature maps based on the impervious surface index with the TsHARP method to ASTER satellite data: A case study from the metropolitan Kuala Lumpur, Malaysia. *Measurement*, 125: 262–278. https://doi.org/10.1016/j.measurement.2018.04.092.

Sawabe, Y., T. Matsunaga, S. Rokugawa, and A. Hoyano, 2003. Temperature and emissivity separation for multi-band radiometer and validation ASTER TES algorithm. *J. Remote Sens. Soc. Jap.*, 23: 364–375.

Schaepman-Strub, G., M. E. Schaepman, T. H. Painter, S. Dangel, and J. V. Martonchik, 2006. Reflectance quantities in optical remote sensing-definitions and case studies. *Remote Sens. Environ.*, 103: 27–42.

Scherrer, D., and C. Körner, 2010. Infra-red thermometry of alpine landscapes challenges climatic warming projections. *Glob. Chang. Biol.*, 16: 2602–2613.

Schlerf, M., and C. Atzberger, 2012. Vegetation structure retrieval in beech and spruce forests using spectrodirectional satellite data. *Sel. Top. Appl. Earth Obs. Remote Sens.*, 5: 8–17.

Schmidhuber, J., 2015. Deep learning in neural networks: An overview. *Neural Netw.*, 61: 85–117. https://doi.org/10.1016/j.neunet.2014.09.003.

Schmugge, T. J., S. J. Hook, and C. Coll, 1998. Recovering surface temperature and emissivity from thermal infrared multispectral data. *Remote Sens. Environ.*, 65: 121–131.

Schneider, E. D., and J. J. Kay, 1994a. Life as a manifestation of the second law of thermodynamics. *Math. Comput. Model.*, 19(6–8): 25–48.

Schneider, E. D., and D. Sagan, 2005. *Into the Cool; Energy Flow, Thermodynamics, and Life* (p. 378). The University of Chicago Press.

Schneider, K., and W. Mauser, 1996. Processing and accuracy of Landsat thematic Mapper data for lake surface temperature measurement. *Int. J. Remote Sens.*, 17: 2027–2014.

Schonlau, M., and R. Y. Zou, 2020. The random forest algorithm for statistical learning. *Stata J.*, 20: 3–29.

Schwarz, N., S. Lautenbach, and R. Seppelt, 2011. Exploring indicators for quantifying surface urban heat islands of European cities with MODIS land surface temperatures. *Remote Sens. Environ.*, 115: 3175–3186.

Seielstad, C., and L. Queen, 2009. Thermal remote monitoring of the Norris Geyser Basin, Yellowstone National Park. *Final Report for the National Park Service Cooperative Ecosystem Studies Unit*, Agreement No. H1200040001: 38 p.

Sekertekin, A., and S. Bonafoni, 2020. Land surface temperature retrieval from Landsat 5, 7, and 8 over rural areas: Assessment of different retrieval algorithms and emissivity models and toolbox implementation. *Remote Sens.*, 12(2): 294.

Sepulcre-Cantó, G., P. J. Zarco-Tejada, J. Jiménez-Muñoz, J. Sobrino, E. D. Miguel, and F. Villalobos, 2006. Detection of water stress in an olive orchard with thermal remote sensing imagery. *Agric. For. Meteorol.*, 136: 31–44.

Sharma, K. V., S. Khandelwal, and N. Kaul, 2020. Downscaling of coarse resolution land surface temperature through vegetation indices based regression models. In J. K. Ghosh, and I. da Silva (Eds.), *Applications of Geomatics in Civil Engineering* (p. 33). Lecture Notes in Civil Engineering. https://doi.org/10.1007/978-981-13-7067-0_51.

Sheikhrahimi, A., A. B. Pour, B. Pradhan, and B. Zoheir, 2019. Mapping hydrothermal alteration zones and lineaments associated with orogenic gold mineralization using ASTER data: A case study from the Sanandaj-Sirjan Zone, Iran. *Adv. Space Res.*, 63: 3315–3332.

Shen, H., Y. Jiang, T. Li, Q. Cheng, C. Zeng, and L. Zhang, 2020. Deep learning-based air temperature mapping by fusing remote sensing, station, simulation and socioeconomic data. *Remote Sens. Environ.*, 240: 111692.

Shi, H., G. Xian, R. Auch, K. Gallo, and Q. Zhou, 2021. Urban heat island and its regional impacts using remotely sensed thermal data—a review of recent developments and methodology. *Land*, 10: 867. https://doi.org/10.3390/land10080867.

Shi, X., Z. Chen, H. Wang, D. Y. Yeung, W. K. Wong, and W. C. Woo, 2015. *Convolutional Lstm Network: A Machine Learning Approach for Precipitation Nowcasting arXiv:1506. 04214.* Advances in Neural Information Processing Systems.

Shirazi, A., A. Hezarkhani, A. B. Pour, A. Shirazy, and M. Hashim, 2022. Neuro-fuzzy-AHP (NFAHP) technique for copper exploration using advanced spaceborne thermal emission and reflection radiometer (ASTER) and geological datasets in the Sahlabad Mining Area, East Iran. *Remote Sens.*, 14: 5562. https://doi.org/10.3390/rs14215562.

Shirmard, H., E. Farahbakhsh, E. Heidari, A. Beiranvand Pour, B. Pradhan, D. Müller, and R. Chandra, 2022. A comparative study of convolutional neural networks and conventional machine learning models for lithological mapping using remote sensing data. *Remote Sens.*, 14(4): 819.

Shrestha, A., and A. Mahmood, 2019. Review of deep learning algorithms and architectures. *IEEE Access*, 7: 53040–53065. https://doi.org/10.1109/ACCESS.2019.2912200.

Siebert, S., F. Ewert, E. E. Rezaei, H. Kage, and R. Graß, 2014. Impact of heat stress on crop yield—on the importance of considering canopy temperature. *Environ. Res. Lett.*, 9: 044012.

Singh, A., 1993. Principal components analysis in remote sensing. *Proc. IEEE Int. Geosci. Remote Sens. Symp.*, 4: 1680–1682.

Sismanidis, P., I. Keramitsoglou, B. Bechtel, and C. T. Kiranoudis, 2017. Improving the downscaling of diurnal land surface temperatures using the annual cycle parameters as disaggregation kernels. *Remote Sens.*, 9: 23. https://doi.org/10.3390/rs9010023.

Sismanidis, P., I. Keramitsoglou, and C. T. Kiranoudis, 2015. A satellite-based system for continuous monitoring of surface urban heat islands. *Urban Clim.*, 14: 141–153.

Sismanidis, P., I. Keramitsoglou, C. T. Kiranoudis, and B. Bechtel, 2016. Assessing the capability of a downscaled urban land surface temperature time series to reproduce the spatiotemporal features of the original data. *Remote Sens.*, 8: 274. https://doi.org/10.3390/rs8040274

Skamarock, W. C., J. B. Klemp, J. Dudhia, D. O. Gill, D. M. Baker, W. Wan, et al., 2005. A description of the advanced research WRF version 2. *NCAR Technical Note.* http://www. mmm. ucar. edu/wrf/users/docs/arw_v2. Pdf.

Slater, P. N., 1980. *Remote Sensing, Optics and Optical Systems.* Reading, MA: Addison-Wesley.

Smets, B., C. Wauthier, and N. D'Oreye, 2010. A newmap of the lava flow field of Nyamulagira (D.R. Congo) from satellite imagery. *J. Afr. Earth Sci.*, 58: 778–786.

Smith, W. K., and G. A. Carter, 1988. Shoot structural effects on needle temperatures and photosynthesis in conifers. *Am. J. Bot.*, 75(4): 496–500.

Snyder, W. C., Z. Wan, Y. Zhang, and Y. Z. Feng, 1998. Classification based emissivity for land surface temperature measurement from space. *Int. J. Remote Sens.*, 19: 2753–2774. https://doi.org/10.1080/014311698214497.

Sobrino, J., C. Coll, and V. Caselles, 1991. Atmospheric correction for land surface temperature using NOAA-11 AVHRR channels 4 and 5. *Remote Sens. Environ.*, 38(1): 19–34.

Sobrino, J. A., J. C. Jiménez-Muñoz, and L. Paolini, 2004. Land surface temperature retrieval from LANDSAT TM 5. *Remote Sens. Environ.*, 90(4): 434–440.

Sobrino, J. A., and N. Raissouni, 2000. Toward remote sensing methods for land cover dynamic monitoring: Application to Morocco. *Int. J. Remote Sens.*, 21(2): 353–366.

Sobrino, J. A., R. Oltra-Carrió, G. Sòria, R. Bianchi, and M. Paganini, 2012. Impact of spatial resolution and satellite overpass time on evaluation of the surface urban heat island effects. *Remote Sens. Environ.*, 117: 50–56.

Sobrino, J. A., Z. L. Li, M. P. Stoll, and F. Becker, 1996. Multi-channel and multi-angle algorithms for estimating sea and land surface temperature with ATSR data. *Int. J. Remote Sens.*, 17(11): 2089–2114.

Sodha, M. S., P. K. Bansal, and S. C. Kaushik, 1980. Reduction of heat flux by a flowing water layer over an insulated roof. *Build. Environ.*, 15(2): 133–140.

Soltanifard, H., A. Kashki, and M. Karami, 2022. Analysis of spatially varying relationships between urban environment factors and land surface temperature in Mashhad city, Iran. *Egypt. J. Remote Sens. Space Sci.*, 25: 987–999.

Sontag, E. D., 1992. Feedback stabilization using two-hidden-layer nets. *IEEE Trans. Neural Netw.*, 3: 981–990.

Spronken-Smith, R. A., T. R. Oke, and W. P. Lowry, 2000. Advection and the surface energy balance across an irrigated urban park. *Int. J. Climatol.*, 20: 1033–1047.

Stathopoulou, M., and C. Cartalis, 2009. Downscaling AVHRR land surface temperatures for improved surface urban heat island intensity estimation. *Remote Sens. Environ.*, 112: 2592–2605.

Stavros, E. N., D. Schimel, R. Pavlick, S. Serbin, A. Swann, L. Duncanson, J. B. Fisher, F. Fassnacht, S. Ustin, R. Dubayah, et al., 2017. ISS observations offer insights into plant function. *Nat. Ecol. Evol.*, 1(7): 0194. https://doi.org/10.1038/s41559-017-0194.

Steffke, A. M., and A. J. L. Harris, 2011. A review of algorithms for detecting volcanic hot spots in satellite infrared data. *Bull. Volcanol.*, 73: 1109–1137.

Still, C., R. Powell, D. Aubrecht, Y. Kim, B. Helliker, D. Roberts, A. D. Richardson, and M. Goulden. 2019. Thermal imaging in plant and ecosystem ecology: Applications and challenges. *Ecosphere*, 10(6): e02768. https://doi.org/10.1002/ecs2.2768.

Stobbelaar, P., E. Neinavaz, and P. Nyktas, 2022. Prediction of leaf area index using thermal infrared data acquired by UAS over a mixed temperate forest. *Int. J. Appl. Earth Obs. Geoinf.*, 114: 103049.

Stott, P., 2000. Combustion in tropical biomass fires: A critical review. *Prog. Phys. Geogr.*, 24(3): 355–377.

Su, H., M. F. Mccabe, E. F. Wood, Z. Su, and J. H. Prueger, 2005. Modeling evapotranspiration during SMACEX: Comparing two approaches for local- and regional-scale prediction. *J. Hydrometeorol.*, 5–6: 910–922.

Su, Y., L. Liu, J. Liao, J. Wu, P. Ciais, J. Liao, X. He, X. Liu, X. Chen, and W. Yuan, 2020. Phenology acts as a primary control of urban vegetation cooling and warming: A synthetic analysis of global site observations. *Agric. For. Meteorol.*, 280: 107765.

Su, Z., 2002. The Surface Energy Balance System (SEBS) for estimation of turbulent heat fluxes. *Hydrol. Earth Syst. Sci.*, 6(1): 85–100.

Sun, D., and R. T. Pinker, 2003. Estimation of land surface temperature from a Geostationary Operational Environmental Satellite (GOES-8). *J. Geophys. Res.: Atmos.*, 108(D11).

Sun, W., and Q. Du, 2019. Hyperspectral band selection: A review. *IEEE Geosci. Remote Sens. Mag.*, 7: 118–139.

Sun, X., and P. Shu, 2011. An airborne thematic thermal infrared and electro-optical imaging system. In *International Symposium on Photoelectronic Detection and Imaging 2011: Advances in Infrared Imaging and Applications* (Vol. 8193, pp. 1281–1287). SPIE.

Szymanowski, M., and M. Kryza, 2011. Application of remotely sensed data for spatial approximation of urban heat island in the city of Wrocław, Poland. In *Proceedings of the 2011 Joint Urban Remote Sensing Event, JURSE 2011—Proceedings*, Munich, Germany, 11–13 April, pp. 353–356.

Taha, H., S. Konopacki, and S. Gabersek, 1999. Impacts of large scale modifications on meteorological conditions and energy use: A 10-region modeling study. *Theor. Appl. Climatol.*, 62: 175–185.

Takebayashi, H., and M. Moriyama, 2007. Surface heat budget on green roof and high reflection roof for mitigation of urban heat island. *Build. Environ.*, 42(8): 2971–2979.

Takebayashi, H., and M. Moriyama, 2012. Relationships between the properties of an urban street canyon and its radiant environment: Introduction of appropriate urban heat island mitigation technologies. *Sol. Energy*, 86: 2255–2262.

Tang, H., and Z. L. Li, 2013. *Quantitative Remote Sensing in Thermal Infrared: Theory and Applications*. Springer Science & Business Media.

Thanh Noi, P., and M. Kappas, 2017. Comparison of random forest, k-nearest neighbor, and support vector machine classifiers for land cover classification using Sentinel-2 imagery. *Sensors*, 18(1): 18.

Thermtest Instruments, 2022. *Materials Thermal Properties Database*. https://thermtest.com/thermal-resources/materials-database, accessed on May 10, 2022.

Tikhonov, V. I., 1963. Peaks of random processes. *Sov. Phys. Usp.*, 5: 594–611.

Tiwari, et al., 1982. Cooling by water evaporation over roof-review. *Energy Conv. Manage.*, 22(2): 143–153.

Tonooka, H., 2001. An atmospheric correction algorithm for thermal infrared multispectral data over land—a water-vapor scaling method. *IEEE Trans. Geosci. Remote Sens.*, 39: 682–692.

Tonooka, H., 2005a. Accurate atmospheric correction of ASTER thermal infrared imagery using the WVS method. *IEEE Trans. Geosci. Remote Sens.*, 43: 2778–2792.

Tonooka, H., 2005b, Atmospheric correction of MODIS thermal infrared bands by water vapor scaling method. *IEEE Trans. Geosci. Remote Sens.*, 5979: 152–163.

Townshend, J. R. G., C. O. Justice, D. Skole, J. P. Malingreau, J. Cihlar, P. Teillet, et al., 1994. The 1 km resolution global data set: Needs of the international geosphere biosphere programme. *Int. J. Remote Sens.*, 15: 3417–3441.

Tramutoli, V., 1998. Robust AVHRR techniques (RAT) for environmental monitoring: Theory and applications. *Proc. SPIE*, 3496: 101–113.

Tramutoli, V., P. Claps, M. Marella, N. Pergola, and C. Sileo, 2000. Feasibility of hydrological application of thermal inertia from remote sensing. In *2nd Plinius Conference on Mediterranean Storms, Siena, Italy* (pp. 16–18).

Tran, H., D. Uchihama, S. Ochi, and Y. Yasuoka, 2006. Assessment with satellite data of the urban heat island effects in Asian mega cities. *Int. J. Appl. Earth Obs. Geoinf.*, 8(1), 34–48.

Transfer Learning, 2023. *What Is Transfer Learning? A Guide for Deep Learning*. https://builtin.com/data-science/transfer-learning, accessed on July 29, 2023.

Tuia, D., F. Pacifici, M. Kanevski, and W. J. Emery, 2009. Classification of very high spatial resolution imagery using mathematical morphology and support vector machines. *IEEE Trans. Geosci. Rem. Sens.*, 47: 3866–3879. https://doi.org/10.1109/TGRS.2009.2027895.

Tyralis, H., G. Papacharalampous, A. Burnetas, and A. Langousis, 2019. Hydrological post-processing using stacked generalization of quantile regression algorithms: Large-scale application over Conus. *J. Hydrol.*, 577: 123957.

Udelhoven, T., M. Schlerf, K. Segl, K. Mallick, C. Bossung, R. Retzla, G. Rock, P. Fischer, A. Müller, and T. Storch, 2017. A satellite-based imaging instrumentation concept for hyperspectral thermal remote sensing. *Sensors*, 17: 1542.

Ulivieri, C., and G. Cannizzaro, 1985. Land surface temperature retrievals from satellite measurements. *Acta Astronaut.*, 12(12): 977–985.

Ulivieri, C. M. M. A., M. M. Castronuovo, R. Francioni, and A. Cardillo, 1994. A split window algorithm for estimating land surface temperature from satellites. *Adv. Space Res.*, 14(3): 59–65.

Ullah, S., T. A. Groen, M. Schlerf, A. K. Skidmore, W. Nieuwenhuis, and C. Vaiphasa, 2012a. Using a genetic algorithm as an optimal band selector in the mid and thermal infrared (2.5–14 μm) to discriminate vegetation species. *Sensors*, 12: 8755–8769.

Ullah, S., M. Schlerf, A. K. Skidmore, and C. Hecker, 2012b. Identifying plant species using mid-wave infrared (2.5–6μm) and thermal infrared (8–14μm) emissivity spectra. *Remote Sens. Environ.*, 118: 95–102.

Ullah, S., A. K. Skidmore, A. Ramoelo, T. A. Groen, M. Naeem, and A. Ali, 2014. Retrieval of leaf water content spanning the visible to thermal infrared spectra. *ISPRS J. Photogramm. Remote Sens.*, 93: 56–64.

Urano, A., T. Ichinose, and K. Hanaki, 1999. Thermal environment simulation for three dimensional replacement of urban activity. *J. Wind Eng. Ind. Aerodyn.*, 81: 197–210.

USGS, 2023. *US Geological Survey Urban Heat Islands*. https://www.usgs.gov/media/images/urban-heat-islands, accessed on April 6, 2023.

Valor, E., and V. Caselles, 1996. Mapping land surface emissivity from NDVI: Application to European, African, and South American areas. *Remote Sens. Environ.*, 57: 167–184.

Van de Griend, A. A., and M. Owe, 1993. On the relationship between thermal emissivity and the normalized difference vegetation index for natural surfaces. *Int. J. Remote Sens.*, 14: 1119–1131.

Van den Hurk, B., 2001. Energy balance based surface flux estimation from satellite data, and its application for surface moisture assimilation. *Meteorol. Atmos. Phys.*, 76: 43–52.

Van der Meer, F., C. Hecker, F. Ruitenbeek, H. van der Werff, C. D. Wijkerslooth, and C. Wechsler, 2014. Geologic remote sensing for geothermal exploration: A review. *Int. J. Appl. Earth Obs. Geoinf.*, 33: 255–269.

Van der Meer, F. D., H. M. A. Van der Werff, F. J. A. Van Ruitenbeek, C. A. Hecker, W. H. Bakker, M. F. Noomen, M. Van der Meijde, E. J. M. Carranza, J. B. Smeth, and T. Woldai, 2012. Multi-and hyperspectral geologic remote sensing: A review. *Int. J. Appl. Earth Obs. Geoinf.*, 14: 112–128.

Van Leeuwen, W. J. D., and A. R. Huete, 1996. Effects of standing litter on the biophysical interpretation of plant canopies with spectral indices. *Remote Sens. Environ.*, 55(2): 123–138.

Vaughan, R. G., L. P. Keszthelyi, H. Heasler, C. Jaworowski, J. B. Lowenstern, and D. J. Schneider, 2009. Thermal infrared remote sensing of the yellowstone geothermal system. In *AGU Fall Meeting Abstracts* (Vol. 2009, pp. V21B–1993). https://ui.adsabs.harvard.edu/abs/2009AGUFM.V21B1993V/abstract

Velliangiri, S., S. Alagumuthukrishnan, and S. I. T. Hankumar Joseph, 2019. A review of dimensionality reduction techniques for efficient computation. *Procedia Comput. Sci.*, 165: 104–111. https://doi.org/10.1016/j.procs.2020.01.079.

Veraverbeke, S., S. Harris, and S. Hook, 2011. Evaluating spectral indices for burned area discrimination using MODIS/ASTER (MASTER) airborne simulator data. *Remote Sens. Environ.*, 115: 2702–2709.

Verhoef, A., 2004. Remote estimation of thermal inertia and soil heat flux for bare soil. *Agric. For. Meteorol.*, 123: 221–236. https://doi.org/10.1016/j.agrformet.2003.11.005.

Vidal, A., 1991. Atmospheric and emissivity correction of land surface temperature measured from satellite using ground measurements or satellite data. *Remote Sens.*, 12(12): 2449–2460.

Voogt, J., 2002. Urban heat island. In T. Munn (Ed.), *Encyclopedia of Global Environmental Change* (Vol. 3, pp. 660–666). New York: Wiley.

Voogt, J. A., and T. R. Oke, 2003. Thermal remote sensing of urban climates. *Remote Sens. Environ.*, 86: 370–384.

Wambo, J. D. T., A. B. Pour, S. Ganno, P. D. Asimowc, B. Zoheird, R. R. Sallese, J. P. Nzentia, B. Pradhanf, and A. M. Muslimb, 2020. Identifying high potential zones of gold mineralization in a sub-tropical region using Landsat-8 and ASTER remote sensing data: A case study of the Ngoura-Colomines goldfield, eastern Cameroon. *Ore Geol. Rev.*, 122: 103530.

Wan, Z., 2014. New refinements and validation of the collection-6 MODIS land-surface temperature/emissivity product. *Remote Sens. Environ.*, 140: 36–45.

Wan, Z., and J. Dozier, 1996. A generalized split-window algorithm for retrieving land-surface temperature from space. *IEEE Trans. Geosci. Remote Sens.*, 34(4): 892–905.

Wan, Z., and Z.-L. Li, 1997. A physics-based algorithm for retrieving land-surface emissivity and temperature from EOS/MODIS data, *IEEE Trans. Geosci. Remote Sens.*, 35: 980–996.

Wan, Z., Y. Zhang, Q. Zhang, and Z.-L. Li, 2004. Quality assessment and validation of the MODIS global land surface temperature. *Int. J. Remote Sens.*, 25(1): 261–274.

Wang, H., and D. Y. Yeung, 2020. A survey on bayesian deep learning. *ACM Comput. Surv.*, 53: 1–38. https://doi.org/10.1145/3409383.

Wang, J., M. Kuffer, R. Sliuzas, and D. Kohli, 2019. The exposure of slums to high temperature: Morphology-based local scale thermal patterns. *Sci. Total Environ.*, 650: 1805–1817.

Wang, J., O. Schmitz, M. Lu, and D. Karssenberg, 2020. Thermal unmixing based downscaling for fine resolution diurnal land surface temperature analysis. *ISPRS J. Photogramm. Remote Sens.*, 161: 76–89.

Wang, J., W. Song, W. Wei, Y. Zhang, and S. Liu, 2011. A new algorithm for forest fire smoke detection based on MODIS data in Heilongjiang province. In *Proceedings of the 2011 International Conference on Remote Sensing, Environment and Transportation Engineering*, Nanjing, China, 24–26 June, pp. 5–8.

Wang, J. W., W. T. L. Chow, and Y.-C. Wang, 2020c. A global regression method for thermal sharpening of urban land surface temperatures from MODIS and Landsat. *Int. J. Remote Sens.*, 41(8): 2986–3009. https://doi.org/10.1080/01431161.2019.1697009

Wang, K., and R. E. Dickinson, 2012. A review of global terrestrial evapotranspiration: Observation, modeling, climatology, and climatic variability. *Rev. Geophys.*, 50: RG2005.

Wang, K., S. Jiang, J. Wang, C. Zhou, X. Wang, and X. Lee, 2017a. Comparing the diurnal and seasonal variabilities of atmospheric and surface urban heat islands based on the Beijing urban meteorological network. *J. Geophys. Res. Atmos.*, 122: 2131–2154.

Wang, K., P. Wang, Z. Li, M. Cribb, and M. Sparrow, 2007. A simple method to estimate actual evapotranspiration from a combination of net radiation, vegetation index, and temperature. *J. Geophys. Res.* [Atmospheres], 112(D15): D15107.

Wang, L., and J. J. Qu, 2007. NMDI: A normalized multi-band drought index for monitoring soil and vegetation moisture with satellite remote sensing. *Geophys. Res. Lett.*, 34, L20405. https://doi.org/10.1029/2007GL031021.

Wang, M., G. Heb, Z. Zhang, G. Wang, Z. Wang, R. Yin, S. Cui, Z. Wu, and X. Cao, 2019. A radiance-based split-window algorithm for land surface temperature retrieval: Theory and application to MODIS data. *Int. J. Appl. Earth Obs. Geoinf.*, 76: 204–217.

Wang, N., B. H. Tang, C. Li, and Z.-L. Li, 2010. A generalized neural network for simultaneous retrieval of atmospheric profiles and surface temperature from hyperspectral thermal infrared data. In *IEEE International Geoscience and Remote Sensing Symposium* (pp. 1055–1058). Honolulu: IEEE.

Wang, R., W. Gao, and W. Peng, 2020b. Downscale MODIS land surface temperature based on three different models to analyze surface urban heat island: A case study of Hangzhou. *Remote Sens.*, 12: 2134. https://doi.org/10.3390/rs12132134.

Wang, W., J. Qu, X. Hao, Y. Liu, and W. Sommers, 2007. An improved algorithm for small and cool fire detection using MODIS data: A preliminary study in the southeastern United States. *Remote Sens. Environ.*, 108: 163–170.

Wang, Z., C. B. Schaaf, Q. Sun, J. Kim, A. M. Erb, F. Gao, M. O. Román, Y. Yang, S. Petroy, J. R. Taylor, J. G. Masek, J. T. Morisette, X. Zhang, and S. A. Papuga, 2017b. Monitoring land surface albedo and vegetation dynamics using high spatial and temporal resolution synthetic time series from landsat and the MODIS BRDF/NBAR/albedo product. *Int. J. Appl. Earth Obs. Geoinf.*, 59: 104–117.

Wang, Z., Y. Sun, H. Ren, Q. Qin, and G. Han, 2017c. Downscaling research of remotely sensed land surface temperature. *IGARSS 2017*, IEEE Xplore.

Watson, F. G. R., R. E. Lockwood, W. B. Newman, T. N. Anderson, and R. A. Garrott, 2008. Development and comparison of Landsat radiometric and snowpack model inversion techniques for estimating geothermal heat flux. *Remote Sens. Environ.*,112(2): 471–481.

Welch, R., T. Jordan, H. Lang, and H. Murakami, 1998. ASTER as a source for topographic data in the late 1990's. *IEEE Trans. Geosci. Remote Sens.*, 36: 1282–1289.

Weng, Q., 2009. Thermal infrared remote sensing for urban climate and environmental studies: Methods, applications, and trends. *ISPRS J. Photogramm. Remote Sens.*, 64(4): 335–344.

Weng, Q., 2012. Remote sensing of impervious surfaces in the urban areas: Requirements, methods, and trends. *Remote Sens. Environ.*, 117: 34–49.

Weng, Q., and P. Fu, 2014. Modeling diurnal land temperature cycles over Los Angeles using downscaled GOES imagery. *ISPRS J. Photogramm. Remote Sens.*, 97: 78–88. http://doi.org/10.1016/j.isprsjprs.2014.08.009.

Weng, Q., P. Fu, and F. Gao, 2014. Generating daily land surface temperature at Landsat resolution by fusing Landsat and MODIS data. *Remote Sens. Environ.*, 145: 55–67.

Weng, Q., H. Liu, B. Liang, and D. Lu, 2008. The spatial variations of urban land surface temperatures: Pertinent factors, zoning effect, and seasonal variability. *IEEE Geosci. Remote Sens. Soc.*, 1(2): 154–166. https://doi.org/10.1109/JSTARS.2008.917869.

Weng, Q., H. Liu, and D. Lu, 2007. Assessing the effects of land use and land cover patterns on thermal conditions using landscape metrics in city of Indianapolis, United States. *Urban Ecosyst.*, 10(2): 203–219. https://doi.org/10.1007/s11252-007-0020-0.

Weng, Q., D. Lu, and B. Liang, 2006. Urban surface biophysical descriptors an land surface temperature variations. *Photogramm. Eng. Remote Sens.*, 72(11): 1275–1286.

Weng, Q., D. Lu, and J. Schubring, 2004. Estimation of land surface temperature-vegetation abundance relationship for urban heat island studies. *Remote Sens. Environ.*, 89(4): 467–483. https://doi.org/10.1016/j.rse.2003.11.005.

Wessels, R. L., R. G. Vaughan, M. R. Patrick, and M. L. Coombs, 2013. High-resolution satellite and airborne thermal infrared imaging of precursory unrest and 2009 eruption at redoubt volcano, Alaska. *J. Volcanol. Geotherm. Res.*, 259: 248–269.

Wikipedia, 2022. *Emissivity*. https://en.wikipedia.org/wiki/Emissivity, accessed on April 27, 2022.

Winter, M. E., P. G. Lucey, T. Williams, and M. Wood, 2004. Calibration of the University of Hawaii's airborne hyperspectral imager. In *Proceedings of SPIE 5159, Imaging Spectrometry IX*, 7 January. https://doi.org/10.1117/12.510198.

Wong, C. L., and W. R. Blevin, 1967. Infrared reflectances of plant leaves. *Aust. J. Biol. Sci.*, 20(3): 501–508. https://doi.org/10.1071/BI9670501.

Wood, E. F., H. Su, M. McCabe, and B. Su, 2003. Estimating evaporation from satellite remote sensing. In *Proceedings of the 2003 IEEE International Geoscience and Remote Sensing Symposium (IGARSS '03)*, Toulouse, France, 21–25 July, 2: 1163–1165.

Wooster, M. J., G. Roberts, G. L. W. Perry, and Y. J. Kaufman, 2005. Retrieval of biomass combustion rates and totals from fire radiative power observations: FRP derivation and calibration relationships between biomass consumption and fire radiative energy release. *J. Geophys. Res.*, 110(D24), D24 311.

Wooster, M. J., G. Roberts, A. M. Smith, J. Johnston, P. Freeborn, S. Amici, and A. T. Hudak, 2013. Thermal remote sensing of active vegetation fires and biomass burning events. In *Thermal Infrared Remote Sensing: Sensors, Methods, Applications*, Springer (pp. 347–390).

Wooster, M. J., B. Zhukov, and D. Oertel, 2003. Fire radiative energy for quantitative study of biomass burning: Derivation from the BIRD experimental satellite and comparison MODIS fire products. *Remote Sens. Environ.*, 86(1): 83–107.

Wright, R., S. Carn, and L. P. Flynn, 2005. A satellite chronology of the May–June 2003 eruption of Anatahan volcano. *J. Volcanol. Geotherm. Res.*, 146: 102–116.

Wright, R., S. D. Cruz-Reyna, L. Flynn, A. Harris, and J. J. Gomez-Palacios, 2002b. Integrated geophysical monitoring at Popocatepetl, Mexico: Explosive activity and dome growth processes. *J. Geophys. Res.*, 107(B8). https://doi.org/10.1029/2000JB000125.

Wright, R., and E. Pilger, 2008a. Satellite observations reveal little inter-annual variability in the radiant flux from the Mount Erebus lava lake. *J. Volcanol. Geotherm. Res.*, 177(3): 687–694.

Wright, R., L. Flynn, H. Garbeil, A. Harris, and E. Pilger, 2002a. Automated Volcanic eruption detection using MODIS. *Remote Sens. Environ.*, 82: 135–155.

Wright, R., and L. P. Flynn, 2003. On the retrieval of lava-flow surface temperatures from infrared satellite data. *Geology*, 31: 893–896.

Wright, R., L. P. Flynn, H. Garbeil, A. J. L. Harris, and E. Pilger, 2004. MODVOLC: Near-real-time thermal monitoring of global volcanism. *J. Volcanol. Geotherm. Res.*, 135: 29–49.

Wright, R., and E. Pilger, 2008b. Radiant flux from Earth's subaerially erupting volcanoes. *Int. J. Remote Sens.*, 29: 6443–6466.

Wu, C. D., S. C. C. Lung, and J. F. Jan, 2013. Development of a 3-D urbanization index using digital terrain models for surface urban heat island effects. *ISPRS J. Photogramm. Remote Sens.*, 81: 1–11.

Wu, H., and W. Li, 2019. Downscaling land surface temperatures using a random forest regression model with multitype predictor variables. *IEEE Access*, 7: 21904–21916. https://doi.org/10.1109/ACCESS.2019.2896241.

Wu, J., B. Zhong, S. Tian, A. Yang, and J. Wu, 2019b. Downscaling of urban land surface temperature based on multi-factor geographically weighted regression. *J. Sel. Top. Appl. Earth Obs. Remote Sens.*, 12(8): 2897–2911. 10.1109/JSTARS.2019.2919936.

Wu, M., H. Li, W. Huang, Z. Niua, and C. Wang, 2015. Generating daily high spatial land surface temperatures by combining ASTER and MODIS land surface temperature products for environmental process monitoring. *Environ. Sci. Processes Impacts*, 17: 1396–1404. https://doi.org/10.1039/c5em00254k.

Wu, M. Q., Z. Niu, C. Y. Wang, C. Y. Wu, and L. Wang, 2012. Use of MODIS and Landsat time series data to generate high-resolution temporal synthetic Landsat data using a spatial and temporal reflectance fusion model. *J. Appl. Remote Sens.*, 6(1): 063507.

Wu, P., Z. Yin, H. Yang, Y. Wu, and X. Ma, 2019a. Reconstructing geostationary satellite land surface temperature imagery based on a multiscale feature connected convolutional neural network. *Remote Sens.*, 11(3): 300. https://doi.org/10.3390/RS11030300.

Wu, X., Q. Xiao, J. Wen, D. You, and A. Hueni, 2019c. Advances in quantitative remote sensing product validation: Overview and current status. *Earth Sci. Rev.*, 196: 102875. https://doi.org/10.1016/j.earscirev.2019.102875.

Xia, H., Y. Chen, Y. Li, and J. Quan, 2019. Combining kernel-driven and fusion-based methods to generate daily high spatial resolution land surface temperatures. *Remote Sens. Environ.*, 224: 259–274.

Xia, H., Y. Chen, Y. Zhao, and Z. Chen. 2018. Regression-then-fusion" or "fusion-then-regression"? A theoretical analysis for generating high spatiotemporal resolution land surface temperatures. *Remote Sens.*, 10: 1382. https://doi.org/10.3390/rs10091382.

Xian, G., H. Shi, R. Auch, K. P. Gallo, Q. Zhou, Z. Wu, and M. Kolian, 2021. The effects of urban land cover dynamics on urban heat island intensity and temporal trends. *Gisci. Remote Sens.*, 58(4): 501–515. https://doi.org/10.1080/15481603.2021.1903282.

Xiao, R. B., Q. H. Weng, Z. Y. Ouyang, W. F. Li, W. S. Erich, and Z. M. Zhang, 2008. Land surface temperature variation and major factors in Beijing. China. *Photogramm. Eng. Remote Sens.*, 74(4): 451–461.

Xiao, Y., P. Liu, Y. Liang, S. Stolte, P. Sanelli, A. Gupta, J. Ivanidze, and R. Fang, 2019. STIR-net: Deep spatial-temporal image restoration net for radiation reduction in CT perfusion. *Front. Neurol.*, 10: 647.

Xie, Y., J. J. Qu, X. Xiong, X. Hao, N. Che, and W, Sommers, 2007. Smoke plume detection in the eastern United States using MODIS. *Int. J. Remote Sens.*, 28: 2367–2374.

Xu, H., 2008. A new index for delineating built-up land features in satellite imagery. *Int. J. Remote Sens.*, 29(14): 4269–4276.

Xu, J., F. Zhang, H. Jiang, H. Hu, K. Zhong, W. Jing, J. Yang, and B. Jia, 2020. Downscaling ASTER land surface temperature over urban areas with machine learning-based area-to-point regression kriging. *Remote Sens.*, 12: 1082. https://doi.org/10.3390/rs12071082.

Xu, L. Y., X. D. Xie, and S. Li, 2013. Correlation analysis of the urban heat island effect and the spatial and temporal distribution of atmospheric particulates using TM images in Beijing. *Environ. Pollut.*, 178: 102–114.

Xu, S., Q. Zhao, K. Yin, G. He, Z. Zhang, G. Wang, M. Wen, and N. Zhang, 2021. Spatial downscaling of land surface temperature based on a multi-factor geographically weighted machine learning model. *Remote Sens.*, 13: 1186. https://doi.org/10.3390/rs13061186.

Xu, Y. M., Z. H. Qin, and H. X. Wan, 2010. Spatial and temporal dynamics of urban heat island and their relationship with land cover changes in urbanization process: A case study in Suzhou, China. *J. Indian Soc. Remote Sens.*, 38: 654–663.

Xue, J., Y. Leung, and T. Fung, 2019. An unmixing-based Bayesian model for spatio-temporal satellite image fusion in heterogeneous landscapes. *Remote Sens.*, 11: 324.

Xue, Y., and A. P. Cracknell, 1995. Advanced thermal inertia modelling. *Int. J. Remote Sens.*, 16(3): 431–446. https://doi.org/10.1080/01431169508954411.

Yamaguchi, Y., A. B. Kahle, H. Tsu, T. Kawakami, and M. Pniel, 1998. Overview of advanced spaceborne thermal emission and reflection radiometer (ASTER). *IEEE Trans. Geosci. Remote Sens.*, 36(4): 1062–1071.

Yamamoto, S., Y. Hosoya, N. Tsumura, and K. Ogawa-Ochiai, 2012. Principal component analysis for dental shade color. *Dental Mater.*, 28(7): 736–742.

Yan, L., W. Jia, and S. Zhao, 2021. The cooling effect of urban green spaces in Metacities: A case study of Beijing, China's capital. *Remote Sens.*, 13: 4601. https://doi.org/10.3390/rs13224601.

Yang, C., Q. Zhan, Y. Lv, and H. Liu, 2019. Downscaling land surface temperature using multiscale geographically weighted regression over heterogeneous landscapes in Wuhan,

China. *J. Sel. Top. Appl. Earth Obs. Remote Sens.*, 12(12): 5213–5222. https://doi.org/10.1109/JSTARS.2019.2955551.

Yang, G. J., R. L. Pu, C. J. Zhao, W. J. Huang, and J. H. Wang, 2011. Estimation of subpixel land surface temperature using an endmember index based technique: A case examination on ASTER and MODIS temperature products over a heterogeneous area. *Remote Sens. Environ.*, 115(5): 1202–1219.

Yang, J., Y. Yao, Y. Wei, Y. Zhang, K. Jia, X. Zhang, K. Shang, X. Bei, and X. Guo, 2020. A robust method for generating high-spatiotemporal-resolution surface reflectance by fusing MODIS and Landsat data. *Remote Sens.*, 12: 2312.

Yang, Y., C. Cao, X. Li, and X. Zhu, 2017. Downscaling land surface temperature in an arid area by using multiple remote sensing indices with random forest regression. *Remote Sens.*, 9: 789. https://doi.org/10.3390/rs9080789.

Yang, Y., H. Guan, D. Long, B. Liu, G. Qin, J. Qin, and O. Batelaan, 2015. Estimation of surface soil moisture from thermal infrared remote sensing using an improved trapezoid method. *Remote Sens.*, 7: 8250–8270; https://doi.org/10.3390/rs70708250.

Yao, Y., C. Chang, F. Ndayisaba, and S. Wang, 2020. A new approach for surface urban heat island monitoring based on machine learning algorithm and spatiotemporal fusion model. *IEEE Access*, 164268–164281. https://doi.org/10.1109/ACCESS.2020.3022047.

Yao, Y., Y. Luo, J. Huang, and Z. Zhao, 2013. Comparison of monthly temperature extremes simulated by CMIP3 and CMIP5 models. *J. Clim.*, 26(19): 7692–7707.

Yao, Y., S. Liang, S. Zhao, Y. Zhang, Q, Qin, J, Cheng, K. Jia, X. Xie, N. Zhang, and M. Liu, 2014. Validation and application of the modified satellite-based priestley-taylor algorithm for mapping terrestrial evapotranspiration. *Remote Sens.*, 6: 880–904. https://doi.org/10.3390/rs6010880.

Yin, Z., P. Wu, G. M. Foody, Y. Wu, Z. Liu, Y. Du, and F. Ling, 2021. Spatiotemporal fusion of land surface temperature based on a convolutional neural network. *Trans. Geosci. Remote Sens.*, 59(2): 1808–1822. https://doi.org/10.1109/TGRS.2020.2999943.

Yoo, C., J. Im, S. Park, and D. Cho, 2020. Spatial downscaling of MODIS land surface temperature: Recent research trends, challenges, and future directions. *Korean J. Remote Sens.*, 36(4): 609–626. https://doi.org/10.7780/kjrs.2020.36.4.9

Yu, L., A. Porwal, E.-J. Holden, and M. C. Dentith, 2012. Towards automatic lithological classification from remote sensing data using support vector machines. *Comput. Geosci.*, 45: 229–239.

Yu, X., X. Guo, and Z. Wu, 2014. Land surface temperature retrieval from Landsat 8 TIRS—Comparison between radiative transfer equation-based algorithm, split window algorithm and single channel algorithm. *Remote Sens.*, 6(10): 9829–9852.

Yu, Z., S. Xu, Y. Zhang, G. Jørgensen, and H. Vejre, 2018. Strong contributions of local background climate to the cooling effect of urban green vegetation. *Sci. Rep.*, 8: 6798.

Yuan, F., and M. E. Bauer, 2007. Comparison of impervious surface area and normalized difference vegetation index as indicators of surface urban heat island effects in Landsat imagery. *Remote Sens. Environ.*, 106(3): 375–386. https://doi.org/10.1016/j.rse.2006.09.003.

Yue, W., Y. Liu, P. Fan, X. Ye, and C. Wu, 2012. Assessing spatial pattern of urban thermal environment in Shanghai, China. *Stoch. Environ. Res. Risk Assess.*, 26: 899–911. https://doi.org/10.1007/s00477-012-0638-1.

Yuhaniz, S. S., and T. Vladimirova, 2009. An onboard automatic change detection system for disaster monitoring. *Int. J. Remote Sens.*, 30(23): 6129–6139.

Zadeh, L. A., 1965. Fuzzy sets. *Inf. Control.*, 8(3): 338–353. https://doi.org/10.1016/s0019-9958(65)90241.

Zakšek, K., and K. Oštir, 2012. Downscaling land surface temperature for urban heat island diurnal cycle analysis. *Remote Sens. Environ.*, 117: 114–124.

Zawadzka, J., R. Corstanje, J. Harris, and I. Truckell, 2020. Downscaling Landsat-8 land surface temperature maps in diverse urban landscapes using multivariate adaptive regression splines and very high resolution auxiliary data. *Int. J. Digit. Earth*, 13(8): 899–914. https://doi.org/10.1080/17538947.2019.1593527.

Zha, Y., J. Gao, and S. Ni, 2003. Use of normalized difference built-up index in automatically mapping urban areas from TM imagery. *Int. J. Remote Sens.*, 24(3): 583–594.

Zhai, H., F. Huang, and H. Qi, 2020. Generating high resolution LAI based on a modified FSDAF model. *Remote Sens.*, 12: 150.

Zhan, W., Y. Chen, J. Zhou, J. Wang, W. Liu, J. Voogt, X. Zhu, J. Quan, and J. Li, 2013. Disaggregation of remotely sensed land surface temperature: Literature survey, taxonomy, issues, and caveats. *Remote Sens. Environ.*, 131: 119–139. https://doi.org/10. 1016/j.rse.2012.12.014.

Zhang, H., Y. Song, C. Han, and L. Zhang, 2021a. Remote sensing image spatiotemporal fusion using a generative adversarial network. *Trans. Geosci. Remote Sens.*, 59(5): 4273–4286. https://doi.org/10.1109/TGRS.2020.3010530

Zhang, M., M. Zhang, H. Yang, Y. Jin, X. Zhang, and H. Liu, 2021b. Mapping regional soil organic matter based on sentinel-2a and MODIS imagery using machine learning algorithms and google earth engine. *Remote Sens.*, 13: 2934.

Zhang, N., Y. Hong, Q. Qin, and L. Liu, 2013. VSDI: A visible and shortwave infrared drought index for monitoring soil and vegetation moisture based on optical remote sensing. *Int. J. Remote Sens.*, 34(13): 4585–4609.

Zhang, P., M. L. Imhoff, R. E. Wolfe, and L. Bounoua, 2010. Characterizing urban heat islands of global settlements using MODIS and nighttime lights products. *Can. J. Remote Sens.*, 36: 185–196.

Zhang, T., X. Huang, D. Wen, and J. Li, 2017. Urban building density estimation from high- resolution imagery using multiple features and support vector regression. *IEEE J. Sel. Top. Appl. Earth Obs. Remote Sens.*, 10: 3265–3280. https://doi.org/10.1109/JSTARS.2017.2669217.

Zhang, X., H. Zhao, and J. Yang, 2019. Spatial downscaling of land surface temperature in combination with TVDI and elevation. *Int. J. Remote Sens.*, 40(5–6): 1875–1886. https://doi.org/10.1080/01431161.2018.1489164.

Zhang, Y. Q., F. H. S. Chiew, L. Zhang, R. Leuning, and H. A. Cleugh, 2008. Estimating catchment evaporation and runoff using MODIS leaf area index and the Penman-Monteith equation. *Water Res. Res.*, 44: W10420.

Zhang, Y.-X., X. Li, M. Zang, Q. Kang, W. Weiw, X.-B. Zheng, and Y, Zhang, 2020. On-orbit radiometric calibration for thermal infrared band of FY3D/MERSI-II satellite remote sensor based on Qinghai Lake radiation calibration test-site. *Acta Photonica Sin.* (in Chinese), 49(5): 0528002-1-9. https://doi.org/10.3788/gzxb2020 4905.0528002.

Zhao, H., K. Deng, N. Li, Z. Wang, and W. Wei, 2020. Hierarchical spatial-spectral feature extraction with long short term memory (LSTM) for mineral identification using hyperspectral imagery. *Sensors*, 20: 6854. https://doi.org/10.3390/s20236854.

Zhao, W., and A. Li, 2015. A review on land surface processes modelling over complex terrain. *Adv. Meteorol.*, 2015: 607181. https://doi.org/10.1155/2015/607181.

Zhou, D., J. Xiao, S. Bonafoni, C. Berger, K. Deilami, Y. Zhou, S. Frolking, Y. Yao, Z. Qiao, and J. A. Sobrino, 2019a. Satellite remote sensing of surface urban heat islands: Progress, challenges, and perspectives. *Remote Sens.*, 11(1): 48. https://doi.org/10.3390/rs11010048.

Zhou, D., L. Zhang, D. Li, D. Huang, and C. Zhu, 2016a. Climate-vegetation control on the diurnal and seasonal variations of surface urban heat islands in China. *Environ. Res. Lett.*, 11: 074009. https://doi.org/10.1088/1748-9326/11/7/074009.

Zhou, D., S. Zhao, L. Zhang, and S. Liu, 2016b. Remotely sensed assessment of urbanization effects on vegetation phenology in China's 32 major cities. *Remote Sens. Environ.*, 176: 272–281.

Zhou, J., M. Li, S. Liu, Z, Jia, and Y. Ma, 2015. Validation and performance evaluations of algorithms for estimating land surface temperatures from ASTER Data in the middle reach of the Heihe River Basin, Northwest China. *Remote Sens.*, 7(6): 7126–7156.

Zhou, J., S. Liang, J. Cheng, Y, Wang, and J. Ma, 2019b. The GLASS land surface temperature product. *IEEE J. Sel. Top. Appl. Earth Obs. Remote Sens.*, 12: 493–507.

Zhou, J., S. Liu, M. Li, W. Zhan, Z. Xu, and T. Xu, 2016c. Quantification of the scale effect in downscaling remotely sensed land surface temperature. *Remote Sens.*, 8: 975. http://doi.org/10.3390/rs8120975.

Zhou, L., M. Goldberg, C. Barnet, Z. Cheng, F. Sun, W. Wolf, T. King, X. Liu, H. Sun, and M. Divakarla, 2008. Regression of surface spectral emissivity from hyperspectral Instruments. *IEEE Trans. Geosci. Remote Sens.*, 46(20): 328–333.

Zhou, Q., G. Xian, and H. Shi, 2020. Gap fill of land surface temperature and reflectance products in landsat analysis ready data. *Remote Sens.*, 12: 1192.

Zhou, Y., Q. Weng, R. K. Gurney, Y. Shuai, and X. Hu, 2012. Estimation of the relationship between remotely sensed anthropogenic heat discharge and building energy use. *ISPRS J. Photogramm. Remote Sens.*, 67: 65–72. https://doi.org/10.1016/j.isprsjprs.2011.10.007.

Zhu, S., Y. Liu, J. Hua, G. Zhang, Y. Zhou, and J. Xiang, 2018b. Monitoring spatio-temporal variance of an extreme heat event using multiple-source remote sensing data. *Chin. Geogr. Sci.*, 28(5), 744–757. https://doi.org/10.1007/s11769-018-0989-8

Zhu, X., J. Chen, F. Gao, X. Chen, and J. G. Masek, 2010. An enhanced spatial and temporal adaptive reflectance fusion model for complex heterogeneous regions. *Remote Sens. Environ.*, 114(11): 2610–2623.

Zhukov, B., E. Lorenz, D. Oertel, M. J. Wooster, and G. Roberts, 2005. Experience of detection and quantitative characterization of fires during the experimental small satellite mission BIRD. *DLR-Forschungsbericht*, 1434–8454, p. 96. https://www.semanticscholar.org/paper/Experience-of-detection-and-quantitative-of-fires-Zhukov-Lorenz/1765e24ac37db1d41cbe15328b1c21b8fa2ad27f.

Zhukov, B., E. Lorenz, D. Oertel, M. J. Wooster, and G. Roberts, 2006. Spaceborne detection and characterization of fires during the bi-spectral infrared detection (BIRD) experimental small satellite mission (2001–2004). *Remote Sens. Environ.*, 100(1): 29–51. https://doi.org/10.1016/j.rse.2005.09.019.

Zimmermann, H. J., and P. Zysno, 1980. Latent connectives in human decision making. *Fuzzy Sets Syst.*, 4(1): 37–51.

Zinzi, M., 2010. Cool materials and cool roofs: Potentialities in Mediterranean buildings. *Adv. Build. Energy Res.*, 4: 201–266.

Index

N

O

P

R

S

T

U

V

W